INCLUSION ASPECTS OF MEMBRANE CHEMISTRY

TOPICS IN INCLUSION SCIENCE

VOLUME 2

The titles published in this series are listed at the end of this volume.

INCLUSION ASPECTS OF MEMBRANE CHEMISTRY

Edited by

T. OSA
Tohoku University, Sendai, Japan

and

J. L. ATWOOD
University of Alabama, U.S.A.

KLUWER ACADEMIC PUBLISHERS
DORDRECHT / BOSTON / LONDON

ISBN 0-7923-1123-X

Published by Kluwer Academic Publishers,
P.O. Box 17, 3300 AA Dordrecht, The Netherlands.

Kluwer Academic Publishers incorporates
the publishing programmes of
D. Reidel, Martinus Nijhoff, Dr W. Junk and MTP Press.

Sold and distributed in the U.S.A. and Canada
by Kluwer Academic Publishers,
101 Philip Drive, Norwell, MA 02061, U.S.A.

In all other countries, sold and distributed
by Kluwer Academic Publishers Group,
P.O. Box 322, 3300 AH Dordrecht, The Netherlands.

Printed on acid-free paper

Printed in the Netherlands

TABLE OF CONTENTS

Ion Extraction and Transport by Proton-Ionizable Crown Ethers

Philip R. Brown and Richard A. Bartsch

Department of Chemistry and Biochemistry, Texas Tech University
Lubbock, TX 79409-4260

Table of Contents

1. Introduction.

Interest in macrocyclic polyethers as complexing agents has grown exponentially since Pedersen's original description [1] of the ability of these compounds to lipophilize metal and ammonium cations. Such ligands have been used to complex about half the metals on the periodic table [2], to act as selective agents for metal extractions [3], to serve as phase transfer catalysts [4], to carry metal ions across hydrophobic barriers [5], and to emulate biological binding sites [6]. In view of their widespread applications, macrocycles have

T. Osa and J. L. Atwood (eds.), Inclusion Aspects of Membrane Chemistry, 1–57.

been synthesized which provide a variety of ligand donor atoms for the ionophoric cavity [2]. Many structural variations have also appeared, so that macrocycles, macropolycycles (cryptands) [7], speleands [8], spherands [9], and torands [10] are frequently mentioned in the chemical literature. These variations on the theme of lipophilic host molecules form the basis for an ever-widening circle of interest in host-guest chemistry and separation science using synthetically prepared ionophoric molecules.

The use of macrocycles as extraction agents has also led to the development of molecules in which the lipophilicity has been increased by the addition of aryl or alkyl substituents to the outer surfaces of the macrocyclic ring structures [1,11]. A natural extension of such structural modification was the addition of extra ligand sites in side chains [11] in an effort to increase the stability and selectivity of complexes formed by these molecules. A further outgrowth was the addition of proton-ionizable moieties [12] in the side chains to increase their complexing abilities by providing the requisite anion(s) for complexation.

The volume of macrocyclic polyether research conducted during the past two decades has been so great that several substantive reviews on macrocycles [2], cryptands [7] and lariat ethers [11] are available. These reviews cover topics as widely varied as synthesis [13], complexation chemistry [14], extraction [15] and transport behavior [5, 16, 17], analytical applications [18, 19] and complexation with neutral guest molecules [20]. Only in a very limited number of these compilations [2, 5, 7, 18, 19] are proton-ionizable crown ethers identified as an important class of macrocyclic polyethers. Since this class of compounds is now quite extensive and has unique advantages for applications in separation science, this review will concentrate on the synthesis, complexation behavior, use in solvent extraction, and utilization in membrane transport of proton-ionizable macrocyclic polyethers. The constraints of time and space dictate a restriction to only those compounds which have oxygen atoms as the most numerous electron donor atoms in the ring structure of the macrocycle. Thus, this review will not mention the extensive research being conducted with cyclams, lactams, porphyrins, and thiacrowns. Even within this restricted field, there is a wealth of valuable and interesting chemistry to be understood and applied to practical separation problems.

2. Syntheses of Proton-Ionizable Macrocyclic Compounds.

The publication frequency of papers dealing with the synthesis of crown ether compounds which contain ionizable moieties has been steadily increasing since the first such paper appeared in 1973 (Figure 1).

The topography of a molecule synthesized by a particular

researcher is usually designed in direct response to a particular interest of that worker. This is particularly evident among the ligands designed with pendant side arms. Two major motivations for molecular design are the construction of ionophores which will; (1) mimic the catalytic or ion-specific sites of a biological molecule; and, (2) aid in the selective separation of metal ions from aqueous solutions.

A review of the literature reveals that three primary methods for incorporation of proton-ionizable moieties in macrocycles have evolved. These are:

(1) attachment of a pendant side arm which includes or is terminated by an ionizable entity [7,12,21-66];

(2) inclusion within the crown ether framework of a subunit which projects an ionizable entity, so that both the proton and the atom which bears it are within the cavity [67-75]; and

(3) inclusion, as part of the crown ether framework, of a unit which places a heteroatom in a ligating position, with an attached proton projecting into the ring cavity [76-84].

The macrocyclic compounds in these three classes will be designated as macrocycles with ionizable pendant groups, ionizable intra-annular groups and ionizable ring atoms, respectively, throughout this chapter.

Figure 1: A histogram of publication frequency for journal articles dealing with proton-ionizable crown ether synthesis.

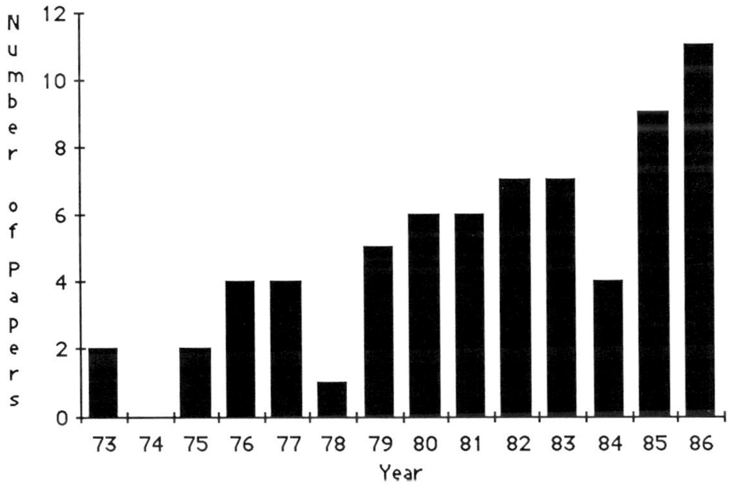

2.1 Ionizable Pendant Groups.

Side arms have been attached to the perimeters of macrocyclic ring structures by substitution on the aryl groups of benzo (**I -XVIII**), biphenyl (**XIX - XXVII**) and binaphthyl (**XXVIII - XXXIII**) crown ethers, by the addition of substituents directly to carbon atoms of the macrocyclic framework (**XXXIV - CXLVII**), and by substitution at the nitrogen atom of a monoazacrown (**CXLVIII - CLVIII**) or both nitrogen atoms of a diazacrown (**CLIX-CLXXVIII**). The resultant side arms include ionizable sites either to provide a specific binding site for biomimetic experiments, or to eliminate the need for inorganic anions to accompany extracted cations during solvent extraction and membrane transport procedures.

When crown ethers with pendant side arms are considered, the intended application of a compound strongly influences the type of molecule that is constructed. Biomimetic molecules have side arms which, by virtue of either the length or the structure of the side arm (e.g.,**XXVIII-XXXIII,CXXXIII, CXXXIV**), maintain relatively fixed steric locations in relation to the macrocyclic cavity. This is desirable in hosts which are designed to produce high chiral or structural recognition of proposed guest molecules. The success of the technique is illustrated by the ability of Cram and co-workers to separate racemates with suitable macrocyclic hosts [85].

In contrast with these more rigid structures, molecules designed for metal ion separations tend to have side arms which are more flexible, to allow for arrangement of the molecule so that the charged unit of the ionized moiety can more easily approach a desirable ion-pairing position in relation to the cation sequestered within the cavity (e.g., **XXXIV-LVIII**). Other variations of side arms designed for metal ion complexation include the addition of chromophores for the colorimetric detection of complexed metals (e.g.,**LXII, LXIII, LXXIV–LXXXIII**) [42, 57], and the addition of photo-responsive bonds [39] (**I-III**) which allow the use of light-switching to control the efficiency and selectivity of metal ion solvent extraction and liquid membrane transport.

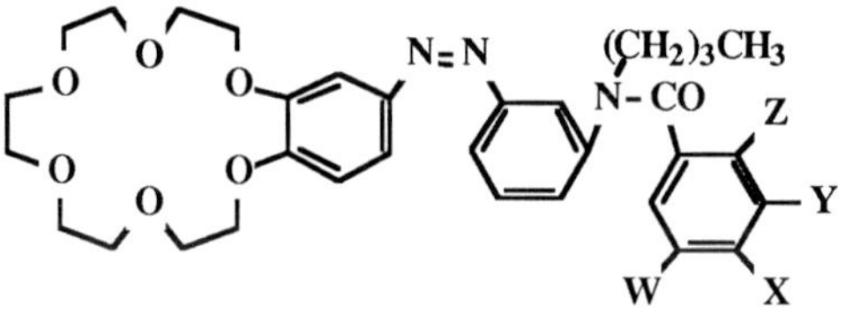

Cmpd	Ref	W	X	Y	Z
I	[39]	H	H	OH	H
II	[39]	H	OH	H	H
III	[39]	OH	H	H	NO_2

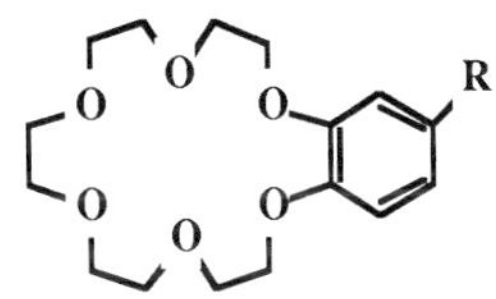

Cmpd	Ref	R
IV	[31]	N=N-2-hydroxy-1-naphthyl
V	[31]	N=N-4-hydroxyphenyl
VI	[31]	N=N-3-t-butyl-4-hydroxyphenyl
VII	[31]	N=N-3-chloro-4-hydroxyphenyl
VIII	[22]	$CONH(CH_2)_5CO_2H$
IX	[31]	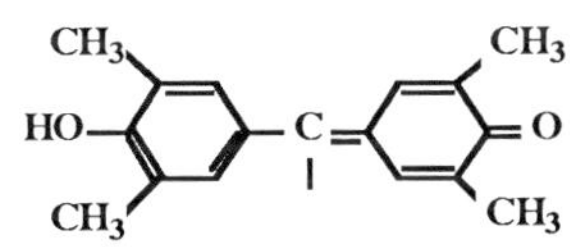

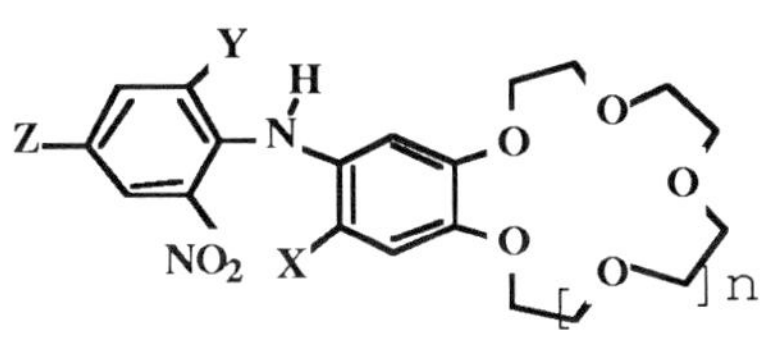

Cmpd	Ref	n	X	Y	Z
X	[31]	2	H	NO_2	NO_2
XI	[31]	2	NO_2	NO_2	NO_2
XII	[25]	1	H	NO_2	NO_2
XIII	[25]	1	Br	NO_2	NO_2
XIV	[25]	1	NO_2	NO_2	NO_2
XV	[32]	1	H	NO_2	CN
XVI	[34]	1	H	NO_2	CF_3
XVII	[34]	1	H	CF_3	NO_2

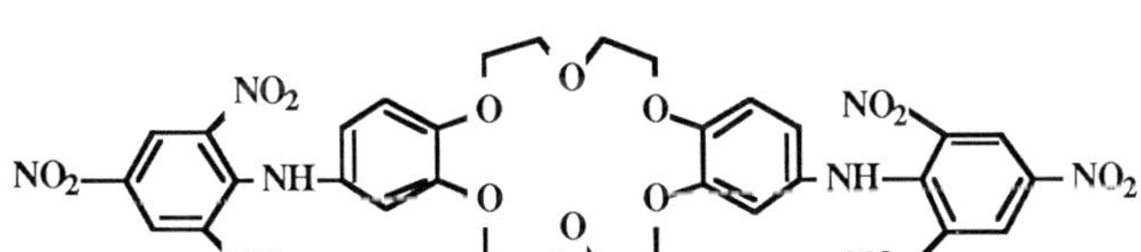

Cmpd	Ref
XVIII	[31]

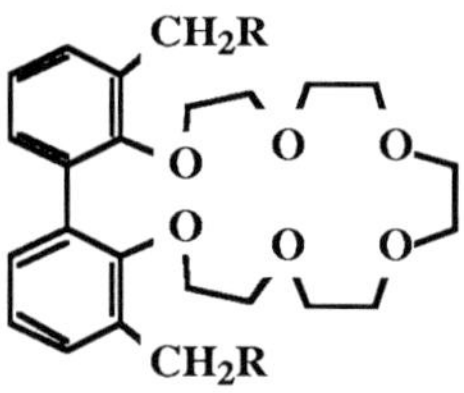

Cmpd	Ref	R	Cmpd	Ref	R
XIX	[36]	SCH_2CO_2H	**XXI**	[36]	SO_3M
XX	[36]	OSO_3M	**XXII**	[36]	$PO(OH)_2$

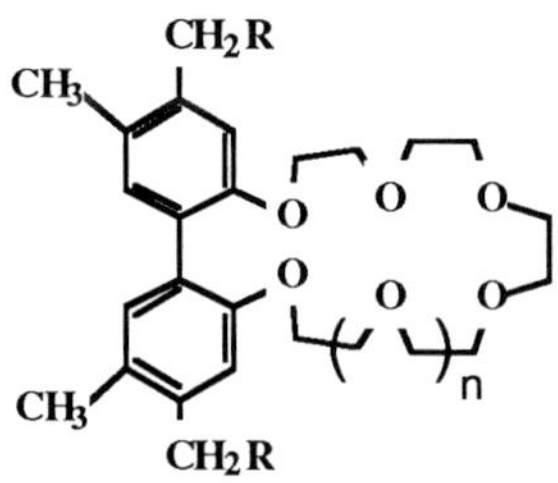

Cmpd	Ref	n	R	Cmpd	Ref	n	R
XXIII	[36]	1	SCH_2CO_2H	**XXVI**	[36]	2	$S(CH_2)_3CO_2H$
XXIV	[36]	1	$S(CH_2)_2CO_2H$	**XXVII**	[36]	1	$O(CH_2)_2OSO_3M$
XXV	[36]	2	$S(CH_2)_2CO_2H$				

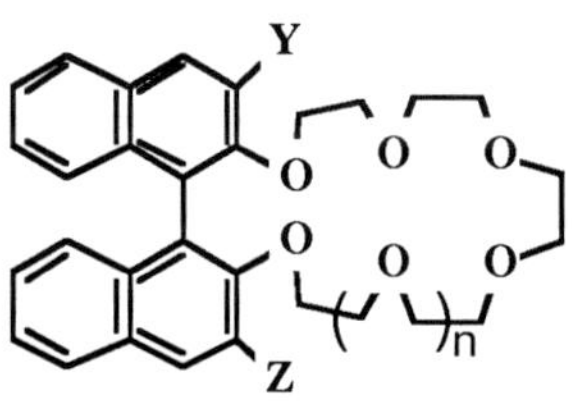

Cmpd	Ref	n	Y	Z
XXVIII	[12]	1	H	$CH_2OCH_2CO_2H$
XXIX	[12]	1	$CH_2OCH_2CO_2H$	$CH_2OCH_2CO_2H$
XXX	[12]	0	$CH_2OCH_2CO_2H$	$CH_2OCH_2CO_2H$
XXXI	[12]	2	$CH_2OCH_2CO_2H$	$CH_2OCH_2CO_2H$
XXXII	[24]	1	CO_2H	CO_2H
XXXIII	[24]	2	$PO(OEt)_2$	$PO(OEt)_2$

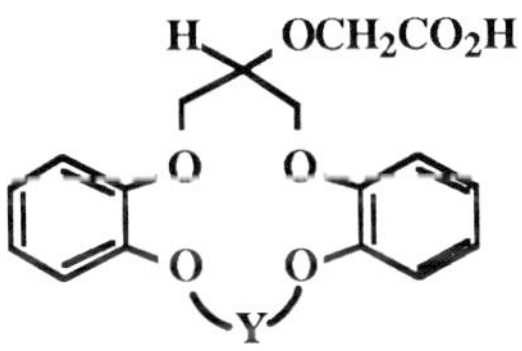

Cmpd	Ref	Y	Cmpd	Ref	Y
XXXIV	[38]	CH_2CH_2	**XXXVI**	[38]	$CH_2CH_2OCH_2CH_2$
XXXV	[38]	$CH_2CH_2CH_2$	**XXXVII**	[38]	$CH_2(CH_2OCH_2)_2CH_2$

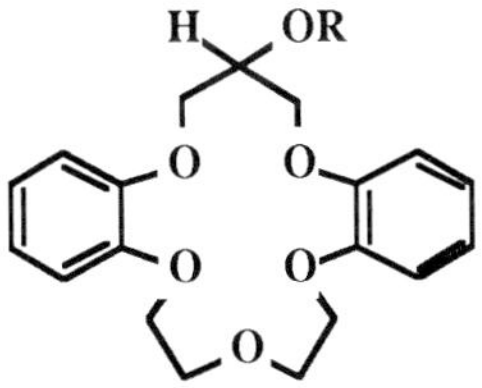

Cmpd	Ref	R	Cmpd	Ref	R
XXXVIII	[38]	$CH(C_2H_5)CO_2H$	**XLII**	[38]	$(CH_2)_2CO_2H$
XXXIX	[38]	$CH(C_4H_9)CO_2H$	**XLIII**	[38]	$(CH_2)_4CO_2H$
XL	[38]	$CH(C_6H_{13})CO_2H$	**XLIV**	[61]	$(CH_2)_3SO_3H$
XLI	[38]	$CH(C_8H_{17})CO_2H$	**XLV**	[61]	$(CH_2)PO(OH)(OEt)$

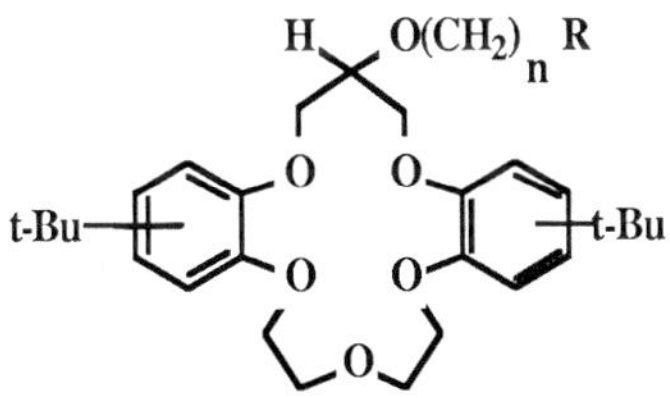

Cmpd	Ref	n	R
XLVI	[62]	1	PO(OH)(OEt)
XLVII	[62]	2	PO(OH)(OEt)
XLVIII	[62]	3	PO(OH)(OEt)
IL	[62]	4	PO(OH)(OEt)
L	[40]	1	CO_2H

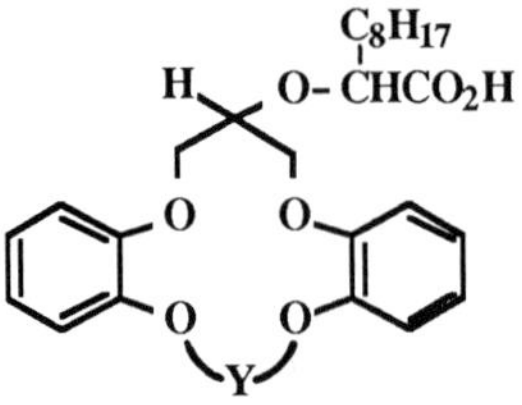

Cmpd	Ref	Y
LI	[46]	$CH_2CH_2CH_2$
LII	[46]	$CH_2(CH_2OCH_2)_2CH_2$
LIII	[46]	$CH_2(CH_2OCH_2)_3CH_2$

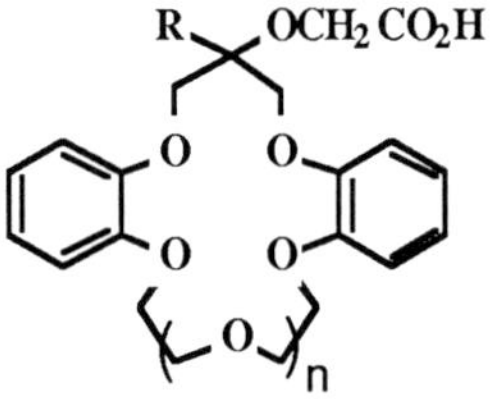

Cmpd	Ref	n	R
LIV	[46]	1	C_4H_9
LV	[46]	1	C_8H_{17}
LVI	[46]	1	$C_{14}H_{29}$
LVII	[46]	2	C_8H_{17}

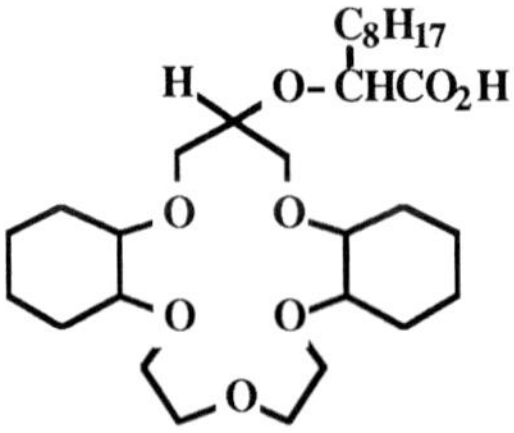

Cmpd	Ref
LVIII	[46]

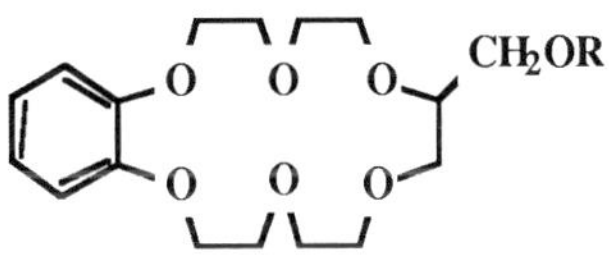

Cmpd	Ref	R
LIX	[48]	$-\underset{}{CH}(C_8H_{17})CO_2H$
LX	[48]	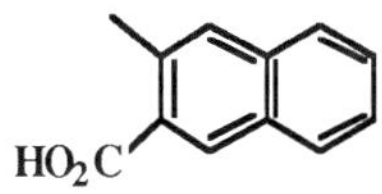

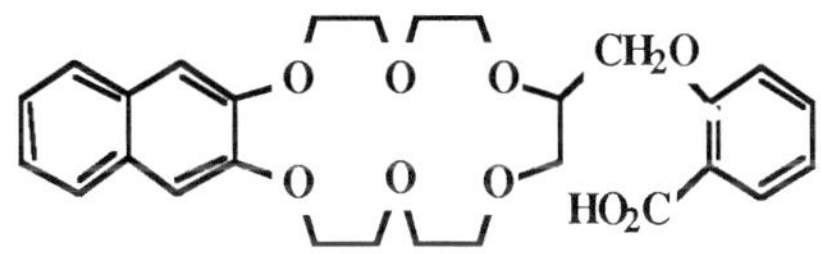

Cmpd	Ref
LXI	[47]

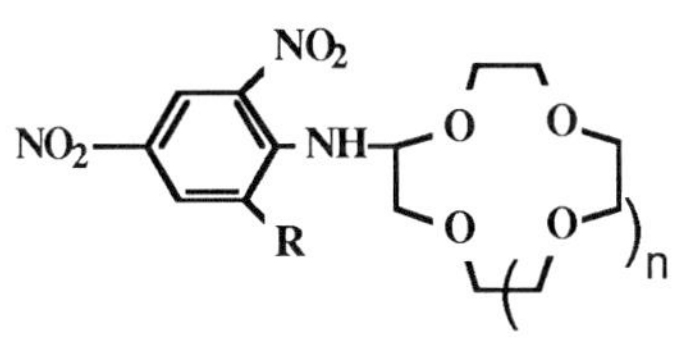

Cmpd	Ref	n	R
LXII	[54]	1	CF_3
LXIII	[54]	2	NO_2

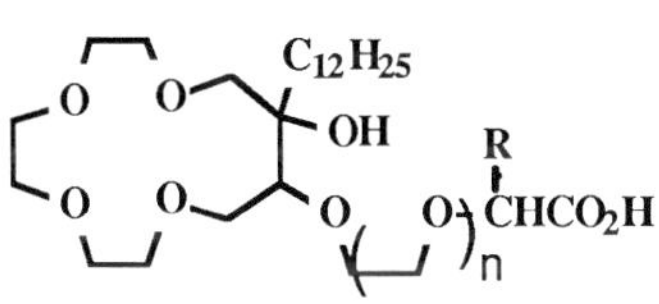

Cmpd	Ref	n	R
LXIV	[55]	0	H
LXV	[55]	1	H
LXVI	[55]	2	H
LXVII	[55]	3	H
LXVIII	[55]	1	C_6H_{13}
LXIX	[55]	2	C_6H_{13}
LXX	[55]	3	C_6H_{13}

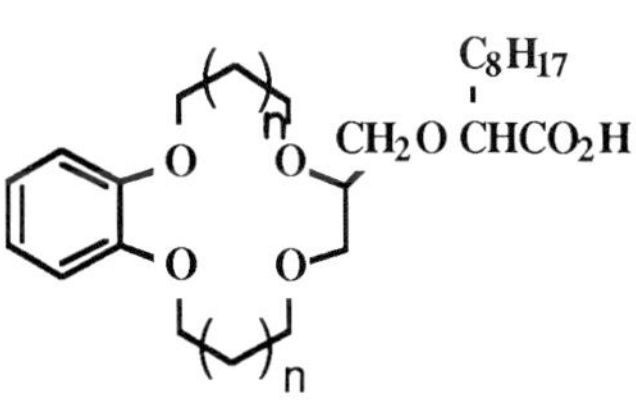

Cmpd	Ref	n
LXXI	[60]	0
LXXII	[60]	1

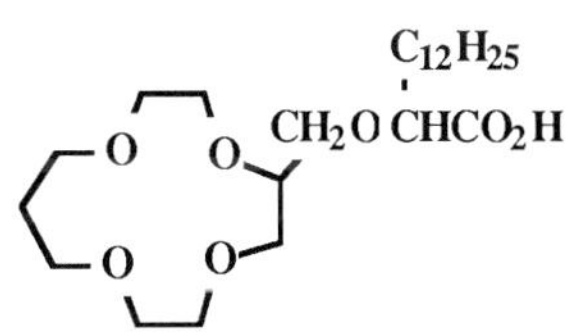

Cmpd	Ref
LXXIII	[60]

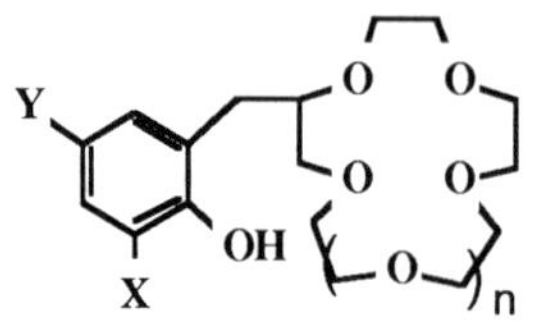

Cmpd	Ref	n	X	Y
LXXIV	[57]	1	NO_2	NO_2
LXXV	[57]	2	NO_2	NO_2
LXXVI	[57]	1	H	p-nitrophenylazo
LXXVII	[57]	2	H	p-nitrophenylazo
LXXVIII	[57]	1	p-nitrophenylazo	p-nitrophenylazo

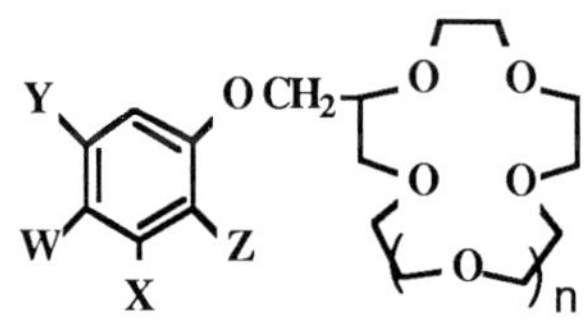

Cmpd	Ref	n	W	X	Y	Z
LXXIX	[57]	0	H	NO_2	NO_2	OH
LXXX	[42]	1	H	NO_2	NO_2	OH
LXXXI	[42]	2	H	NO_2	NO_2	OH
LXXXII	[42]	1	H	H	p-nitrophenylazo	OH
LXXXIII	[42]	2	H	H	p-nitrophenylazo	OH
LXXXIV	[51]	1	$C_{10}H_{21}$	H	H	PO(OH)(OEt)
LXXXV	[51]	2	$C_{10}H_{21}$	H	H	PO(OH)(OEt)
LXXXVI	[51]	3	$C_{10}H_{21}$	H	H	PO(OH)(OEt)
LXXXVII	[51]	4	$C_{10}H_{21}$	H	H	PO(OH)(OEt)
LXXXVIII	[59]	1	t-Bu	H	H	PO(OH)(OEt)

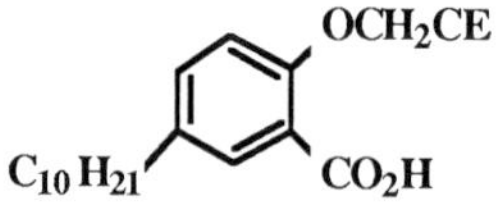

Cmpd	Ref	CE	Cmpd	Ref	CE
LXXXIX	[60]	12C4(2)*	**XCVI**	[60]	16C5(3)
XC	[60]	13C4(2)	**XCVII**	[60]	18C6
XCI	[60]	13C4(3)	**XCVIII**	[60]	19C6(2)
XCII	[60]	14C4(2)	**IC**	[60]	21C7
XCIII	[60]	14C4(3)	**C**	[60]	24C8
XCIV	[60]	15C4(3)	**CI**	[60]	27C9
XCV	[60]	15C5	**CII**	[60]	30C10

* (2) indicates the crown ether is attached to one of the carbon atoms in a two atom bridge; (3) indicates attachment is at the central carbon atom of a three carbon bridge.

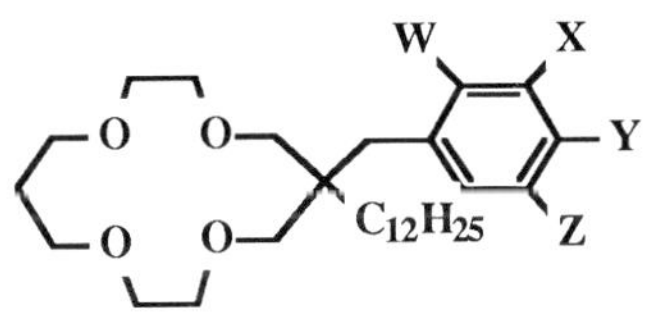

Cmpd	Ref	W	X	Y	Z
CIII	[56]	OH	H	H	H
CIV	[56]	OH	H	H	NO_2
CV	[56]	OH	NO_2	H	H
CVI	[56]	OH	NO_2	H	NO_2
CVII	[66]	H	NO_2	OH	NO_2
CVIII	[66]	OH	H	H	p-(N,N-dimethyl-sulfonamido)phenylazo
CIX	[66]	OH	H	H	p-nitrophenylazo
CX	[66]	OH	p-nitrophenylazo	H	H
CXI	[66]	OH	p-nitrophenylazo	H	p-nitrophenylazo
CXII	[66]	OH	NO_2	H	p-nitrophenylazo

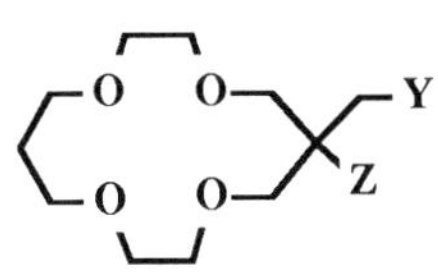

Cmpd	Ref	Y	Z
CXIII	[66]	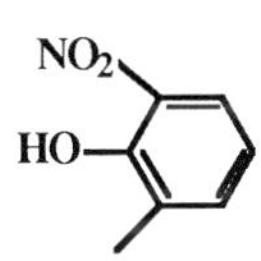	H
CXIV	[66]	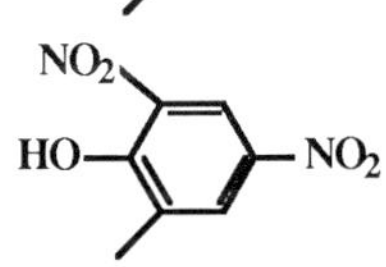	H
CXV	[66]	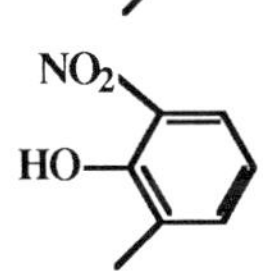	CH_3
CXVI	[66]	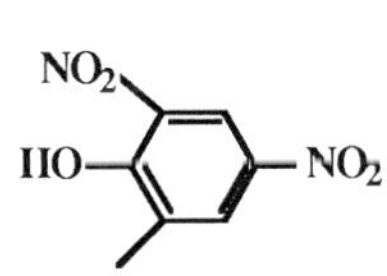	CH_3

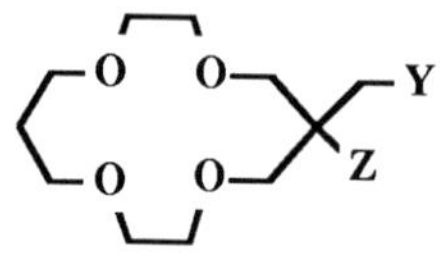

Cmpd	Ref	Y	Z
CXVII	[66]	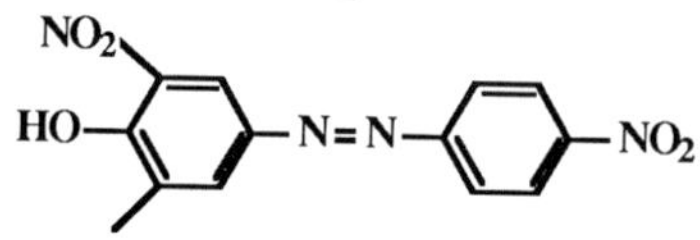	CH_3
CXVIII	[66]	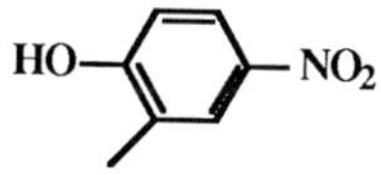	CH_2Ph
CXIX	[66]	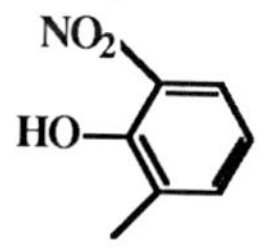	CH_2Ph
CXX	[66]	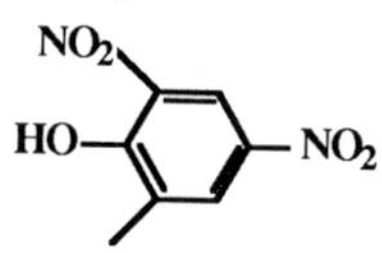	CH_2Ph
CXXI	[66]	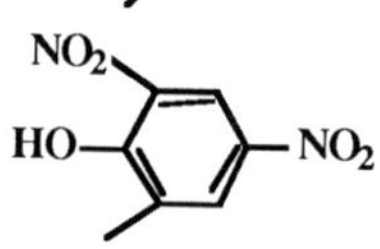	$CH_2C_6H_4$-p-NO_2

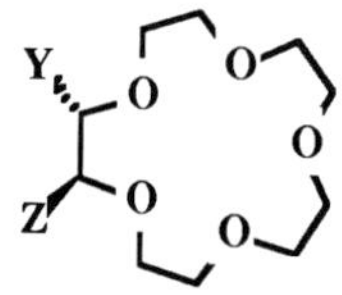

Cmpd	Ref	Y	Z
CXXII	[28]	CO_2H	CO_2H
CXXIII	[50]	CO_2H	$CON[(CH_2)_7CH_3]_2$
CXXIV	[29]	CO_2H	$CONH(CH_2)_7CH_3$
CXXV	[29]	CO_2H	$CONH(CH_2)_{13}CH_3$
CXXVI	[29]	CO_2H	$CONH(CH_2)_{17}CH_3$

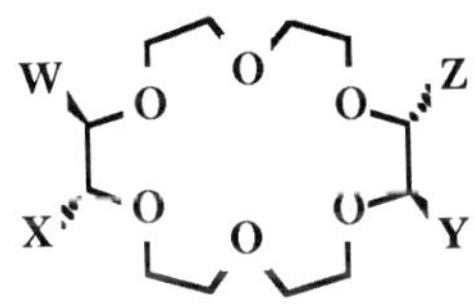

Cmpd	Ref	W	X	Y	Z
CXXVII	[119]	CO_2H	CO_2H	$C_{10}H_{21}$	$C_{10}H_{21}$
CXXVIII	[119]	CO_2H	$CONHC_8H_{17}$	CO_2H	$CONHC_8H_{17}$
CXXIX	[119]	CO_2H	$CONHC_8H_{17}$	$CONHC_8H_{17}$	CO_2H
CXXX	[21]	CO_2H	CO_2H	CO_2H	CO_2H
CXXXI	[28]	CH_2CO_2H	CH_2CO_2H	CH_2CO_2H	CH_2CO_2H
CXXXII	[23]	All 4; L-$CONHCH(CO_2N(CH_3)_4)CH_2$(3-indole)			
CXXXIII	[23]	All 4; L-$CONHCH_2CO_2(CH_3)_4$			
CXXXIV	[33]	CO_2H	CONHPh	CONHPh	CO_2H
CXXXV	[33]	CONHPh	CO_2H	CONHPh	CO_2H
CXXXVI	[43]	CO_2H	Nitroxide[a]	CO_2H	Nitroxide[a]
CXXXVII	[43]	CO_2H	Nitroxide[a]	Nitroxide[a]	CO_2H
CXXXVIII	[58]	Bridge[b]	CO_2H	Bridge[b]	CO_2H
CXXXIX	[58]	CO_2H	Bridge[b]	Bridge[b]	CO_2H
CXL	[29]	CO_2H	$CONH(CH_2)_4CH_3$	H	H
CXLI	[29]	CO_2H	$CONH(CH_2)_7CH_3$	H	H
CXLII	[29]	CO_2H	$CONH(CH_2)_{13}CH_3$	H	H
CXLIII	[29]	CO_2H	$CONH(CH_2)_{17}CH_3$	H	H

a. Nitroxide = 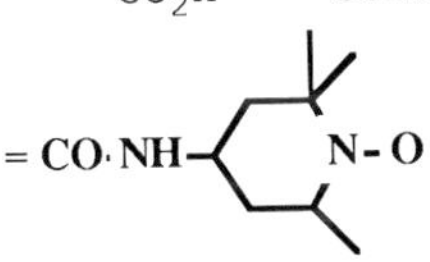

b. Bridge = 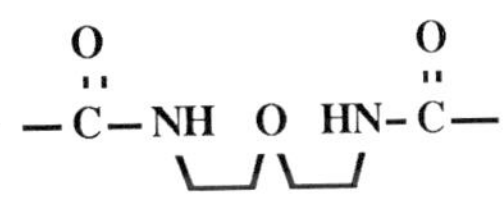

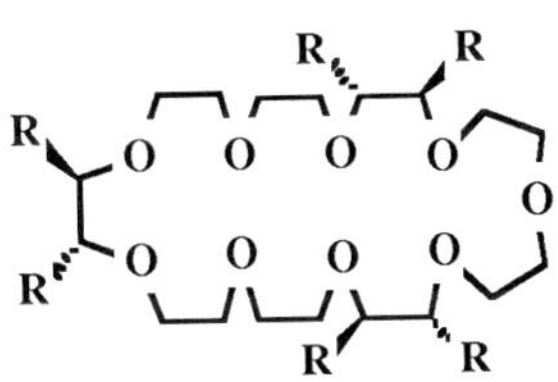

Cmpd	Ref	R
CXLIV	[28]	$CO_2NH(CH_2CH_2OH)_3$

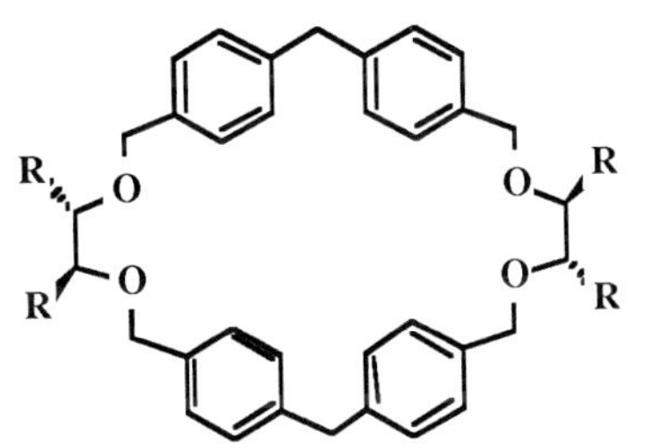

Cmpd	Ref	R
CXLV	[53]	CO_2H
CXLVI	[53]	$CONHCH_3$

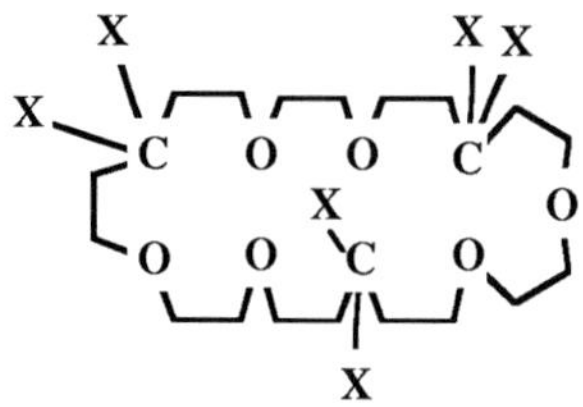

Cmpd	Ref	X
CXLVII	[30]	CO_2H

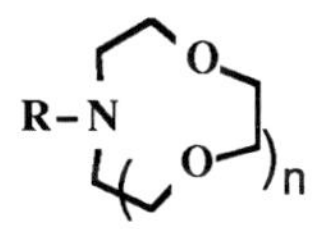

Cmpd	Ref	n	R
CXLVIII	[57]	1	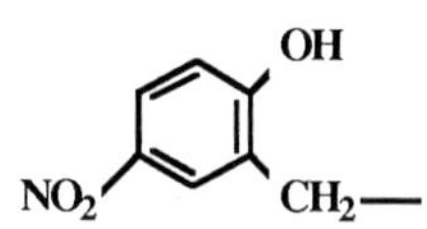
CIL	[57]	2	
CL	[57]	3	
CLI	[57]	4	
CLII	[57]	2	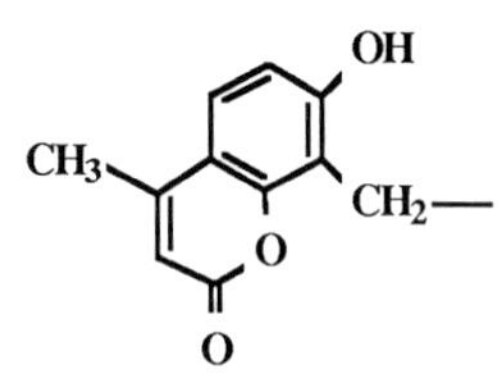
CLIII	[57]	3	
CLIV	[57]	4	
CLV	[57]	3	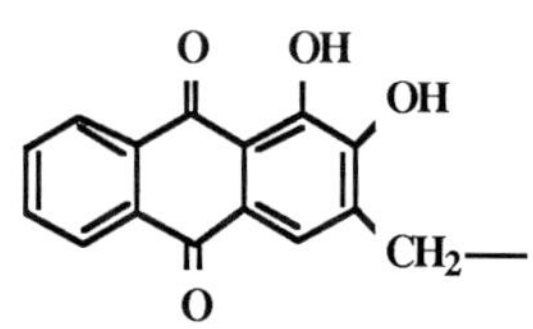
CLVI	[57]	4	

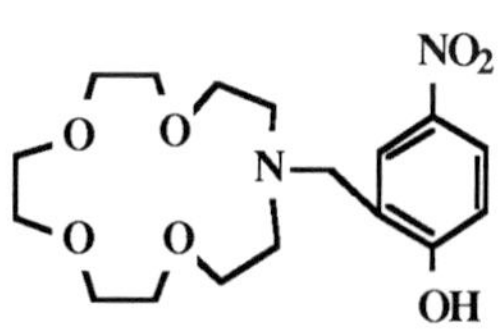

Cmpd	Ref
CLVII	[65]

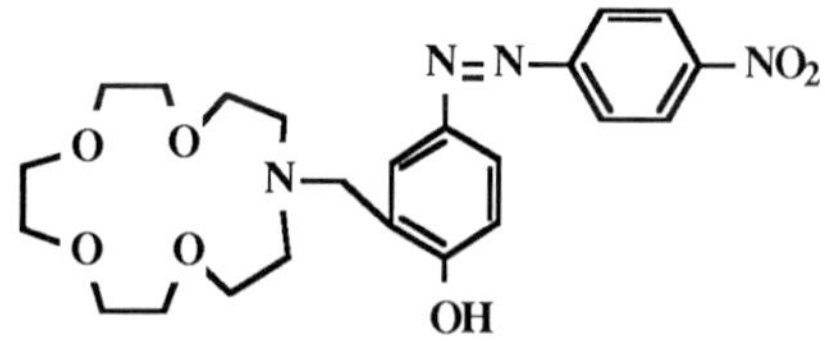

Cmpd	Ref
CLVIII	[65]

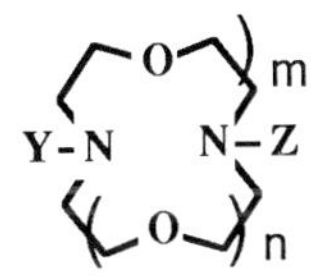

Cmpd	Ref	m	n	Y	Z
CLIX	[45]	2	2	CH_2CO_2H	$CO(CH_2)_{10}CH_3$
CLX	[26]	2	2	CH_2CONH_2	CH_2CONH_2
CLXI	[44]	1	1	CH_2CO_2H	CH_2CO_2H
CLXII	[44]	1	2	CH_2CO_2H	CH_2CO_2H
CLXIII	[26]	2	2	CH_2CO_2H	CH_2CO_2H
CLXIV	[44]	2	2	$CH(CH_3)CO_2H$	$CH(CH_3)CO_2H$
CLXV	[65]	2	1	Y = Z = $CH_2C_6H_4$-P-NH[2,4,6$(NO_2)_3C_6H_2$]	
CLXVI	[65]	2	2	Y = Z = $CH_2C_6H_4$-P-NH[2,4,6$(NO_2)_3C_6H_2$]	
CLXVII	[65]	2	2	Y = Z = 2-(4'-Nitrophenol)	

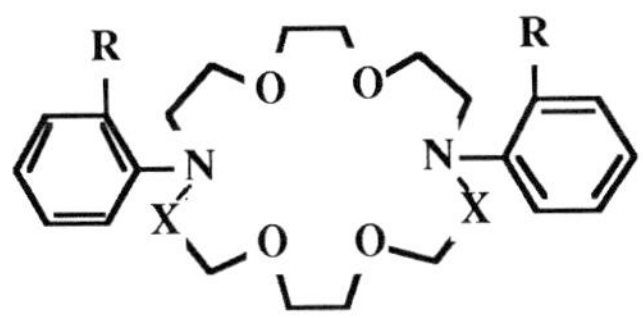

Cmpd	Ref	X	R	Cmpd	Ref	X	R
CLXVIII	[52]	CO	NHTs	**CLXX**	[52]	CH_2	NH_2
CLXIX	[52]	CH_2	NHTs	**CLXXI**	[52]	CH_2	$NHCOCH_2CH_2CO_2H$

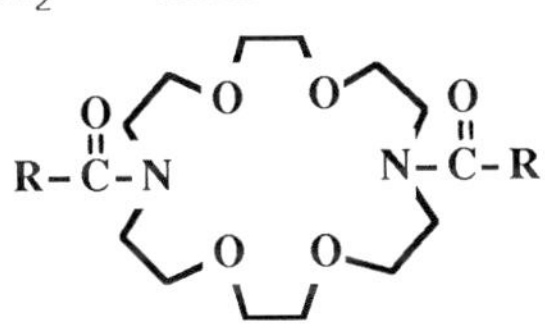

Cmpd	Ref	R
CLXXII	[31]	2-Phenol
CLXXIII	[31]	2-(p-Nitrophenol)
CLXXIV	[31]	$-(CH_2)_2$-2-Phenol
CLXXV	[31]	$-(CH_2)_2$-2-(p-Nitrophenol)
CLXXVI	[31]	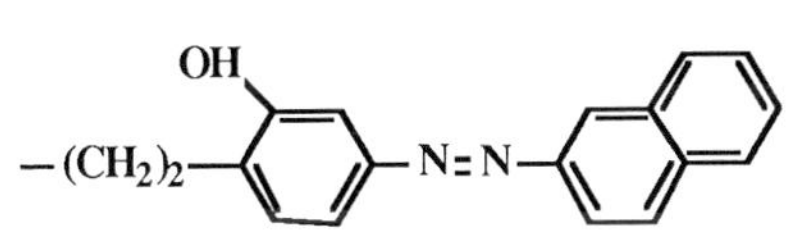

CLXVII [31]

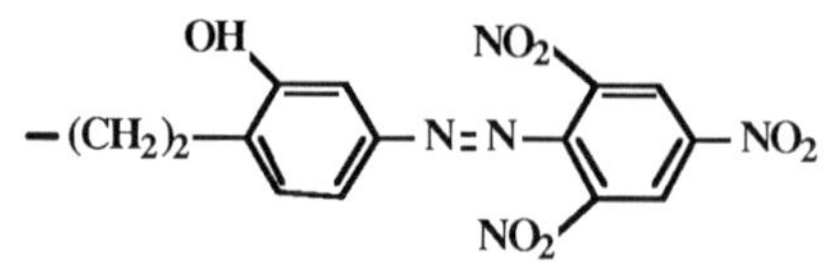

CLXVIII [31]

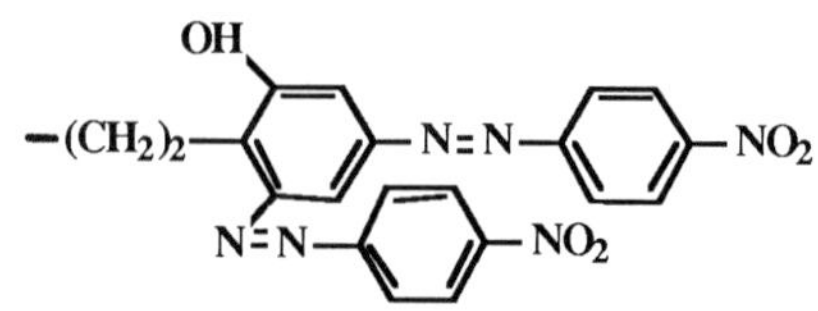

2.2 Ionizable Intra-annular Groups.

Several papers have appeared [67-74] which report the syntheses of macrocycles with built-in molecular subunits which project into the cavity of the macrocycle. Among such molecules, the only ionizable subunits used have been either phenolic -OH groups (**CLXXIX-CLXXXVIII** and **CXCIII-CXCV**) or -COOH groups on a benzene ring (**CLXXXIX-CXCII, CXCVI, CXCVII**). Variations in these derivatives center around the addition of substituents at the para position of the benzene ring which bears the ionizable unit, to make the pKa values for the ionization more appropriate for the separation scheme being pursued. A notable exception has been the synthesis [75] of a series of macrocyclic phosphate esters (**CXCVIII-CCIII**) in which the phosphorus atom forms part of the ring with an ionizable -OH group projecting into the ring cavity.

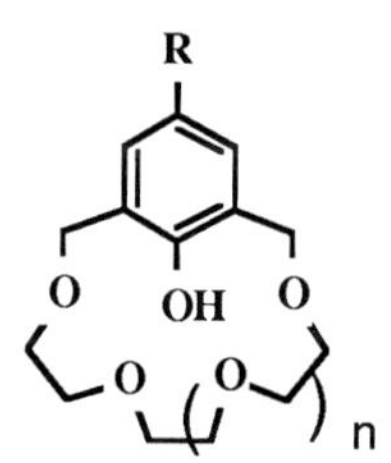

Cmpd	Ref	n	R	Cmpd	Ref	n	R
CLXXIX	[69]	1	H	**CLXXXIV**	[69]	2	NO_2
CLXXX	[69]	1	NO_2	**CLXXXV**	[71]	2	OH
CLXXXI	[71]	1	OH	**CLXXXVI**	[72]	2	AzoDNB[a]
CLXXXII	[72]	1	AzoDNB[a]	**CLXXXVII**	[74]	3	H
CLXXXIII	[69]	2	H	**CLXXXVIII**	[74]	3	NO_2

a. AzoDNB =

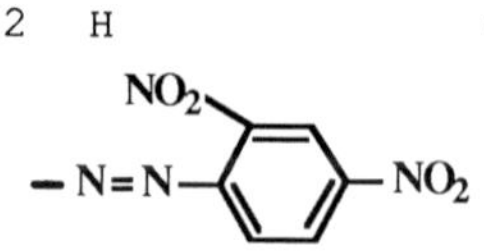

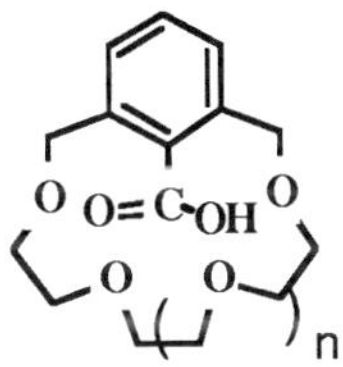

Cmpd	Ref	n	Cmpd	Ref	n
CLXXXIX	[67]	1	**CXCI**	[67]	3
CXC	[67]	2	**CXCII**	[67]	6

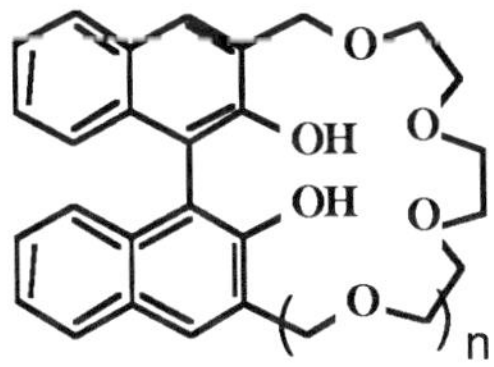

Cmpd	Ref	n
CXCIII	[68]	1
CXCIV	[68]	2
CXCV	[68]	3

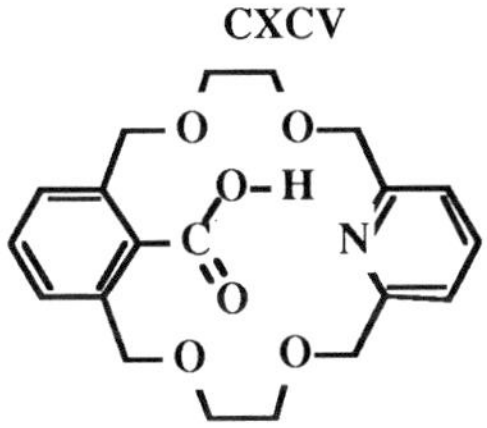

Cmpd	Ref
CXCVI	[73]

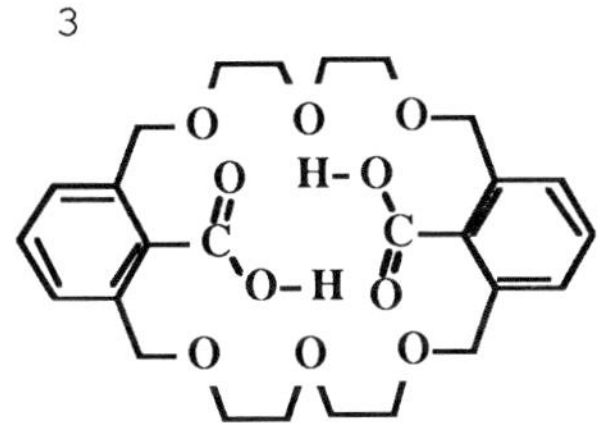

Cmpd	Ref
CXCVII	[73]

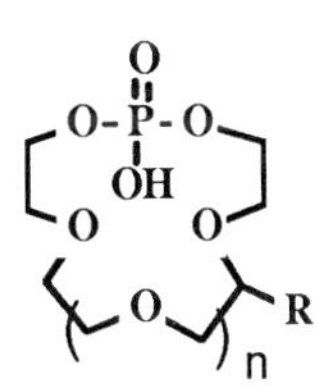

Cmpd	Ref	n	R	Cmpd	Ref	n	R
CXCVIII	[75]	1	H	**CCI**	[75]	2	C_8H_{17}
CIC	[75]	1	C_8H_{17}	**CCII**	[75]	2	$C_{10}H_{21}$
CC	[75]	2	H	**CCIII**	[75]	2	$C_{16}H_{33}$

2.3 Ionizable Ring Atoms.

The third method of providing a site that bears a removable proton has been to construct a macrocycle with a structural subunit which positions a proton-bearing atom at the location of one of the oxygen atoms in an equivalent polyether. To date all attempts to do this have made use of nitrogen atoms in hydroxypyridine or pyridone (**CCIV-CCXX**) [76, 78, 80, 82, 84], triazole (**CCXXI-CCXXXII**) [77, 81], or sulfonamide (**CCXXXIII-CCXLII**) [79, 83] subunits.

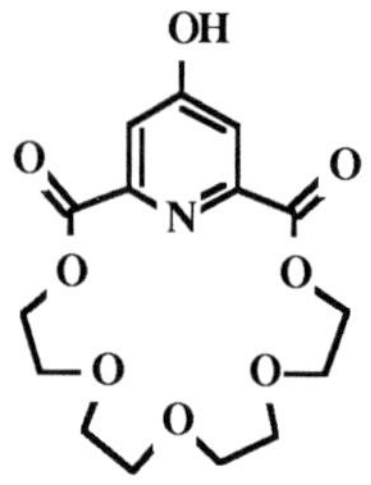

Cmpd	Ref
CCIV	[78]

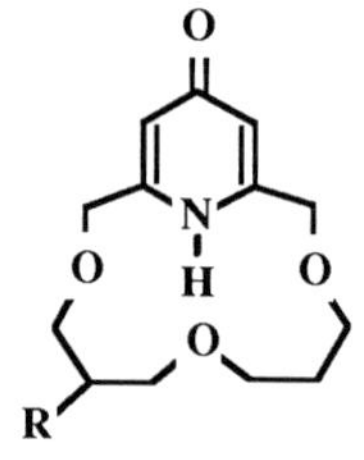

Cmpd	Ref	R
CCV	[84]	H
CCVI	[84]	C_6H_5
CCVII	[84]	C_8H_{15}

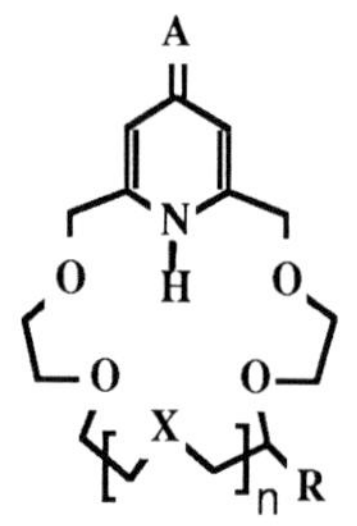

Cmpd	Ref	n	A	X	R
CCVIII	[80]	0	O	O	H
CCIX	[78]	1	O	O	H
CCX	[80]	2	O	O	H
CCXI	[82]	0	O	O	C_8H_{17}
CCXII	[80]	1	O	O	C_8H_{17}
CCXIII	[82]	2	O	O	C_8H_{17}
CCXIV	[82]	1	O	O	C_6H_5
CCXV	[80]	1	O	S	H
CCXVI	[82]	0	S	O	H
CCXVII	[82]	0	S	O	C_8H_{17}
CCXVIII	[82]	1	S	O	H
CCXIX	[82]	1	S	O	C_8H_{17}
CCXX	[82]	1	S	O	C_6H_5

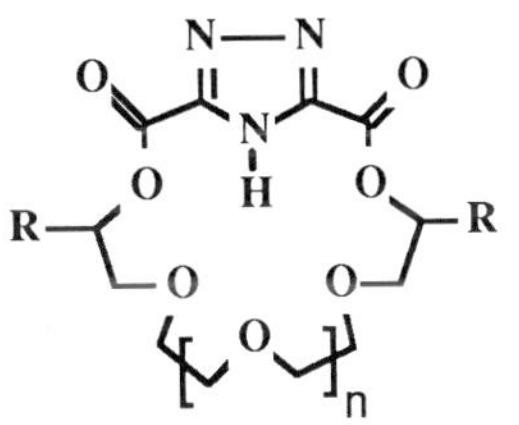

Cmpd	Ref	n	R
CCXXI	[77]	1	H
CCXXII	[77]	2	H
CCXXIII	[77]	1	CH_3 (S,S)

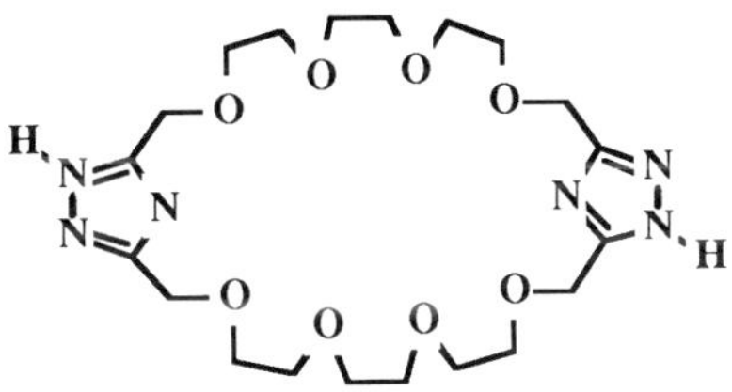

Cmpd	Ref
CCXXIV	[83]

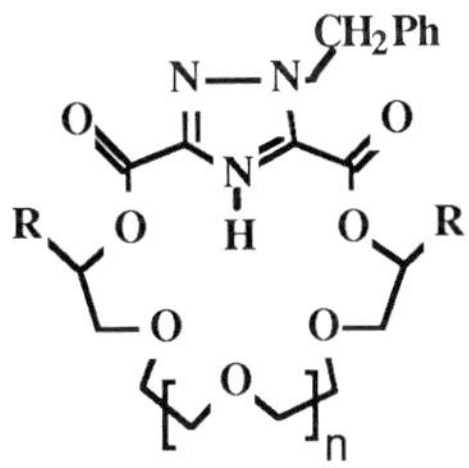

Cmpd	Ref	n	R
CCXXV	[77]	1	H
CCXXVI	[77]	2	H
CCXXVII	[77]	1	CH_3 (S,S)

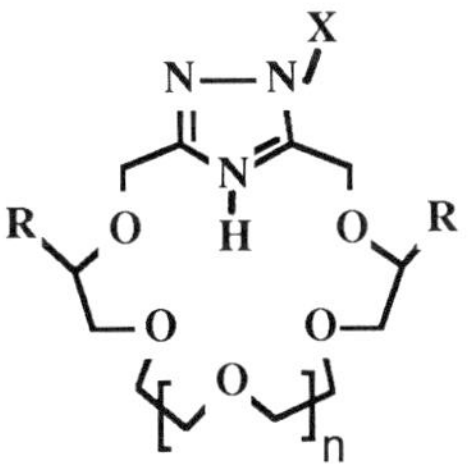

Cmpd	Ref	n	R	X
CCXXVIII	[81]	0	H	H
CCXXIX	[81]	1	H	H
CCXXX	[81]	2	H	H
CCXXXI	[81]	1	CH_3 (S,S)	H
CCXXXII	[81]	1	H	Allyl

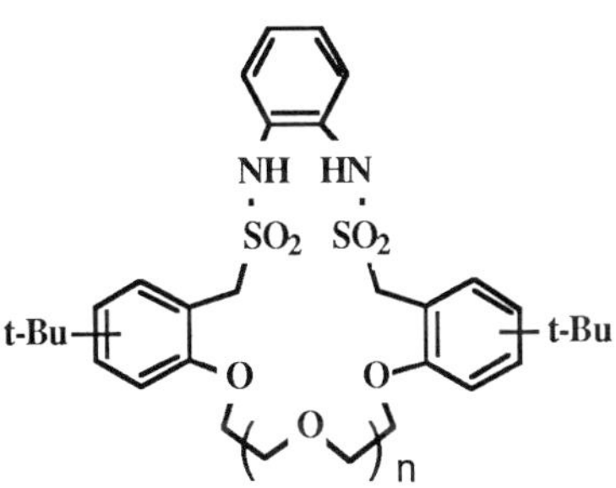

Cmpd	Ref	n
CCXXXIII	[79]	1
CCXXXIV	[79]	2

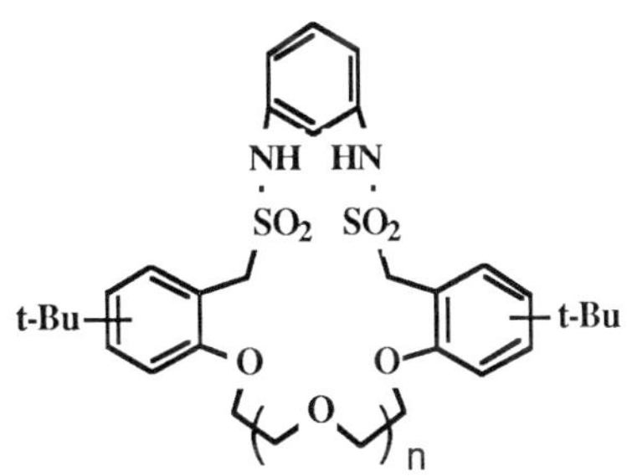

Cmpd	Ref	n
CCXXXV	[83]	1
CCXXXVI	[83]	2

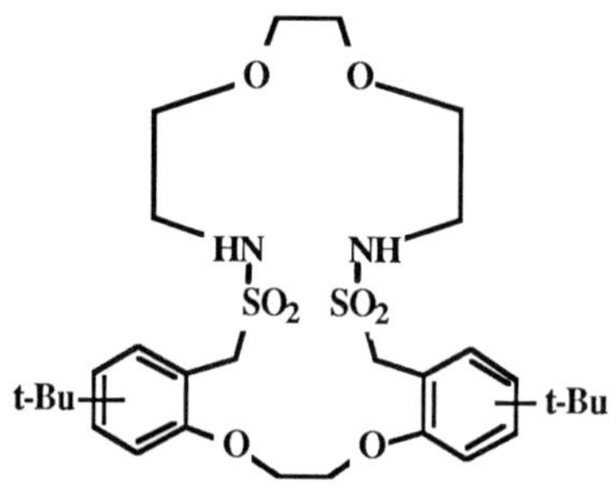

Cmpd	Ref
CCXXXVII	[83]

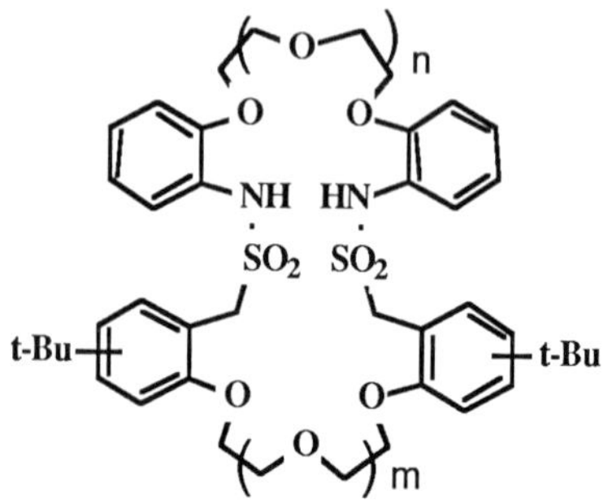

Cmpd	Ref	m	n
CCXXXVIII	[83]	0	0
CCXXXIX	[83]	1	0
CCXL	[83]	0	1
CCXLI	[83]	1	1

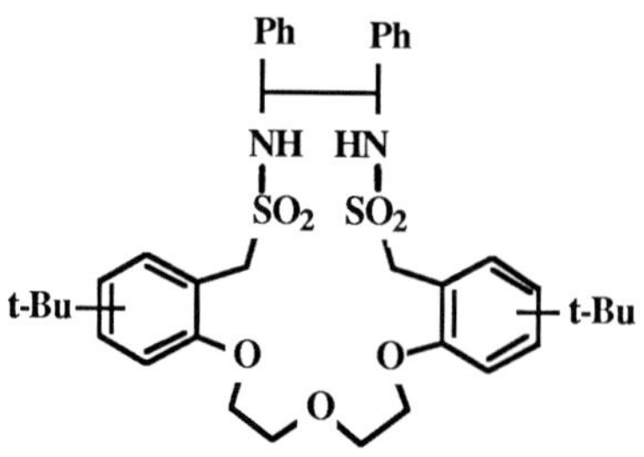

Cmpd	Ref
CCXLII	[83]

3. Complexation Behavior of Proton-Ionizable Macrocycles.

As was noted above, the motives for designing particular molecules have been varied. In like manner, studies of complexation behavior have also differed. Researchers interested in biomimetic chemistry have studied the complexation behavior of crown ethers with ammonium, alkylammonium, and amino acid compounds, usually with only a passing reference to metal ion complexation behavior. Those motivated by interest in metal ion separations report a thorough investigation of complexation behavior with a range of metal cations. It is encouraging to note that many of the molecules tested and reported are highly successful at meeting the needs for which they were designed, indicating that current understanding of macrocyclic chemistry is quite good.

3.1 Acid Dissociation Constants.

The advantage of proton-ionizable crown ethers for separation chemistry is the formation of electroneutral, lipophilic complexes by reaction between a cationic species of interest and the anionic form of the ligand. The degree of dissociation of the ionizable crown ether in a given medium has a direct bearing on the efficiency of the desired complexation reactions. It is therefore of fundamental interest to know the pKa values of proton-ionizable crown ethers as a function of substituent, ring size, proton location, and solvent. The pKa values for most of the proton-ionizable crown ethers studied have been determined in water or methanol-water solutions, although some have been determined in aqueous dioxane [42, 87]. Investigators who use these data for planning solvent extraction or membrane transport experiments should remember that complexation generally occurs in the organic-aqueous interfacial region so effective pKa values will be somewhat higher than those determined in water.

Several researchers have found that pKa values of proton-ionizable crown ethers decrease when a metal cation which will complex efficiently with the macrocycle is introduced into the solution. This results from competition between the metal ions and protons for the anionic form of the crown molecules. A complete reaction scheme has been proposed (Scheme I) [88] which describes all of the competitive reactions possible, regardless of set pH conditions, when both metal ions and an ionizable crown ether are present.

Scheme I. The ionic equilibria possible in a solution containing a proton-ionizable crown ether and competing metal ions.

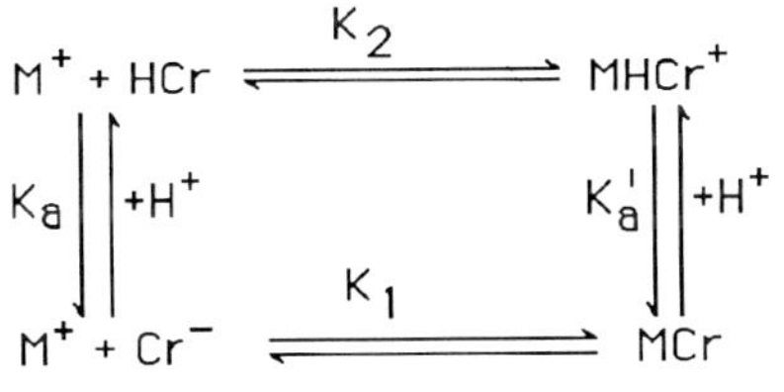

If solution conditions are adjusted so that the ionizable crown ether, HCr, is completely deprotonated, then only the equilibria characterized by K_a and K_1 in Scheme I apply. The addition of a second liquid phase, as in solvent extraction or liquid membrane transport experiments, further complicates the cycle because the phase transfers of the various crown ether species also must be considered. However, the basic cycle still applies.

Consideration of this cycle reveals that metal ion association with the crown ether will have pH-dependent free-energy considerations for the complexation reaction with the polyether binding site and for the Coulombic interaction with the anionic site of the ionized crown ether. In an interesting correlation based on this observation, Adamic et.al. [88] were able to estimate the increased stability of complexation due to the combination of both interactions in one molecule by use of Equation 1,

$$(\Delta G^{o}_{total})^2 = (\Delta G^{o}_{cavity})^2 + 2\lambda\Delta G^{o}_{cavity}\Delta G^{o}_{anion} + (\lambda\Delta G^{o}_{anion})^2 \quad (1)$$

where λ is a factor representing the amount of interaction between the cation and the anionic portion of the molecule ($\lambda = 0$ for no interaction, $\lambda = 1$ for maximum interaction). Thus, $(\Delta G^{o}cavity)^2$ is the contribution to the free energy of metal ion complexation with the unionized macrocycle, $(\lambda\Delta G^{o}anion)^2$ is the contribution of the purely Coulombic interaction between the cation and the anionic site, and $2\lambda\Delta G^{o}cavity\Delta G^{o}anion$ is the term which reflects the increased stability due to the inclusion of both polyether and anionic binding sites in one molecule. Data from several sources analyzed in Reference 88 show that a reasonably constant proportion (30-50%) of the total free energy change on complexation is due to this middle term in Equation 1, while the free energy changes due to the other two terms may vary much more significantly.

From further consideration of the details of Scheme I, it is seen that values of log K determined for the metal cation-ionizable crown ether interaction will be pH dependent. Thus, it is important to consider the pH conditions under which log K(association) data were determined when planning solution conditions for extraction or membrane experiments.

3.1.1 Ionizable Pendant Groups.

The proton-ionizable crown ethers with side arms that have been synthesized thus far have utilized largely carboxylic acid units (e.g., **XXIII- XXVI**) [12,22,23,26,28,30,33,36,38,40,43-48,50,53,55,58,60], phenolic OH groups (e.g., **LXXIV- LXXXIII**) [31,35,39,42,56,57,65,66] or amine nitrogens (e.g., **X-XVIII**) [25,31,32,34,52,54,65] to provide the acidic proton. A few have been synthesized with phosphonic acid functions (e.g., **XLV-IL**) [24,36,51,59,61,62] and others have a sulfonic acid group (**XXI, XLIV**) [36,61]. As a general rule, it appears that the pKa value for a crown ether derivative of this type is not appreciably different from that of the parent acid group, unless there is a significant degree of hydrogen-bonding possible between the ionizable proton and the oxygen atoms in the macrocyclic cavity. Thus for crown phenols **I**, **II**, and **III**, pKa values of 9.57, 9.06 and 6.33, respectively, are reported [39]. These values do not vary greatly from those for phenol (pKa = 9.89) and p-nitrophenol (pKa = 7.15).

Similarly, crown carboxylic acids **XXXIV - XXXVII** [91] have pKa values that are comparable with those of methoxy- and ethoxyacetic acid (Table 1), indicating that the proximity of the crown ether does not markedly affect the proton equilibrium.

Table 1: pKa Values for some crown ether carboxylic acids and model compounds.

Compound	pKa (H_2O)[a]
XXXIV	3.69
XXXV	4.36
XXXVI	4.59
XXXVII	3.89
methoxyacetic acid	3.57
ethoxyacetic acid	3.87

a. From Reference 91.

In contrast, crown carboxylic acids **XXXVIII - XLIII** show quite marked variations in pKa (Table 2). Compounds **XXXVIII - XLI** (Table 2) show enhanced pKa values which can be explained by considering that the long alkyl side chain forces the carboxylic group into a position

Table 2: pKa Values for dibenzo-16-crown-5 carboxylic acids substituted with various alkyl groups on the acid-bearing side arm.

Compound	pKa (H_2O)[a]
XXXVI	4.59
XXXVIII	5.30
XXXIX	5.90
XL	7.19
XLI	7.35
XLII	4.89
XLIII	6.41

a. From Reference 91.

favorable for the formation of intramolecular hydrogen bonds with the oxygens in the polyether ring. In a similar fashion, it is envisioned that the longer chains in compounds **XLII** and **XLIII** (Table 2) enable the carboxylate group to approach the polyether rings more closely than in compound **XXXVI**, thus enhancing the degree of intra-annular hydrogen bonding that occurs and raising the pKa value.

Nakamura, Takagi and Ueno [57] have synthesized several phenolic crown ether compounds among which pairs of molecules can be identified that differ only in the presence or absence of an oxygen atom in the side arm (e.g., **LXXIV** and **LXXX**). The pKa values for such pairs of crown phenols are compared in Table 3. The inclusion of the ether linkage causes a decrease of about 0.5 units in the pKa. The change in pKa is apparently due to a change in electron density at the acidic proton, caused by the inclusion of the oxygen atom. A similar decrease in pKa is observed when comparing 2-methylphenol (pKa = 10.31) with 2-methoxyphenol (pKa = 9.99).

Of particular interest for metals separations are the pKa variations caused by changing the identity of the ionizable group on a molecule. The molecule sym-(dibenzo-16-crown-5)-propanesulfonic acid

Table 3: pKa Values for pairs of similar phenolic crown ethers.

with O	pKa	without O	pKa
LXXX	3.16	**LXXIV**	3.97
LXXXI	3.27	**LXXV**	3.92
LXXXII	7.51	**LXXVI**	7.97
LXXXIII	7.54	**LXXVII**	8.11

a. From Reference 57.

(**XLIV**) has a pKa value in 80% methanol/water mixtures of 2.6 [61], while a similar crown with a phosphonic acid monoethyl ester (**XLV**) moiety has a pKa of 3.57 [61] in the same solvent. In contrast, crown carboxylic acid molecules **XXXVI** and **XLII** have pKa values in 80% methanol/water mixtures of 5.90 and 6.72 [61], respectively. The crown phosphonic acid monoethyl ester has been found to transport metal ions rapidly across a chloroform membrane even with a source phase pH ("basic" side, see following discussion on membranes) as low as pH 6 [124].

A further note of interest is raised in connection with azacrown ethers such as compounds **CXLVIII-CLVIII** in which ionizable side-arms have been attached at the nitrogen atom. All such nitrogen containing crowns have two pKa values, one for the protonation of the nitrogen to form a cationic species, and the other for the deprotonation of the acid group to form the anionic species. Several researchers have found [45,57,94] that such molecules can exist in the zwitterionic form. Compound **CLIX** has been shown to be successful in membrane transport experiments as the zwitterion [45], when K^+ ion was transported best at pH=5.9, which is a pH level commonly found in industrial applications.

3.1.2 Ionizable Intra-annular Groups.

A number of crown ethers have been synthesized which have phenolic (**CLXXIX-CLXXXVIII, CXCIII-CXCV**) [68,69,71,72,74] or benzoic acid (**CLXXXIX- CXCII**) [67,70,73] subunits included in the macrocyclic ring framework, so that the ionizable moiety is projected into the cavity of the macrocycle. These crowns offer maximum ability to form electrostatic ion pairs during the complexation of a desired cationic species. It seems reasonable, however, to expect that hydrogen bond interactions within the cavity might affect the acidity of the molecules. Examination of the pKa values for compounds **CLXXXIX-CXCII**

Table 4: pKa Values for compounds which have a COOH unit projecting into the macrocyclic cavity .

Compound	pKa (22°C) [a]
CLXXXIX	4.8
CXC	4.8
CXCI	3.8
CXCII	3.4
1,3-di(methoxymethyl)benzoic acid	3.3

a. From Reference 64.

(Table 4) [67] shows a ring-size effect on pKa compared with that of the model compound 1,3-di(methoxymethyl)benzoic acid when the ionizable group is the relatively bulky COOH group. However, pKa values for compounds **CLXXIX**, **CLXXX**, **CLXXXIII- CLXXXV** and **CLXXXVIII** (Table 5) [74] show very little hydrogen bond interaction for the smaller phenolic group.

The significance and extent of intra-annular hydrogen bonding in crown carboxylic acids was probed by Cram and co-workers [73]. Several compounds were synthesized and studied to determine structural effects on pKa values. Evidence gathered during the study indicates that **CXC** has a hydrogen bond to the oxygen atom of the macrocyclic ring directly

Table 5: pKa values for macrocyclic compounds with a phenolic OH projecting into the cavity.

Compound	pKa (20°C) [a]
CLXXIX	10.8
CLXXXIII	10.6
CLXXXVII	10.5
CLXXX	6.8
CLXXXIV	6.6
CLXXXVIII	6.5
phenol	10.0
2,6-dimethylphenol	10.7
p-nitrophenol	7.2

a. From Reference 74.

opposite the carboxylic acid group. Compound **CXC** has a pKa value of 4.8, while the acyclic model 1,3-di(methoxymethyl)benzoic acid has a pKa value of 3.3. The pyridino-carboxylic acid analog **CXCVI** has a pKa of 5.8 and the existence of a hydrogen bond from the carboxylic acid to the nitrogen atom was established in an X-ray crystal structure. If the pyridine nitrogen of **CXCVI** is protonated, the pKa value of the carboxylic acid drops to 3.6, almost the same as that for the model benzoic acid, rather than that of **CXC**. This demonstrates the existence of a hydrogen bonding interaction in **CXC**. In the same study, it was shown that hydrogen bonding to the macrocyclic ring oxygens on either side of an intra-annular ionizable group does not readily occur since the pKa values of 1,3-pyridino-18-crown-6 and 1,3-di(methoxymethyl)pyridine are virtually identical, at 4.9 and 4.8, respectively.

3.1.3 Ionizable Ring Atoms.

As previously noted, all of the crown ethers in this class use nitrogen atoms as the ionizable site. The methods of placing the nitrogen atom in the ring framework include the incorporation of hydroxypyridine or pyridone rings (**CCIV-CCXX**) [76, 78, 80, 82, 84], triazole subunits (**CCXXI-CCXXXII**) [77, 81] or sulfonamide groups (**CCXXXIII-CCXLII**) [79, 83] as part of the macrocycic framework. Hydroxypyridine and pyridone crown ethers are effective complexing agents for alkylammonium and amino acid moieties. This should make these compounds of considerable interest for use in separations of such compounds and for biomimetic chemistry.

Bradshaw's group have synthesized [77,78,81] a number of nitrogen containing macrocycles which have retained the ability to form both an anionic and a cationic form, according to the reactions

$$H_2L^+ = H^+ + HL \quad (2)$$

$$HL = H^+ + L^- \quad (3)$$

Values for both pKa's have been determined for four of these compounds, two of which (**CCIV,CCIX**) contain 4-hydroxypyridine or pyridone subunits [78] and two (**CCXXI,CCXXIX**) that have triazole subunits [77, 81]. In all cases the pKa values for formation of the anionic form of the ligand are higher than 8 so that extraction and membrane transport procedures require feed solutions with high pH values.

The only other compounds with ionizable heteroatoms in the ring are the crown sulfonamides **CCXXXIII-CCXLII**. The pKa values for these compounds had not been determined at the time of this review.

3.2 Biomimetic Complexation Studies.

Pedersen [1] found that macrocyclic polyethers, particularly those with the 18-crown-6 cavity, were capable of complexing ammonium ions. Cram [6] pointed out that the addition of extra units could give such compounds more shape and additional binding sites. Since then a number of researchers have prepared host molecules which contain a combination of shaped and charged side units in efforts to duplicate the chiral and electrostatic recognition of guest molecules afforded in natural systems by enzymes and proteins. Consequently, several investigations of the complexation of bioorganic molecules or their mimics by both neutral and proton-ionizable crown ether molecules have been conducted.

The potential of ionizable crown ethers for complexation and separation of biomimetic molecules is perhaps best illustrated by the work of Cram and coworkers [85]. In this research, a chiral binaphthyl crown ether (**XXIX**) with two pendant carboxylic acid groups located over a 20-crown-6 cavity was used successfully to completely resolve a racemic mixture of R- and S-valine, an α-amino acid. Lehn and co-workers [23] have shown that a given macrocyclic host will have a range of stability constants for complexation of related, similar bioorganic molecules, or for alkyl substituent variation within an ammonium salt, in the same way that a given macrocycle has a range of association constant values for various metal cations. Thus it would appear feasible to design systems that would be highly selective for a given bioorganic molecular species.

Of equal interest are reports [95-97] of the use of chiral compounds derived from proton-ionizable crown ethers to provide highly selective sites for molecular catalysis. In this instance, crown carboxylic acids were modified to produce molecules which present a reactive substituent on the host macrocycle to a complexed guest molecule in such a fashion as to catalyse a transacylation or hydrogen transfer and thereby provide high chiral selectivity.

3.3 Metal Complexation Studies.

Metal cation complexation properties for most of the ionizable crown ethers have been studied to a greater or lesser extent. The following list of desirable structural characteristics has emerged.
(i) The macrocyclic cavity should be of an appropriate size to match the radius of the cation to be complexed.
(ii) The anionic moiety formed on the crown ether should be positioned in the molecular structure such that the charge center can approach a desirable ion-pairing distance from the sequestered cation.
(iii) The anionic charge on the deprotonated crown ether should match the cation charge to form the most stable 1:1 complex.

The extra structural features of this class of ligand necessitates additional consideration for the interpretation of complexation chemistry. There have been some unusual and unexpected observations. The first factor is that both the neutral and ionized forms of a given ligand are capable of forming complexes with the metal cation. This is neither unusual nor unexpected behavior, yet only a few papers [75,77,88,91] record complexation constants for both types of association. In such cases, the association constant is always lower for the neutral ligand than for the ionized form, suggesting that competition occurs between the proton and the metal ion for the ligand site. An extension of this phenomenon was noted by Fyles et.al. [119] for crown ethers which have two ionizable groups attached to a single polyether framework. It was found that alkaline earth cations form stable complexes with the crown diacids in which the ligand:metal: proton ratio is 1:1:1 if the carboxylic acid groups are on the same side of the framework or if the diacids extend over the same face of the cavity (**CXXVII,CXXVIII**). This suggests that divalent cations are complexed more strongly by electrostatic attraction to the anion than by interaction with the cavity oxygens. An isomer, diacid **CXXIX** in which the carboxylic groups are on opposite sides and opposite faces of the crown cavity, forms 1:1 ligand:metal complexes readily, presumably because the divalent cation can fit into the cavity and still coordinate strongly with both anionic sites.

Addition of ionizable groups at sites remote from the cavity adds the possibility of complexation equilibria involving factors other than simple macrocycle-cation interaction. Robison and Bartsch [59] used ^{1}H NMR to show that the dominant coordination site for **LXXXVIII** is the anionic charge center, rather than the macrocyclic cavity, if the cation:cavity size matching is not appropriate. While investigating this type of complexation behavior, Takagi and co-workers [65] performed an extensive analysis of structural effects on bonding. The authors proposed the existence of both "chelate" and "intramolecular ion-pair" complexes, with a range of structures from "loose", or solvent-separated, ion pairs to "tight", or contact, ion pairs.

In an interesting study using crowns **CLXVIII-CLXXV** with two ionizable side arms attached at the nitrogen atoms of a diazacrown, Sonveaux [52] suggested that a Cu(II) ion is able to complex in two different locations in the cavity, depending on whether one or two of the charged side arms are involved in coordination with the metal. If only one side arm is ionized, the Cu(II) ion coordinates close to the nitrogen atom on the same side of the cavity as the ionized side arm, thus being complexed off-center in the cavity. When the second side arm is ionized, it can also coordinate with the Cu(II) ion, thus forcing the cation into the center of the cavity. The two copper complexes are sufficiently different that a color change is associated with the variation in coordination.

Several ionizable crown ethers have been synthesized which have benzo-group subunits as part of the molecular framework. These aromatic rings are derivatized to project acid protons into the macrocyclic cavity. McKervey et.al. have shown [74] that the anions of phenolic compounds of this type participate in extensive ion-pairing with complexed cations, even at very low concentrations. It was also shown [74] that the exact location of the smaller cations in the cavity is affected by the strength of the electrostatic interaction within the ion pair. A corollary to this observation is the notion that remote substitution on the parent aromatic ring could have significant effects on the strength of the electrostatic interactions between the coordinating site in the cavity and the sequestered cation. In an attempt to study the extent of this effect, Cram and co-workers [99] investigated the effect of remote ring substituents on non-ionizable 1',3'-xylyl-18-crown-5 compounds. It was shown that such substitutions do have pronounced effects on the association constant for a cation which is appropriately sized to coordinate with all available oxygen atoms in the crown ether. Electron-withdrawing substituents caused a decrease in the complexation constant for the metal-macrocycle interaction.

An unusual protonation effect was observed during K^+ complexation by an 18-crown-6 macrocycle bearing a pyridone subunit (**CCIX**). In a crystal structure study, Bradshaw [80] observed that the proton which is normally attached to the nitrogen atom and found in the cavity of the uncomplexed ligand translocates across the heterocyclic ring and becomes attached to the oxygen atom at the para-position after K^+ has been complexed. Thus the macrocycle remains an electroneutral species and an inorganic anion must accompany the complexed cation.

Frequently macrocyclic polyethers have been found to complex both alkali metal and alkaline earth metal cations. Therefore, it is of interest to note that the triazolo-18-crown-6 **CCXXIX** [81] forms stable complexes with the alkaline-earth cations, but not with the alkali metal ions, lithium and potassium.

3.3.1 Association Constants.

A comprehensive compilation of thermodynamic data for cation-crown ether interactions has been published recently [2]. The compilation includes, almost exclusively, association constant data for neutral macrocycles. Accordingly, the data which have become available for association of metal cations with proton-ionizable crown ethers are compiled and presented in the Appendix to this chapter.

4. Reactions at the Aqueous-Organic Interface.

Interpretation of solvent extraction and membrane transport processes relies heavily on understanding the chemistry at organic-aqueous interfaces [100-102]. Vandegrift and Delphin [103] showed that even neutral macrocyclic polyethers are surface-active agents which preferentially adsorb at the interface. Interfacial tension measurements were utilized to determine the crown concentrations that were necessary to saturate the surface monolayer of an interface. The authors were able to determine the surface area for individual molecules at the interface, which suggested the probable alignments of the molecules in relation to the surface. The presence of ionizable moieties in proton-ionizable crown ethers makes these molecules even more surface active. Fyles has used interfacial tension measurements, kinctic data and transport results to propose [102] a comprehensive model of the interfaces of a liquid membrane containing a proton-ionizable crown ether. This model is represented schematically in Figure 2 for a liquid membrane transport experiment. The points raised in the following discussion would apply to solvent extraction before solution equilibrium was attained if the source phase chemistry is considered to be the extraction step and the receiving phase chemistry to be the stripping step.

The interface between two immiscible liquids is a region of unknown dimensions and physical properties. It is formed by the partial mixing of the two bulk phases and, as such, is non-uniform in nature. However, the chemistry involved in the movement of dissolved species through the interface can be characterized in terms of an average interfacial chemistry. Using this philosophy, the following picture of a membrane can be proposed (Figure 2).

The symbols are defined as follows; Square brackets indicate molar concentrations; $M^{+}i$ is the ith metal cation species; subscripts As and Ar identify the aqueous source and receiving phases, respectively; subscripts Is and Ir designate the interfaces on the source and receiving phase sides of the membrane, respectively; and the subscript O is for the bulk organic phase. HCr is the protonated form, and MCr is the metal-complexed form of the ligand, while Cr^{-} is the ionized form of the uncomplexed ligand.

Figure 2: A schematic representation of the important parts of a liquid membrane which contains a proton-ionizable crown ether as a selective metal transporting agent.

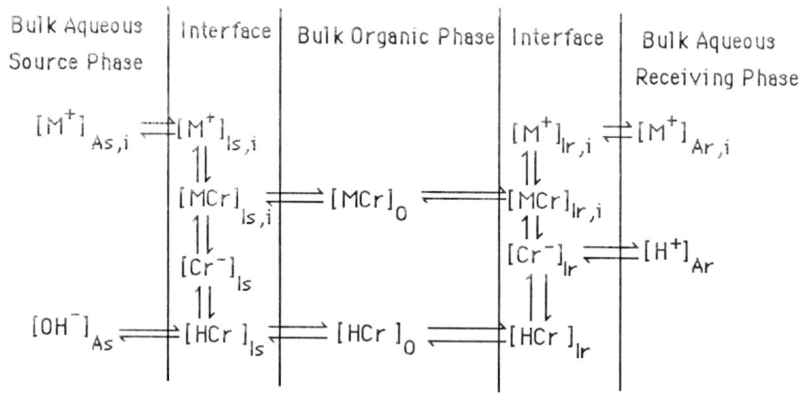

In such a model, the interesting reactions are described in detail in the equations given below:

$$M^+_{As,i} \underset{k_{-1}}{\overset{k_1}{\rightleftharpoons}} M^+_{Is,i} \tag{4}$$

$$HCr_{Is} \underset{k_{-2}}{\overset{k_2}{\rightleftharpoons}} H^+_{As} + Cr^-_{Is} \tag{5}$$

$$M^+_{Is,i} + Cr^-_{Is} \underset{k_{-3}}{\overset{k_3}{\rightleftharpoons}} MCr_{Is,i} \tag{6}$$

$$HCr_O \underset{k_{-4}}{\overset{k_4}{\rightleftharpoons}} HCr_{Is} \tag{7}$$

$$MCr_{Is,i} \underset{k_{-5}}{\overset{k_5}{\rightleftharpoons}} MCr_O \tag{8}$$

$$MCr_O \underset{k_{-6}}{\overset{k_6}{\rightleftharpoons}} MCr_{Ir,i} \tag{9}$$

$$HCr_O \underset{k_{-7}}{\overset{k_7}{\rightleftharpoons}} HCr_{Ir} \tag{10}$$

$$HCr_{Ir} \underset{k_{-8}}{\overset{k_8}{\rightleftharpoons}} H^+_{Ar} + Cr^-_{Ir} \tag{11}$$

$$MCr_{Ir,i} \underset{k_{-9}}{\overset{k_9}{\rightleftharpoons}} M^+_{Ir,i} + Cr^-_{Ir} \tag{12}$$

$$M^+_{Ir,i} \underset{k_{-10}}{\overset{k_{10}}{\rightleftharpoons}} M^+_{Ar,i} \tag{13}$$

Equations 4-8 apply to the extraction of a species on the source phase side of the membrane and equations 9-13 refer to the stripping of that species on the receiving phase side. The quotients of the forward and reverse rate constants for each equilibrium can be identified as various equilibrium or extraction constants in the following way. The constants for equations 4, 7, 8, 9, 10, and 13 are all extraction constants for the distribution of various species from one solution to another which has a different dielectric constant. The quotients for equations 5 and 11 are the acid dissociation constants for the ionizable crown ether in the interface. The quotients for equations 6 and 12 are the equilibrium constants for the complexation reaction between the metal ion and the ionized macrocycle in the dielectric constant of the interface. Although several of these constants appear to be the same on either side of the membrane (e.g., those arising from equations 5 and 11), the differences in structure of the interfaces may mean that all 10 constants are unique.

The use of a simpler view of interfacial reactions, as applied in the work of Vandegrift and Horwitz [104] to solvent extractions, indicates that the rate at which a number of moles, n_i, of the ith species enters the organic phase is described by the equation

$$dn_i/dt = A_s(k_{ao}[M^+]_{as,i} - k_{oa}[M^+]_o) \quad (14)$$

where A_s is the area of the source phase interface, and the subscripts ao and oa indicate the rate constants for movement of ions from the aqueous phase to the organic phase or vice-versa, respectively. A linear plot of the function $\ln\{1 - ([M^+]_o / [M^+]_o^\infty)\}$ vs. time will have a slope given by the equation

$$\text{Slope} = -A_s/V_o \ (1 + \{V_o/V_a\}K_d)k_{oa} \quad (15)$$

and

$$k_{ao} = K_d k_{oa} \quad (16)$$

when K_d is the distribution ratio of the ith species at equilibrium such that

$$K_d = ([M^+]_o^\infty / [M^+]_a^\infty) \quad (17)$$

In all of these equations, the superscript ∞ indicates the concentration at infinite time, or equilibrium for a solvent extraction.

On the receiving phase side of the membrane, stripping of the ith species would be given by the equation

$$-dn_i/dt = A_r(k_{ao}[M^+]_{as,i} - k_{oa}[M^+]_o) \quad (18)$$

where A_r is the surface area of the receiving phase interface in cm^2. Equations 15-17 should then hold true for the stripping phase also, except that the sign of the slope will be reversed. The mass transfer coefficients k_{ao} and k_{oa} should coincide in magnitude with the smallest of the mass transfer rate constants given in the more detailed view of equations 4-13.

Considering the model proposed for the membrane, and the kinds of equilibria and distributions that are involved, it is expected that the overall chemistry of extractions or membranes will depend on such variables as $[H^+]_A$, $[M^+]_A$, $[HCr]_O$, and the interfacial activity of each crown species. The Gibbs equation in the form

$$F = (-1/kT)(d\Gamma/d\ \ln[Cr])max \qquad (19)$$

can be used to estimate the concentration at which a surface active species saturates an interface with a monolayer of the species. The quantity F is known as the surface excess of the molecule, and is defined as the concentration of the molecule at the surface less the concentration of the molecule in the bulk phase. Measurements of F from interfacial tension data will thus give indications of macrocycle concentrations in the equilibria represented in equations 5,6,7,10,11, and 12. Since the concentration of crown in the interface is that in a monolayer, it is possible to calculate the number of molecules, and hence the number of complexation sites, available at the interface. Fyles [102] has pointed out that the number of complexation sites available in the interface can be modified by the addition of non-complexing surfactants to the solutions, resulting in the replacement of some of the surfactant crown molecules. Further, the addition of cationic or anionic non-complexing surfactants would encourage the attraction or repulsion of, say, divalent over monovalent cations at the interface, thus adding a possibility for enhancing a desired selectivity. Selectivity can also be modified by adding a complexing agent to the aqueous source phase which would preferentially bind some of the cations. This would alter the distribution coefficients for the cations so that the desired species would be more available, and undesired ones less available, for complexation in the interface.

Information obtained by plotting the interfacial tension between two solutions versus concentration of the crown species in the organic phase can also be used to estimate the surface area per molecule of the species that is saturating the interface. This area is calculated simply as $10^{16}/F$ square angstroms. The surface area of the molecule exposed to the interface can be compared with surface areas measured from CPK models [104] to suggest possible physical alignments of the crown molecules in the interface. Thus, it would appear possible to physically model the interfacial region quite accurately with some relatively simple measurements. To date, no record has been found of attempts to do so to this level of detail.

5. Solvent Extraction Studies.

Detailed solvent extraction studies have been performed with about half of the ionizable crown ethers that have been synthesized. The presence of the ionizable moiety leads to complex equilibrium and distribution chemistry considerations [86] for this class of extractant unless the aqueous solution is made very basic. Indeed, at neutral to acidic pH values, Nakamura et.al. [86] and Bartsch et.al. [105] observed the presence of 2:1 ligand to metal complexes, and Bartsch's group postulated a 3:1 ligand to metal complex stoichiometry to account for certain extraction results with Li^+, Na^+, K^+, and Rb^+. However, in this latter study [105] only 1:1 complexes with Cs^+ were found. A detailed discussion of all the equilibrium constants that may be involved in studying species distributions during an extraction with a proton- ionizable crown ether can be found in the papers by Nakamura and co-workers [25, 86].

A variety of organic solvents have been utilized for the extraction studies. The most common solvents are chloroform, dichloromethane, 1,2-dichloroethane, toluene, and dichlorobenzene. Comparison of results obtained with a given macrocycle to extract alkali metal cations into two different organic solvents [105, 106] showed a large variation in results; the variance depending primarily on the choice of solvent. In an effort to probe this solvent effect, Charewicz, Walkowiak and Bartsch [107] performed a study using six different organic solvents with a common crown. Effectiveness of the solvents was found to be chloroform > 1,1,1-trichloroethane > tetrahydronaphthalene > benzene > toluene > p-xylene. Three of the solvents, the alkyl-substituted benzenes, all formed a second organic phase at high pH conditions. The extraction results were found to correlate with E_T, the empirical Dimroth solvent polarity parameter.

Of interest in separation science are the effectiveness of an extractant both in terms of selectivity and efficiency (loading). Meaningful comparison of solvent extraction results may be made difficult by the several different reporting methods employed. Reported selectivities for a variety of proton-ionizable crown ethers are summarized in Table 6. The selectivities are those reported for extraction studies in competitive systems or those extrapolated from the results of single-species extraction experiments. Several workers [17, 125] have shown that selectivities extrapolated from single-species experiments may prove to be erroneous in either order or magnitude when the selectivity is tested in a competitive extraction system.

Table 6: Selectivity orders for several proton-ionizable crown ethers in solvent extraction of metal cations from aqueous solution.

Ligand	Solvent	Order	T (ºC)	Ref.
CLXXXIX	dichloromethane	Na>K=Li>Ca	22	67
CXC	"	Na>Li>K>Ca	"	"
CXCI	"	Na>K>Li>Ca	"	"
CXCII	"	K>Na>Li>Ca	"	"
I,II	o-dichlorobenzene (dark)	K>Cs>Rb>>Na	30	39
"	" (UV)	Rb>Cs>K>>Na	"	"
III	o-dichlorobenzene-	Rb>K>Cs>Na	"	"
"	butanol (9:1) (dark)	Rb>Cs>K>Na	"	"
"	"	Rb>Cs>K>Na	"	"
"	"	Cs>Rb>K>Na	"	"
"	"	Ca>Ba>Sr	"	"
"	" (UV)	Ca>Ba>Sr	"	"

Ligand	Solvent	Order	T(°C)	Ref.
X	chloroform	K>Rb>Cs>Na	25	86
XI	"	K>Rb>Cs>Na	"	"
XII	"	K>Rb>Na>Cs	"	"
XIII	"	K>Rb>Na>Cs	"	"
XIV	"	K>Rb>Na>Cs	"	"
LXXX	1,2-dichloro-	Na>K>Rb>Cs>Li	25	57
LXXXI	ethane	K>Rb>Cs>Na>Li	"	"
LXXXII	"	Na>K>Rb>Li>Cs	"	"
LXXXIII	"	K>Rb>Na>Li>Cs	"	"
CL	"	Li>Na>K	"	"
CLI	"	K>Na>Rb>Li>Cs	"	"
CLIII	"	Li>Na>K(no Rb, Cs)	"	"
CLIV	"	K>Na≈Rb>Li	"	"
CLV	"	Li>Na>K	"	"
CLVI	"	K>Na>Rb>Li	"	"
LXXXIV	"	K>Na>Rb	"	"
LXXV	"	K>Rb>Na	"	"
LXXVI	"	K>Rb>Na>Cs	"	"
LXXVII	"	K>Cs>Na	"	"
LXXXIV	chloroform	Na>K>Rb>Li≈Cs	25	51
LXXXV	"	K>Rb>Cs(no Na, Li)	"	"
LXXXVI	"	Rb≈Cs>K>Na>Li	"	"
LXXXVII	"	K≈Na>Cs>Rb>Li	"	"
LXXXIX	chloroform	Li>Na>K>Rb>Cs	25	49
XCV	"	Na>K>Rb>Li>Cs	"	"
XCVII	"	K>Rb>Li>Cs>Na	"	"
IC	"	Cs>Rb>K>Li>Na	"	"
XXXIV	chloroform	K>Rb>Cs>Na>Li	25	108
XXXV	"	K>Na>Rb>Li≈Cs	"	"
XXXVI	"	Na>K>Rb≈Cs>Li	"	105
XXXVII	"	K>Rb≥Na≈Cs>Li	"	108
XXXIV	"	Ba>Ca>Sr>Mg	"	109
XXXVI	"	Ba>Ca>Sr>Mg	"	"
XXXVII	"	Ba>Ca>Sr>Mg	"	"
XXXVIII	chloroform	Na>>K>Rb>Cs	25	92
XXXIX	"	Na>>K>Li>Rb>Cs	"	"
XL	"	Na>>K>Li>Rb≈Cs	"	"
XLI	"	Na>>K>Li>Rb≈Cs	"	"
L	chloroform	Na>>K>Rb≈Cs	25	40
L	toluene	Na>K>LiRb>Cs	"	106
XLI	"	Na>K≈Li>Rb>Cs	"	"
XLI	"	Na≈K>Li>Rb>Cs	"	110
LXXI	chloroform	Li>Na>K>Rb>Cs	"	111
LXXII	"	Li>Na>K>Rb (no Cs)	"	"
LI	"	Na>Li>K>Rb>Cs	"	"
LXXXIX	"	Li>Na>K>Rb>Cs	"	"
XC	"	Li>Na>K>Rb>Cs	"	"
XCI	"	Li>Na>K>Rb,Cs	"	"

Ligand	Solvent	Order	T(°C)	Ref.
XCII	chloroform	Li>>Na(no K, Rb, Cs)	25	111
XCIII	"	Li>>Na(no K, Rb, Cs)	"	"
XCIV	"	Li>Na>K,Cs>Rb	"	"
XXXV	chloroform(pH12)	Mg>Ca>Li	25	64
LI	" (pH6)	Ba>Ca>Mg>Li	"	"
XLI	chloroform(pH8.7)	Na>>Li=K>Rb>Cs	"	107
"	1,1,1-trichloro-ethane	Na>Li=K>Rb=Cs	"	"
"	tetrahydro-naphthalene	Na>Li=K>Rb>Cs	"	"
"	benzene	Na>Li=K>Rb=Cs	"	"
"	toluene	Na>Li=K>Rb=Cs	"	"
"	p-xylene	Na>Li=K=Rb>Cs	"	"
XXXVI	chloroform-heptanol(4:1)	Lu>Yb>Er≈Tb>Eu>Sm>Pr>La	25	112
XLVI	chloroform	Na>Li>K>Rb,Cs	"	69
XLVII	"	Na>>K>Li>Cs>Rb	"	"
XLVIII	"	Li>Na>K>Rb,Cs	"	"
IL	"	Li>Na>K>Rb,Cs	"	"
LXIV	chloroform	Li>Na>K	25	55
LXV	"	Na>>Li≈K	"	"
LXVI	"	Na>>K>Li	"	"
LXVIII	"	Na>>Li>K	"	"
LXIX	"	Na>K>Li	"	"
LXX	"	K>Na≈Li	"	"

The available information concerning the extraction efficiencies of proton-ionizable crown ethers as extractants is summarized in Table 7. The percentage of metal cation which is extracted from the aqueous phase into the organic medium is recorded.

Table 7: Efficiency of proton-ionizable crown ethers as extractants where reports indicate % of metal extracted (at 25°C unless otherwise noted).

Ligand	pHaq	Solvent	Cation	%	Ref.
XXXVI	4.5	chloroform-	Lu	20	112
"	5.0	heptanol(4:1)*	"	50	"
"	5.5	"	"	81	"
"	6.0	"	"	95	"
"	6.5	"	"	99	"
"	7.5	"	"	91	"
LXXXIV	9.0	chloroform	Li	5	51
"	"	"	Na	50	"
"	"	"	K	21	"
"	"	"	Rb	14	"

Ligand	pHaq	Solvent	Cation	%	Ref.
LXXXIV	9.0	chloroform	Cs	5	51
LXXXV	"	"	Li	0	"
"	"	"	Na	0	"
"	"	"	K	57	"
"	"	"	Rb	24	"
"	"	"	Cs	5	"
LXXXVI	"	"	Li	10	"
"	"	"	Na	12	"
"	"	"	K	17	"
"	"	"	Rb	33	"
"	"	"	Cs	30	"
LXXXVIII	"	"	Li	11	"
"	"	"	Na	22	"
"	"	"	K	23	"
"	"	"	Rb	18	"
"	"	"	Cs	20	"
CLXXXIX	>11	dichloromethane	Li	1.4	70
"	"	"	Na	1.5	"
"	"	"	K	1.4	"
"	"	"	Ca	1.1	"
CXC	"	"	Li	7.2	"
"	"	"	Na	7.9	"
"	"	"	K	6.7	"
"	"	"	Ca	4.8	"
CXCI	"	"	Li	6.1	"
"	"	"	Na	8.7	"
"	"	"	K	6.8	"
"	"	"	Ca	1.8	"
CXCII	"	"	Li	3.4	"
"	"	"	Na	5.2	"
"	"	"	K	8.0	"
"	"	"	Ca	2.9	"
III**	[MOH]=0.5	o-dichlorobenzene-butanol (9:1)	Na	3.6	39
"	"		K	85.6	"
"	"	"	Rb	74.4	"
"	"	"	Cs	75.1	"
"	"	"	Ca	95.7	"
"	"	"	Sr	90.7	"
"	"	"	Ba	95.0	"
LXIV	13	chloroform	Li	24	55
"	"	"	Na	15	"
"	"	"	K	7	"
LXV	"	"	Li	4	"
"	"	"	Na	20	"
"	"	"	K	3	"
LXVI	"	"	Li	2	"
"	"	"	Na	25	"
"	"	"	K	8	"
LXVIII	"	"	Li	8	"

Ligand	pHaq	Solvent	Cation	%	Ref.
LXVIII	13	chloroform	Na	38	55
"	"	"	K	2	"
LXIX	"	"	Li	5	"
"	"	"	Na	29	"
"	"	"	K	14	"
LXX	"	"	Li	11	"
"	"	"	Na	11	"
"	"	"	K	28	"
XXIII	>8	chloroform- methanol(4:1)*	Ba	82	37
XXIV	"	chloroform	Ba	55	"
"	"	"	Sr	40	"
"	"	"	Ca	<1	"
XXV	"	"	Ba	36	"
XXVI	"	"	Ba	84	"
XXVII	"	"	Ba	82	"
XIX	"	"	Ba	6	"
"	"	"	Sr	74	"
"	"	"	Ca	59	"
"	"	chloroform-methanol(4:1)	Ba	40	"
XXI	"	chloroform	Ba	45	"

* Additional results recorded in reference 112 show that the 4:1 mixture is optimal for these extractions.

** These data were obtained at 30°C and the authors reported the extraction as the total concentration of the crown ether detected in the organic phase.

Extraction efficiencies for a number of proton-ionizable crown ethers are reported as Kex, the extraction constant, in references 39, 57, and 86. The extraction constant is defined as

$$K_{ML}^{ex} = [ML]_{org}[H^+]_{aq}/[HL]_{org}[M^+]_{aq} \qquad (20)$$

The data in Table 7 and in the references 39, 57, and 86 reveals a wide range of metal ion extraction efficiencies. This is a result of many factors. Lipophilicity of the total molecule is an important consideration in the design of extractant molecules. It is evident from the data in Table 7 that macrocycles with very little aryl- or alkyl-substitution on the outer surfaces of the crown ether framework extract only small amounts of metals. A proton-ionizable crown ether extraction agent will be lost from the organic phase upon deprotonation if it has insufficient lipophilicity, even if the crown compound forms stable complexes with the metal being extracted. The structural features that have been used so far to increase lipophilicity suggest that the best scheme for such lipophilization is the addition of long alkyl tails at points on the molecule which are remote from both the ionizable function and the crown ether ring.

When ionizable groups are contained in side arms, the selectivity of the resultant crown anion may be affected by the very structure of that side arm. Features that appear to affect selectivity include the nature of the pivot atom to which the side arm is attached [86], the length of the side arm [62], the presence of heteroatoms along the side arm [86], and the nature of the ionizable group itself [51]. The influence of side arm length is illustrated in a study by Pugia et.al. [62] using dibenzo-16-crown-5 macrocycles with a monoalkyl phosphonic acid monoethyl ester moiety at the end of various length side arms (**XLVI - IL**). The ionizable crown ethers were used in competitive solvent extraction experiments with alkali metal cations at pH=9. When only one methylene unit was included in the side arm (**XLVI**), the crown ether was selective for sodium, but selectivity was poor. The addition of a second methylene group (**XLVII**) apparently increased the freedom for the anion to approach the sequestered cation, and sodium selectivity was greatly enhanced. However, the incorporation of additional methylene groups (**XLVIII, IL**) changed the selectivity completely, and the ligand became selective for lithium over all other alkali metal cations.

The extraction selectivity and efficiency data presented in Tables 6 and 7 are given for single pH(aqueous) values in each case. However this may be misleading. Several studies [96, 105] show that selectivity can change as the pH is raised. For example Bartsch et.al. [105] report that for competitive extractions into chloroform with **VI**, K^+ is selected over Na^+ at pH=6, but that the opposite is true at pH=9. Hence the need to conduct extractions at several aqueous phase pH values is demonstrated.

6. Membrane Transport Studies.

Perhaps the most intriguing aspect of all macrocyclic chemistry is the potential to develop membrane systems that will perform clean, selective, low-energy separations of ionic species in aqueous solutions. Several commercial applications of this kind have already been devised [114, 115], although none of them are currently using macrocyclic polyethers as the selective agent. In a recent publication [116], Izatt and co-workers have clearly outlined the general considerations for designing effective liquid membrane systems. There are three major classes of membranes that are likely to be encountered during the development of new separation schemes. Membranes from each of the three classes have been utilized to perform metal ion separations with proton-ionizable crown ether ligands, so each type will be described briefly.

The simplest, and most extensively used, type of membrane system is the bulk liquid membrane. In the most common configuration, a chloroform solution of the proton-ionizable crown ether in the lower portion of a U-tube separates two aqueous solutions in the upper arms. One of the two aqueous phases is designated the source phase and contains the cationic species to be separated and sufficient base to raise the pH above the effective pK_a of the carrier being investigated. The other aqueous phase is termed the receiving phase. This latter aqueous phase may be only water, but more commonly contains a mineral acid or a buffer to decrease the pH of the solution, and sometimes contains metal ions at concentrations equal to the initial metal ion concentrations in the source phase. The matching of ion concentrations is usually done to prove true counter transport as the cation of interest is transported against its own concentration gradient. An alternative cell design is a tube-within-a-beaker, in which the source aqueous phase is contained inside a suspended tube and the receiving aqueous phase is on the outside or vice-versa.

Bulk liquid membranes of this type have little significance in terms of commercial development, but they are extremely useful for laboratory study. The transport experiments can be conducted on a small scale in simple cells which allows rapid screening [17] of large numbers of carriers or testing of a wide range of conditions. Bulk liquid membranes in larger configurations also enable sampling of all three phases as a function of time [121] so that mass balances can be determined, and rates of appearance and disappearance of metal ions in the various phases can be determined. This allows kinetic and mechanistic studies to be performed which reveal the detailed chemistry taking place in a given system. Studies conducted in larger cells have shown some of the chemical differences that exist between separation schemes in which neutral and proton-ionizable crown ethers are used.

The second type of membrane system, liquid emulsion or surfactant membranes, have not been used as extensively. There are only four reports of results [50, 106, 110, 126] obtained with proton-ionizable crown ethers in this type of membrane system. One of these [50] is really the use of phospholipid vesicles as the liquid membrane phase. Liquid emulsion membranes are formed by emulsifying an organic solution of a surfactant and the carrier of interest with an aqueous solution which will act as the receiving phase. This emulsion is then stirred slowly with an aqueous solution that acts as the source phase. Hence the metal ion transport is from an aqueous solution which surrounds the emulsion droplet into an aqueous solution encapsulated within the emulsion droplet. The large surface area of the organic membrane exposed to the source phase and the small distances across the organic barrier make transport across these membranes extremely rapid (10-30 minutes for completion) compared with several days for bulk liquid membrane experiments. However, the emulsions are relatively unstable, have limited lifetimes, and require batch operation. The water-oil- water emulsions are also difficult to characterise physically, which reduces the capability for mathematical modelling. The use of phospholipids as the organic phase does allow the formation of regular vesicles which can be characterised readily (i.e., surface area determinations) and also provides membranes which have many of the characteristics of biological membranes.

The third membrane type is the polymer-supported liquid membrane in which the organic solution of the carrier is absorbed into the pores of a porous polymer in either a sheet or hollow fiber configuration. The polymeric barrier is then used to separate the two source and receiving aqueous phases. Polymer-supported liquid membranes of this type are attractive for industrial development because of their durability. Support of the liquid membrane within the polymeric walls of hollow fibers provides a high membrane surface area in a minimal volume.

Investigations conducted with liquid membranes which contain macrocyclic polyether carriers show that the most important chemistry in the overall transport process occurs in the organic-aqueous interfacial regions. Extensive modelling by Danesi [100], Lehn [101] and Fyles [102] has established the importance of various rate-limiting steps in different membrane situations. Determination of which step will be rate limiting is a function of the type of carrier being used and the chemical conditions in the bulk solutions on either side of the interface. In most cases, the rate limiting step is associated with movement of a metal ion from the aqueous source phase to the macrocycle-metal complex in the bulk organic phase, but the details can vary from diffusion of the ion through the aqueous boundary layer to the rate at which a complex molecule can desorb from the interface into the organic solvent. When the carrier is a neutral macrocycle, the rate limitations in the source phase interface are made even worse by the fact that an anion must accompany the complex into the organic phase. Thus, membranes which use neutral macrocycles as carriers are found to be anion dependent [117]. In 1974, Cussler et al. [118] demonstrated anion-independent, proton-coupled counter transport of a metal ion across a membrane by inclusion of monensin, a proton- ionizable natural ionophore, in a supported membrane. Monensin is a non-cyclic polyether which coordinates with a Na^+ ion in solution by wrapping around the ion in a cyclic configuration. In 1977, McKervey and Mulholland reported the first duplication of this type of experiment [69] using a synthetic ionophore which was a crown ether phenol. Since then, a steadily increasing number of investigations of the ability of proton-ionizable crown ethers to selectively transport metal ions across organic barriers have been reported.

6.1 Bulk Liquid Membrane Transport.

The description of bulk liquid membranes provided above highlights the similarity between membrane transport and solvent extraction. The major difference is that extraction and stripping processes are proceeding simultaneously during the membrane transport process. Therefore, transport processes are subject to the same chemical requirements as are extractions, but with the advantage that conditions on either side of the membrane may be adjusted to achieve separation as a continuous process. Effects of ring size, solvent, pH gradient, metal ion concentration, carrier concentration, identity of the ionizable moiety, and structural variation on both the selectivity and efficiency of transport has been examined for a variety of proton-ionizable crown ether carriers. The majority of these bulk liquid membrane transport studies have concentrated on separation of alkali metal cations. In a few [39, 109, 115], alkaline earth cation transport has been investigated. In one study [45] complexed metal ions were used to assist in the separation of amino acids, while in another the separation of primary, secondary and tertiary amines [71] was probed.

6.1.1 Ring Size Effects.

As is the case for neutral macrocycles, the selectivity of a given proton-ionizable crown ether is governed largely by the relationship between the cation and macrocyclic polyether cavity diameters .

6.1.2 Solvent Effects.

The significant influence of organic solvent identity on the efficiency of solvent extraction has already been noted. Bulk liquid membrane transport with neutral macrocyclic carriers has also been found [120] to be a function of solvent. Hence, it is hardly surprising to find that the organic solvent plays an important role when proton-ionizable carriers are utilized. For two different ionizable crown ethers (**XLI, L**), Bartsch et.al. [40, 106] have shown that selectivity orders are unchanged by replacement of a chloroform membrane with toluene. However, the actual selectivity and the transport efficiency are both affected quite markedly, with the chloroform membrane yielding both higher transport rates and better selectivity. Since these particular systems both have rate limiting steps associated with the extraction of the ions into the organic phase, similarity with the previously reported results for solvent extraction is not unexpected. It is possible that membrane transport systems in which the rate limiting step is part of the stripping processes may not show a similar response to changes in the nature of the solvent.

6.1.3 pH Gradient Effects.

The counter transport of metal ions by proton-ionizable crown ethers is driven by the removal of protons from the crown carrier in the source phase interface and the replacement of those protons in the receiving phase interface. The complexation equilibrium between a cation and the carrier is a function of the acid dissociation constant for the crown ether [88]. Thus the pH of the source phase must be sufficiently high to completely deprotonate the carrier in the interface for maximum complexation of metal cations to occur. Similarly, the pH of the receiving phase must be low enough to force a complete reversal of the complexation-deprotonation balance to encourage maximum stripping. Although it is possible that these equilibria will respond to variations of pH to some extent, it seems likely that rapid, uniform responses will be found at any pH values that exceed these minimum requirements. However, the counter-transport is driven partly by the pH gradient across the membrane, so many membranes show a metal flux response to variations of the pH values on either side of the membrane which maximize the difference between the proton concentrations in each phase. There is not a universal pattern of pH response, however. Fyles and coworkers [121] found that their systems showed only very small responses to pH changes in either aqueous phase. In contrast, Izatt et.al. [116, 122, 123] have found dramatic responses to changing the pH of the receiving phase while using pyridino-crown carriers, and Bartsch and coworkers [93] have found significant responses to changes in the source phase pH when crown ether carboxylic acids are the carriers. Perhaps the most intriguing example of pH effects is the change in selectivity exhibited by a carboxy-18-crown-6 carrier and reported by Hriciga and Lehn [113]. This crown ether changed selectivity completely during competitive transport experiments from preferential transport of K^+ at pH=2 to preferential transport of Ca^{2+} at pH=9. The results were attributed to the differences in complexation behavior for the protonated and deprotonated forms of the ligand involved.

6.1.4 Concentration of the Metal Ions.

The requirement for metal-macrocycle interaction for extraction of metals suggests that membrane transport should exhibit some rate dependence on the initial metal ion concentration or activity in the source phase. This was demonstrated to be the case for neutral macrocyclic carriers. Hence in most studies conducted with proton-ionizable carriers the influence of initial metal cation variation has not been assessed. Fyles et.al. [121] found an apparent first order dependence on metal ion concentration only while the metal ion concentrations are of similar magnitude to or smaller than the crown concentrations. If the metal ions are present in appreciably larger concentrations than the carrier, the extraction becomes apparently zeroth order in metal. There is insufficient data to determine whether this is a universal pattern of behavior for all proton-ionizable crown ethers. Current studies in this laboratory [124] which utilize crown carboxylic acid carriers (e.g., **XXXVIII- XLIII**), and metal ion concentrations higher than those of the crown, yield plots of metal transported vs. time which are typical of the consecutive first-order regime proposed by Fyles [121]. Such behavior is explained in terms of the balance between interfacial surface charge and adsorption [102] discussed in the earlier section on reactions at the aqueous-organic interface. The consecutive first-order regime applies when desorption of the complex from the interface is the rate-limiting step. It would appear that this rate limitation may apply to many membrane transport experiments where the carrier has strong surface activity in both the deprotonated and the complexed forms.

6.1.5 Crown Concentration.

As discussed in Section 4, proton-ionizable crown ethers are preferentially adsorbed at an aqueous interface and can attain saturation concentrations there which are higher than the concentration in the bulk organic phase [104]. This suggests that it should be possible to identify a maximum effective concentration of carrier that needs to be added to the membrane. At present there is insufficient data to support this hypothesis. For very low concentrations of a proton-ionizable crown ether, Fyles and coworkers [121] found that the flux of a metal through a membrane showed a linear dependence on the carrier concentration.

6.1.6 Identity of the Ionizable Moiety.

Bulk liquid membrane transport experiments have been conducted with carriers bearing carboxylic acid , phosphonic acid monoethyl ester, phenol and 4-hydroxypyridine or pyridone moieties. One of the carboxylic acid-bearing crowns (**CLIX**) had two ionizable side arms attached to nitrogen atoms of the diaza-18-crown-6 cycle. (This crown amino acid gave its best transport at a pH value of 5.9, indicating that the zwitterion was the active agent.) As a general rule, the metal ion transport selectivity of proton-ionizable crown ethers is controlled by the cation-cavity match, as discussed above. A change in the ionizable moiety appears to alter only the pH requirements of the basic source phase which are needed to promote rapid transport.

6.1.7 Structural Variations.

Crown ethers which possess a side arm that includes an ionizable moiety are subject to many structural variations. Structural changes with side arms or attachment points should affect transport rates only in so far as they alter stability constants or the lipophilicity of the resulting proton-ionizable crown ether. In a systematic study of side arm variation for a series of (sym-dibenzo-16-crown-5-oxy)-alkanoic acids **XXXVIII-XLI**, Strzelbicki and Bartsch [125] found that varying the length of a side arm, without changing the position of the carboxyl group relative to the cavity, did not affect the selectivity order of the crowns. However, the selectivity coefficients and the fluxes of ions were affected significantly. There was no direct correlation between the side arm length and transport selectivity, but the relationship between flux and length of side arm indicated that the more lipophilic the crown the higher the flux. In the same study, it was found that lengthening the side arm by adding methylene units so that the carboxylic acid group was moved further from the cavity greatly reduced the selectivity and flux even though the lipophilicity was increased. This is consistent with the observation [91] that higher selectivity is achieved when the anionic site can closely approach the cation sequestered in the cavity of the crown.

6.2 Emulsion Liquid Membranes.

Only four studies have been reported in which [50, 106, 110, 126] proton-ionizable crown ethers were used as carriers in water-in-oil-in-water emulsion membranes, and one of these [50] used a phospholipid vesicle as the oil phase. The study reported in reference 126 demonstrates that the transport rate and selectivity of these liquid membrane systems are influenced by the diameter of the cavity of the proton-ionizable crown ether, the attachment site of lipophilic side arms which do not include the ionizable group, and the basicity of the oxygen in the carboxylic acid group.

A study performed with phospholipid vesicles [50] was designed as a deliberate attempt to mimic biological membranes. The carriers tested included the natural ionophores valinomycin (neutral) and nigericin (ionizable), and both an ionizable and a neutral analog of 15-crown-5. The experiment showed that the synthetic crown compounds mimicked the natural ionophoric transport in form, although the synthetic ionizable crown carried less total metal than did nigericin, probably because of lower complex stability. The complexation reactions were shown to occur at the interface, and a pH gradient of only 1 unit was necessary to drive the exchange of Na^+ for H^+. Thus, these membranes appear to be highly successful mimics of natural membranes.

6.3 Supported Liquid Membranes.

Considering their potential usefulness, there have been surprisingly few reports of experiments involving this class of liquid membranes. The viability of proton-coupled transport as a commercial separation scheme has been proven, albeit not with ionizable macrocycles as the selective agents. An excellent review of supported liquid membrane transport techniques, capabilities, and mathematical treatments is available [127]. Much of the work currently being done with ion selective electrodes involves membranes which are similar in principle to supported liquid membranes, and treatises on such electrodes contain a wealth of useful information relevent to this aspect of membrane transport chemistry. One of the few reports about proton-ionizable crown ethers in supported membrane media concerns the attempt to use compounds **CXXV**, **CXXVI**, **CXLI**, and **CXLII** in ion selective electrodes [128].

Two papers [56, 66] have appeared which describe alkali metal cation transport across polymer-supported liquid membranes by 14-crown-4 nitrophenol compounds **CIII-CXXI**. The membranes were found to be highly selective for lithium, with the degree of selectivity being affected by the choice of solvent. The total flux of lithium through a given membrane was also found to be a function of the pH gradient across the membrane. A minimum pH had to be attained in the source phase before transport would commence, and this minimum was a function of the pKa of the particular nitrophenol moiety that was attached to the crown. There was also found to be a pH value for the receiving phase, below which flux did not respond to changes in pH, but above which the flux rapidly declined to zero. Flux of lithium was also found to be a function of lithium concentration in the source phase whenever that concentration was less than or equal to that of the carrier, but once the carrier concentration was exceeded, metal concentration variation did not influence the flux.

7. Concluding Remarks.

Proton-ionizable crown ether ligands are a valuable addition to the arsenal of complexing agents which are available to the separation scientist. Separations are facilitated by elimination of the need to transport an aqueous phase anion into the organic medium. In solvent extraction and membrane separations, selectivity may be controlled by varying the structure of the ligand, changing the organic solvent, mixing non-carrier surfactants with the crown to influence the number of available complexation sites in the interface, and altering the aqueous phase composition to influence the metal ions available for complexation within the interface. At present, insufficient experimental work has been conducted to allow for the development of predictive models which include the influence of these factors. Therefore determination of optimum conditions for a particular separation is still a matter of laboratory testing.

Present results indicate that additional research should be undertaken to probe the responses caused by changing pH and metal ion concentration conditions on either side of liquid membranes containing proton-ionizable crown ethers. The structure and influence of the organic-aqueous interface on metal ion separations should be investigated more thoroughly to enable a reliable predictive model to be developed so that separations could be designed without the time and expense required for protracted laboratory experimentation. Additional research on polymer-supported liquid membranes would speed the appearance of this promising technique as a viable alternative for industrial separations of both metallic and biological cationic species.

8. Acknowledgement.

Financial support by the Division of Basic Chemical Sciences of the United States Department of Energy (Contract DE-AS05-80ER10604 and Grant DE-FG05-88ER13832) for portions of this research which were conducted at Texas Tech University is gratefully acknowledged.

9. References.

1) C. J. Pedersen, *J. Am. Chem. Soc.*, **89**, 2495(1967).
2) R. M. Izatt, J. S. Bradshaw, S. A. Nielsen, J. D. Lamb, J. J. Christensen and D. Sen, *Chem. Rev.*, **85**, 271(1985).
3) e.g., C. J. Pedersen and H. K. Frensdorf, *Angew. Chem., Int. Ed.*, **11**, 16(1972).
4) M. I. Page, *Macromol. Chem.*, **1**, 397(1980).
5) R. M. Izatt, G. A. Clark, J. S. Bradshaw, J. D. Lamb and J. J. Christensen, *Sep. Purif. Methods*, **15**, 21(1986).
6) D. J. Cram and J. M. Cram, *Science*, **183**, 803(1974).
7) J. M. Lehn, *Pure Appl. Chem.*, **50**, 871(1978).
8) J. Canceill, A. Collet, J. Gabard, F. Kotzyba-Hibert and J. M. Lehn, *Helv. Chim. Acta*, **65**, 1894(1982).
9) R. C. Helgeson, J. P. Mazaleyrat and D. J. Cram, *J. Am. Chem. Soc.*, **103**, 3929(1981).
10) T. W. Bell and A. Firestone, *J. Am. Chem. Soc.*, **108**, 8109(1986).
11) D. M. Dishong, C. J. Diamond, M. I. Cinoman, and G. W. Gokel, *J. Am. Chem. Soc.*, **105**, 586(1983).
12) R. C. Helgeson, J. M. Timko, and D. J. Cram, *J. Am. Chem. Soc.*, **95**, 3023(1973).
13) J. S. Bradshaw in "Synthetic Multidentate Macrocyclic Compounds", pp 53-109, Eds., R. M. Izatt and J. J. Christensen, Academic: New York, N.Y., (1978).
14) J. M. Lehn, *Acc. Chem. Res.*, **11**, 49(1978).
15) R. A. Schwind, T.J. Gilligan and E. L. Cussler in "Synthetic Multidentate Macrocyclic Compounds", pp 289-308, Eds., R. M. Izatt and J. J. Christensen, Academic: New York, N.Y., (1978).
16) J. M. Lehn, *Pure Appl. Chem.*, **51**, 979(1979).
17) D. W. McBride, Jr., R. M. Izatt, J. D. Lamb and J. J. Christensen in "Inclusion Compounds", Vol. 3, Ch. 16, (1984).
18) I. M. Kolthoff, *Anal. Chem.*, **51**, 1R(1979).
19) M. Takagi and H. Nakamura, *J. Coord. Chem.*, **15**, 53(1986).
20) P. D. J. Grootenhuis, J. W. H. M. Uiterwijk, D. N. Reinhoudt, C. J. van Staveren, E. J. R. Sudhölter, M. Bos, J. van Erden, W. T.

Klooster, L. Kruise and S. Harkema, *J. Am. Chem. Soc.*, **108**, 780(1986), and references therein

21) J. M. Girodeau, J. M. Lehn and J. P. Sauvage, *Angew. Chem., Int. Ed.*, **14**, 763 (1975).

22) R. Ungaro, B. El Haj and J. Smid, *J. Am. Chem. Soc.*, **98**, 5198(1976).

23) J. P. Behr, J. M. Lehn and P. Vierling, *J. Chem. Soc., Chem. Commun.*, 621(1976).

24) R. C. Helgeson, G. R. Weisman, J. L. Toner, T. L. Tarnowski, Y. Chao, J. M. Mayer and D. J. Cram, *J. Am. Chem. Soc.*, **101**, 4928(1979).

25) H. Nakamura, M. Takagi and K. Ueno, *Talanta*, **26**, 921(1979).

26) S. Külstad and L. A. Målmsten, *Acta Chim. Scand.*, **B33**, 469(1979).

27) J. M. Lehn, P. Vierling and R. C. Hayward, *J. Chem. Soc., Chem. Commun.*, 296(1979).

28) J. P. Behr, J. M. Girodeau, R. C. Hayward, J. M. Lehn and J. P. Sauvage, *Helv. Chim. Acta*, **63**, 2098(1980).

29) L. A. Frederick, T. M. Fyles, N. P. Gurprasad and D. M. Whitfield, *Can. J. Chem.*, **59**, 1724(1980).

30) I. Tabushi, Y. Kobuke, K. Ando, M. Kishimoto and E. Ohara, *J. Am. Chem. Soc.*, **102**, 5948(1980).

31) T. Yamashita, H. Nakamura, M. Takagi and K. Ueno, *Bull. Chem. Soc., Jpn.*, **53**, 1550(1980).

32) G. E. Pacey and B. P. Bubnis, *Anal. Letters*, **13**, 1085(1980).

33) J. P. Behr, J. M. Lehn, D. Moras and J. C. Thierry, *J. Am. Chem. Soc.*, **103**, 701(1981).

34) G. E. Pacey, Y. P. Wu and B. P. Bubnis, *Analyst*, **106**, 636(1981).

35) H. Nakamura, H. Sakka, M. Takagi and K. Ueno, *Chem. Lett.*, 1305 (1981).

36) D. N. Reinhoudt, F. de Jong, and E. M. van de Vondervoort, *Tetrahedron*, **37**, 1753(1981).

37) D. N. Reinhoudt, F. de Jong, and E. M. van de Vondervoort, *Tetrahedron*, **37**, 1985(1981).

38) R. A. Bartsch, G. S. Heo, S. I. Kang, Y. Liu and J. Strzelbicki, *J. Org. Chem.*, **47**, 457(1982).

39) S. Shinkai, T. Minami, Y. Kusano, and O. Manabe, *J. Am. Chem. Soc.*, **104**, 1967(1982).

40) W. A. Charewicz, G. S. Heo and R. A. Bartsch, *Anal. Chem.*, **54**, 2094 (1982).

41) H. Nakamura, H. Nishida, M. Takagi and K. Ueno, *Bunseki Kagaku*, **31**, E131(1982).

42) H. Nakamura, H. Nishida, M. Takagi and K. Ueno, *Anal. Chim. Acta*, **139**, 219(1982).

43) H. Dugas and M. Ptak, *J. Chem. Soc., Chem. Commun.*, 710(1982).
44) R.A. Kolinski and J. Mrozinski, *Polyhedron*, **2**, 1217(1983).
45) S. Shinkai, H. Kinda, Y. Araragi and O. Manabe, *Bull. Chem. Soc., Jpn.*, **56**, 559(1983).
46) R. A. Bartsch, Y. Liu, S. I. Kang, B. Son, G. S. Heo, P. G. Hipes, and L. J. Bills, *J. Org. Chem.*, **48**, 4864(1983).
47) B. Czech, A. Czech and R. A. Bartsch, *Org. Prep. Proced. Int.*, **15**, 349 (1983).
48) B. Czech, S. I. Kang and R. A. Bartsch, *Tetrahedron Lett.*, **24**, 457 (1983).
49) B. Czech, B. Son and R. A. Bartsch, *Tetrahedron Lett.*, **24**, 2923(1983).
50) C. Thomas, C. Sauterey, M. Castaing, C. M. Gary-Bobo, J. M. Lehn and P. Plumere, *Biochem. Biophys. Res. Commun.*, **116**, 981(1983).
51) J. F. Koszuk, B. P. Czech, W. Walkowiak, D. A. Babb and R. A. Bartsch, *J. Chem. Soc., Chem. Commun.*, 1504(1984).
52) E. Sonveaux, *Tetrahedron*, **40**, 793(1984).
53) M. Dhaenens, L. Lacombe, J. M. Lehn and J. P. Vigneron, *J. Chem. Soc., Chem. Commun.*, 1097(1984).
54) B. P. Bubnis and G. E. Pacey, *Tetrahedron Lett.*, **25**, 1107(1984).
55) S. Inokuma, T. Kohno, K. Inoue, K. Yabusa and T. Kuwamura, *J. Chem. Soc. Jpn., Chem. and Ind. Chem.*, **8**, 1585(1985).
56) K. Kimura, H. Sakamoto, S. Kitazawa and T. Shono, *J. Chem. Soc., Chem. Commun.*, 669(1985).
57) Y. Katayama, K. Nita, M. Ueda, H. Nakamura, M. Takagi and K. Ueno, *Anal. Chim. Acta*, **173**, 193(1985).
58) J. P. Behr, C. J. Burrows, R. Heng and J. M. Lehn, *Tetrahedron Lett.*, **26**, 215(1985).
59) T. W. Robison and R. A. Bartsch, *J. Chem. Soc., Chem. Commun.*, 990 (1985).
60) B. P. Czech, A. Czech, B. Son, H. K. Lee and R. A. Bartsch, *J. Heterocyclic Chem.*, **23**, 465 (1985).
61) R. J. Adamic, B. A. Lloyd, E. M. Eyring, S. Petrucci and R. A. Bartsch, *J. Phys. Chem.*, **90**, 1659(1986).
62) M. J. Pugia, G. Ndip, H. K. Lee, I. W. Yang and R. A. Bartsch, *Anal. Chem.*, **58**, 2723(1986).
63) A. Anantanarayan, V. A. Carmichael, P. J. Dutton, T. M. Fyles and M. J. Pitre, *Synth. Commun.*, **16**, 1771(1986).
64) E. Uhlemann, H. Geyer, K. Gloe and P. Mühl, *Anal. Chim. Acta*, **185**, 279 (1986).
65) Y. Katayama, R. Fukuda and M. Takagi, *Anal. Chim. Acta*, **185**, 295 (1986).
66) K. Kimura, M. Tanaka, S. Iketani and T. Shono, *J. Org. Chem.*, **52**, 836(1987); H. Sakamoto, K. Kimura, M. Tanaka and T. Shono, *Bull.*

Chem. Soc. Jpn., **62**, 3394(1989).
67) M. Newcomb and D. J. Cram, *J. Am. Chem. Soc.*, **97**, 1237(1975).
68) K. E. Koenig, R. C. Helgeson and D. J. Cram, *J. Am. Chem. Soc.*, **98**, 4018(1976).
69) M. A. McKervey and D. L. Mulholland, *J. Chem. Soc., Chem. Commun.*, 438 (1977).
70) M. Newcomb, S. S. Moore and D. J. Cram, *J. Am. Chem. Soc.*, **99**, 6405 (1977).
71) K. Sugihara, H. Kamiya, M. Yamaguchi, T. Kaneda and S. Misumi, *Tetrahedron Lett.*, **22**, 1619(1981).
72) T. Kaneda, K. Sugihara, H. Kamiya and S. Misumi, *Tetrahedron Lett.*, **22**, 4407(1981).
73) T. W. Bell, P. G. Chang, M. Newcomb and D. J. Cram, *J. Am. Chem. Soc.*, **104**, 5185(1982).
74) C. M. Browne, G. Ferguson, M. A. McKervey, D. L. Mulholland, T. O'Connor and M. Parvez, *J. Am. Chem. Soc.*, **107**, 2703(1985).
75) J. S. Bradshaw, P. Huszthy and R. M. Izatt, *J. Heterocyclic Chem.*, **23**, 1673(1986).
76) M. Newcomb, J. M. Timko, D. M. Walba and D. J. Cram, *J. Am. Chem. Soc.*, **99**, 6392(1977).
77) J. S. Bradshaw, D. A. Chamberlin, P. E. Harrison, B. E. Wilson, G. Arena, N. K. Dalley, J. D. Lamb and R. M. Izatt, *J. Org. Chem.*, **50**, 3065(1985).
78) Y. Nakatsuji, J. S. Bradshaw, P. K. Tse, G. Arena, B. E. Wilson, N. K. Dalley and R. M. Izatt, *J. Chem. Soc., Chem. Commun.*, 749(1985).
79) J. F. Biernat, J. S. Bradshaw, B. E. Wilson, N. K. Dalley and R. M. Izatt, *J. Heterocyclic Chem.*, **23**, 1667(1986).
80) J. S. Bradshaw, Y. Nakatsuji, P. Huszthy, B. E. Wilson, N. K. Dalley and R. M. Izatt, *J. Heterocyclic Chem.*, **23**, 353(1986).
81) J. S. Bradshaw, R. B. Nielsen, P. K. Tse, G. Arena, B. E. Wilson, N. K. Dalley, J. D. Lamb, J. J. Christensen and R. M. Izatt, *J. Heterocyclic Chem.*, **23**, 361(1986); J. S. Bradshaw, C. W. McDaniel, B. D. Skidmore, R. B. Nielsen, B. E. Wilson, N. K. Dalley and R. M. Izatt, *J. Heterocyclic Chem.*, **24**, 1085(1987).
82) J. S. Bradshaw, P. Huszthy, H. Koyama, S. A. Wood, S. A. Strobel, R. B. Davidson, R. M. Izatt, N. K. Dalley, J. D. Lamb and J. J. Christensen, *J. Heterocyclic Chem.*, **23**, 1837(1986).
83) J. S. Bradshaw, H. Koyama, N. K. Dalley, R. M. Izatt, J. F. Biernat and M. Bochenska, *J. Heterocyclic Chem.*, **24**, 1077(1987).
84) J. S. Bradshaw, J. M. Guynn, S. G. Wood, B. E. Wilson, N. K. Dalley and R. M. Izatt, *J. Heterocyclic Chem.*, **24**, 415(1987).
85) R. C. Helgeson, K. Koga, J. M. Timko and D. J. Cram, *J. Am. Chem. Soc.*, **95**, 3621(1973).

86) H. Nakamura, M. Takagi and K. Ueno, *Anal. Chem.*, **52**, 1668(1980).
87) J. P. Shukla, M. J. Pugia, B. Son and R. A. Bartsch, *Thermochim. Acta*, **111**, 257(1987); J. P. Shulka, E. G. Jeon, B. E. Knudsen, M. J. Pugia, J. S. Bradshaw and R. A. Bartsch, *Thermochim. Acta*, **130**, 103(1988).
88) R. J. Adamic, E. M. Eyring, S. Petrucci and R. A. Bartsch, *J. Phys. Chem.*, **89**, 3752(1985).
89) J. M. Timko, R. C. Helgeson and D. J. Cram, *J. Am. Chem. Soc.*, **100**, 2828(1978).
90) H. Sakamoto, K. Kimura and T. Shono, *Anal. Chem.*, **59**, 1513(1987).
91) R. J. Adamic, B. A. Lloyd, E. M. Eyring, S. Petrucci, R. A. Bartsch, M. J. Pugia, B. E. Knudsen, Y. Liu and D. H. Desai, *J. Phys. Chem.*, **90**, 6571(1986).
92) J. Strzelbicki and R. A. Bartsch, *Anal. Chem.*, **53**, 2251(1981).
93) W. Walkowiak, P. R. Brown, J. P. Shukla and R. A. Bartsch, *J. Membrane Sci.*, **32**, 59,(1987).
94) C. A. Chang and M. E. Rowland, *Inorg. Chem.*, **22**, 3866(1983).
95) Y. Chao and D. J. Cram, *J. Am. Chem. Soc.*, **98**, 1015(1976).
96) J. P. Behr and J. M. Lehn, *J. Chem. Soc., Chem. Commun.*, 143(1978).
97) J. M. Lehn and C. Sirlin, *J. Chem. Soc., Chem. Commun.*, 949(1978).
98) C. A. Chang, J. Twu and R .A. Bartsch, *Inorg. Chem.*, **25**, 396(1986).
99) S. S. Moore, T. L. Tarnowski, M. Newcomb and D. J. Cram, *J. Am. Chem. Soc.*, **99**, 6398(1977).
100) P. R. Danesi, E. P. Horwitz, G. F. VandeGrift and R. Chiarizia, *Sep. Sci. Technol.*, **16**, 201(1981).
101) J. P. Behr, M. Kirch and J. M. Lehn, *J. Am. Chem. Soc.*, **107**, 241 (1985).
102) T. M. Fyles, *J. Chem. Soc., Faraday Trans. 1*, **82**, 617(1986).
103) G. F. Vandegrift and W. H. Delphin, *J. Inorg. Nucl. Chem.*, **42**, 1359 (1980).
104) G. F. Vandegrift and E. P. Horwitz, *J. Inorg. Nucl. Chem.*, **39**, 1425 (1977).
105) J. Strzelbicki and R. A. Bartsch, *Anal. Chem.*, **53**, 1894(1981).
106) W. A. Charewicz and R. A. Bartsch, *Anal. Chem.*, **54**, 2300(1982).
107) W. A. Charewicz, W. Walkowiak and R. A. Bartsch, *Anal. Chem.*, **59**, 494(1987).
108) J. Strzelbicki, G. S. Heo and R. A. Bartsch, *Sep. Sci. Technol.*, **17**, 635(1982).
109) J. Strzelbicki and R. A. Bartsch, *Anal. Chem.*, **53**, 2247(1981).
110) W. A. Charewicz and R. A. Bartsch, *J. Membrane Sci.*, **12**, 323(1983).
111) R. A. Bartsch, B. P. Czech, S. I. Kang, L. E. Stewart, W. Walkowiak, W. A. Charewicz, G. S. Heo and B. Son, *J. Am. Chem. Soc.*, **107**, 4997(1985).

112) J. Tang and C. M. Wai, *Anal. Chem.*, **58**, 3233(1986).
113) A. Hriciga and J. M. Lehn, *Proc. Nat'l. Acad. Sci., U.S.A.*, **80**, 6426(1983).
114) R. W. Baker, M. E. Tuttle, D. J. Kelly and H. K. Lonsdale, *J. Membrane Sci.*, **2**, 213(1977).
115) W. C. Babcock, R. W. Baker, E. D. Lachapelle and K. L. Smith, *J. Membrane Sci.*, **7**, 71(1980).
116) R. M. Izatt, G. C. LindH, R. L. Bruening, J. S. Bradshaw, J. D. Lamb and J. J. Christensen, *Pure Appl. Chem.*, **58**, 1453(1986).
117) J. D. Lamb, J. J. Christensen, S. R. Izatt, K. Bedke, M. S. Astin and R. M. Izatt, *J. Am. Chem. Soc.*, **102**, 3399(1980).
118) E. M. Choy, D. F. Evans and E. L. Cussler, *J. Am. Chem. Soc.*, **96**, 7085(1974).
119) L. M. Dulyea, T. M. Fyles and D. M. Whitfield, *Can. J. Chem.*, **62**, 498(1984); P. J. Dutton, T. M. Fyles and S. J. McDermid, *Can. J. Chem.*, **66**, 1097(1988); P. J. Dutton, T. M. Fyles and S. P. Hansen, *J. Inclusion Phenom.*, **7**, 173(1989).
120) R. M. Izatt, D. W. McBride, Jr., P. R. Brown, J. D. Lamb and J. J. Christensen, *J. Membrane Sci.*, **28**, 69(1986).
121) T. M. Fyles, V. A. Malik-Diemer and D. M. Whitfield, *Can. J. Chem.*, **59**, 1734(1981).
122) R. M. Izatt, G. C. LindH, G. A. Clark, J. S. Bradshaw, Y. Nakatsuji, J. D. Lamb and J. J. Christensen, *J. Chem. Soc., Chem. Commun.*, 1676(1985).
123) R. M. Izatt, G. C. LindH, G. A. Clark, J. S. Bradshaw, Y. Nakatsuji, J. D. Lamb and J. J. Christensen, *J. Membrane Sci.*, in press.
124) E. G. Jeon, P. R. Brown and R. A. Bartsch, unpublished results.
125) J. Strzelbicki and R. A. Bartsch, *J. Membrane Sci.*, **10**, 35(1982).
126) R. A. Bartsch, W. A. Charewicz and S. I. Kang, *J. Membrane Sci.*, **17**, 97(1984).
127) P. R. Danesi, *Sep. Sci. Technol.*, **19**, 857(1985).
128) T. M. Fyles and C. McGavin, *Anal. Chem.*, **54**, 2103(1982).
129) T. M. Fyles and D. M. Whitfield, *Can. J. Chem.*, **62**, 507(1984).
130) C. A. Chang and V. O. Ochaya, *Inorg. Chem.*, **25**, 355(1986).
131) R. Delgado, J. J. R. Frausto da Silva and M. Candida T. A. Vaz, *Polyhedron*, **6**, 29(1987).
132) P. L. Zanonato, P. di Bernardo, A. Cassol and G. Tomat, *Polyhedron*, **6**, 417(1987).

10. Appendix A.

This Appendix contains association constant values at 25°C which have been obtained from the chemical literature.

Ligand	Form[a]	Cation	log K	Medium	Ref.
VIII	HL	Na	3.21	acetone	22
X	L	Na	1.30	water	28
	"	K	1.92	"	"
	"	Rb	1.52	"	"
	"	Cs	1.36	"	"
	HL	Na	1.00	"	"
	"	K	1.62	"	"
	"	Rb	1.20	"	"
	"	Cs	1.08	"	"
XXXII	HL	Li	4.62	chloroform-d	24
	"	Na	6.30	"	"
	"	Rb	5.94	"	"
	"	Cs	5.39	"	"
	"	NH4	5.32	"	"
XXXVI	HL	Na	2.35	80% methanol	61
	L	Na	3.27	"	"
	HL	Na	2.72	99% methanol	"
	L	Na	3.93	"	"
XXXVI	HL(pH3)	K	2.23	80% methanol	88
	HL(pH3)	Ba	2.71	"	"
	HL(pH4)	Na	2.58	"	"
	HL(pH5)	Na	2.71	"	"
	HL+L(pH6)	Na	2.93	"	"
	HL+L(pH7)	Na	3.27	"	"
	L(pH9)	Na	3.33	"	"
	L(pH11)	Li	N.R.	"	"
	"	Na	3.27	"	"
	"	K	3.11	"	"
	"	Rb	2.81	"	"
	"	Cs	2.35	"	"
	"	Ba	5.73	"	"
XXXVI	L	Na	4.02	99% methanol	98
	"	K	3.71	"	"
XXXVI	L	Ca	4.10	"	"
	HL	Na	2.75	"	"
	"	K	2.78	"	"
	"	Ca	2.34	"	"
XLII	HL	Na	2.38	80% methanol	61
	L	Na	2.83	"	"
XLIV	L	Na	2.73	"	"
XLV	HL	Na	2.45	"	"
	L	Na	3.2	"	"

Ligand	Form	Cation	log K	Medium	Ref.
CXXIV	HL	Na	2.7	90% methanol	29
	"	K	2.15	"	"
	"	Rb	1.8	"	"
	L	Na	3.4	"	"
	"	K	2.95	"	"
	"	Rb	2.65	"	"
CXXVI	HL	K	2.1	"	"
	L	K	2.9	"	"
CXXVI	L	K	2.9	"	129
	"	Ca	3.5	"	"
	"	Sr	4.1	"	"
	"	Ba	4.4	"	"
	2L(ß)	Ca	7.1	"	"
	"	Sr	8.5	"	"
	"	Ba	8.2	"	"
	HL	K	2.1	"	"
CXXVII	L^{2-}	K	6.3	"	"
	"	Ca	9.4	"	"
	"	Sr	10.1	"	"
	"	Ba	9.8	"	"
	HL^-	K	4.3	"	"
	"	Ca	6.0	"	"
	"	Sr	6.6	"	"
	"	Ba	6.3	"	"
CXXVII	L^{2-}	K	6.1	"	"
(C=D=H)	"	Ca	9.1	"	"
	"	Sr	9.3	"	"
CXXVII	L^{2-}	Ba	9.2	"	"
(C=D=H)	HL^-	K	4.2	"	"
	"	Ca	5.5	"	"
	"	Sr	5.6	"	"
	"	Ba	5.6	"	"
CXXVIII	L^{2-}	K	5.4	"	"
	"	Ca	8.9	"	"
	"	Sr	8.9	"	"
	"	Ba	8.0	"	"
	HL^-	K	3.6	"	"
	"	Ca	5.3	"	"
	"	Sr	5.4	"	"
	"	Ba	4.3	"	"
CXXIX	L^{2-}	K	6.4	"	"
	"	Ca	10.1	"	"
	"	Sr	10.0	"	"
	"	Ba	9.5	"	"
	HL^-	K	4.8	"	"
	"	Ca	N.R.	"	"
	"	Sr	N.R.	"	"
	"	Ba	N.R.	"	"

Ligand	Form	Cation	log K	Medium	Ref.
CXL	HL	Na	3.2	90% methanol	129
	"	K	4.35	"	"
	"	Rb	3.9	"	"
	L	Na	4.3	"	"
	"	K	5.65	"	"
	"	Rb	4.9	"	"
	"	K	5.8	"	"
	"	Ca	5.8	"	"
	"	Sr	5.5	"	"
	"	Ba	5.5	"	"
	2L(B)	Ca	9.9	"	"
	"	Sr	11.4	"	"
	"	Ba	11.4	"	"
	HL	K	4.5	"	"
CXLI	HL	Na	3.1	"	"
	L	Na	4.2	"	"
CXLII	HL	K	4.4	"	"
CLXII	L	Ni	7.39	water	131
	"	Cu	14.49	"	"
	"	Zn	8.42	"	"
	"	Cd	11.07	"	"
	"	Pb	13.55	"	"
CLXII	H2L	Mg	7.42	"	130
	"	Ca	8.74	"	"
	"	Sr	7.91	"	"
	"	Ba	7.31	"	"
	"	Ni	12.23	"	"
	"	Cu	16.02	"	"
	"	Zn	14.08	"	"
	"	Cd	12.95	"	"
	"	Pb	12.91	"	"
	"	La	10.11	"	"
	"	Ce	10.89	"	"
	"	Pr	11.31	"	"
	"	Nd	11.60	"	"
	"	Sm	11.72	"	"
	"	Eu	11.85	"	"
	"	Gd	11.66	"	"
	"	Tb	11.52	"	"
	"	Dy	11.55	"	"
	"	Ho	11.34	"	"
	"	Er	11.15	"	"
	"	Tm	10.79	"	"
	"	Yb	10.76	"	"
	"	Lu	10.33	"	"
	"	Y	10.85	"	"
	"	Ga	13.96	"	"

Ligand	Form	Cation	log K	Medium	Ref.
CLXII	L	Li	2.14	water	131
	"	Na	2.72	"	"
	"	K	1.69	"	"
	"	Mg	7.53	"	"
	"	Ca	8.68	"	"
	Ca2L	Ca	9.9	"	"
CLXII	L	Sr	8.02	"	"
	Sr2L	Sr	9.99	"	"
	L	Ba	7.41	"	"
	"	Mn	12.11	"	"
	"	Fe	(13.0)	"	"
	"	Co	13.72	"	"
	M2L	Co	16.37	"	"
	L	Ni	12.37	"	"
	M2L	Ni	14.3	"	"
	L	Cu	17.79	"	"
	M2l	Cu	22.79	"	"
	L	Zn	14.44	"	"
	M2L	Zn	17.35	"	"
	L	Cd	13.43	"	"
	M2L	Cd	15.62	"	"
	L	Pb	13.26	"	"
	M2L	Pb	15.69	"	"
CLXIII	L	Na	1.95	"	"
	"	K	3.9	"	"
	"	Mg	<2	"	"
	"	Ca	8.39	"	"
	"	Sr	8.29	"	"
	"	Ba	7.63	"	"
	"	Co	7.00	"	"
CLXIII	H_2L^-	K	3.91	"	94
	"	Ca	8.39	"	"
	"	Sr	8.29	"	"
	"	Ba	7.63	"	"
	"	Ni	7.39	"	"
	"	Cu	14.49	"	"
	"	Zn	8.42	"	"
	"	Cd	11.07	"	"
	"	Pb	13.55	"	"
	"	La	12.21	"	"
	"	Ce	12.23	"	"
	"	Pr	12.22	"	"
	"	Nd	12.21	"	"
CLXIII	H_2L^-	Sm	12.12	"	"
	"	Eu	12.02	"	"
	"	Gd	11.93	"	"
	"	Tb	11.70	"	"
	"	Dy	11.57	"	"

Ligand	Form	Cation	log K	Medium	Ref.
	"	Ho	11.18	water	94
	"	Er	11.30	"	"
	"	Tm	11.10	"	"
	"	Yb	10.90	"	"
	"	Lu	10.84	"	"
CLXXIX	HL	Na	0.8	methanol	132
	"	K	1.26	"	"
	"	Cs	1.30	"	"
CLXXXIV	HL	Na	2.25	"	"
	"	K	3.18	"	"
	"	Cs	2.62	"	"
CLXXXIV	HL	Na	2.51	"	71
	"	K	3.56	"	"
CLXXXIX	HL	Na	1.81	"	"
	"	K	2.82	"	"
CXC	HL	Na	2.68	"	"
	L	Na	4.11	"	"
	HL	Na	2.22	99% methanol	"
	L	Na	5.0	"	"
CXCIII	HL	Li	3.34	chloroform	68
	"	Na	3.90	"	"
	"	K	4.61	"	"
	"	Cs	4.40	"	"
CXCIV	HL	Li	3.85	"	"
	"	Na	4.19	"	"
	"	K	5.37	"	"
	"	Rb	5.35	"	"
	"	Cs	5.30	"	"
CXCV	HL	Li	3.69	"	"
	"	Na	4.49	"	"
	"	K	5.02	"	"
CXCV	HL	Rb	4.94	"	68
	"	Cs	5.35	"	"
CCXXIX	L	H	9.55	water	81
	HL	H	2.36	"	"
	L	Ba	1.7	"	"
	2L	Cu	9.54	"	"
	2H +L	Cu	24.91	"	"
	L	Sr	2.26	"	"

a. L = Deprotonated form of the proton-ionizable crown ether; HL = Protonated form of the ligand; N.R. = No detectable reaction; ß = constant for the two reactions involved in the final product.

PRINCIPLES OF ARTIFICIAL MEMBRANE TRANSPORT OF IONS BY SYNTHETIC IONOPHORES

T.M. Fyles
Department of Chemistry
University of Victoria
Victoria, B.C.
CANADA V8W 2Y2

ABSTRACT. The general principles of ion transport across artificial membranes are examined with emphasis on carrier mediated transport across bulk liquid membranes. The theoretical predictions of a simple diffusion model are compared with key experimental example which are shown to be well correlated by such a model. The structural influences which control ion extraction and transport (ion binding sites, lipophilicity, coextracted ions) are examined. Theoretical and experimental criteria for detecting exceptions to the general behavior are outlined and examples are considered in the light of such criteria. The principles reviewed provide design criteria for the development of carriers and their successful exploitation in selective ion separations.

1. INTRODUCTION

1.1 Overview

It is reasonable to suppose that one of the early events in the evolution of the most primitive forms of life was the sequestering of essential material and reactions from the "primordial soup" by means of a primitive cell membrane. In achieving the separation of internal processes from the environment, the primitive cell was confronted with the problem of how to acquire essential materials from the environment in spite of the membrane barrier. Transport processes thus became essential to the survival of the primitive cell, and through evolution, to the survival of modern organisms.[1] One approach to examining natural transport systems, either modern or primitive, is to examine simple artificial systems which mimic certain functions of biological transport. Such studies can intentionally incorporate structural or functional features of natural transporters (known or presumed) or can involve purely technological objectives such as the development of sensors or practical separations.

T. Osa and J. L. Atwood (eds.), Inclusion Aspects of Membrane Chemistry, 59–110.

Synthetic ionophores such as the crown ethers and cryptands have come to stand at the center of both types of rationale and therefore form the focus of this chapter. The main thrust is to examine the principles of cation transport by these materials and some experimental consequences. This entails an examination of diffusion limited carrier mediated transport (section 2). To begin, we take a larger view of transmembrane forces and fluxes, for which carrier mediated transport is but one alternative. In the subsequent sections the main experimental systems and some key examples are examined in the light of the theory developed. The actual review of the literature is sketchy; the reader is directed to the other contributions in this volume and the literature.[2-11]

1.2 Forces and Fluxes

What is the essential definition of a membrane? A membrane is "a phase, finite in space, which separates two other phases and exhibits individual resistances to the permeation of different species".[3] This definition unites two key ideas; the membrane as a **phase** barrier and the membrane as a **inherently selective** barrier. For the former we deal primarily with liquid membranes in which a non polar liquid phase acts as the barrier between two aqueous phases. The latter idea implicitly gives rise to the range of phenomena associated with membranes: osmosis, electrosmosis, Donnan potential, transport of solutes etc.[3] A exhaustive account[12] examines the various phenomena by considering pairwise, the relationship between the fundamental **driving forces** (concentration gradient, potential gradient, temperature and pressure gradients) and the net **flows** produced (flows of solute, solvent, charge, and heat). If we consider only isothermal processes at constant pressure the major driving forces will be the concentration (activity) gradient and the electrical potential gradient across the membrane. These forces act on the species present to produce net flows which give rise to the different phenomena noted above.

Simply stated, the flux of a species A (J_A) might be given by a relationship such as:

$$J_A = L_{AA}\,\Delta\mu_A + L_{AE}\,\Delta\phi \qquad (1)$$

where $\Delta\mu_A$ is the chemical potential gradient of A across the membrane, $\Delta\phi$ is the transmembrane electrical potential gradient, and L_{AA} and L_{AE} are phenomenological coefficients which describe how A responds to the driving forces. In addition, the flux J_A might be influenced by the fluxes of other species B_1, $B_2 \ldots B_n$ and might also be influenced by some reaction flux J_R. Thus in a more general form (1) becomes:

$$J_A = L_{AA}\Delta\mu_A + L_{AE}\Delta\phi + \sum_{B\neq A} L_{AB}J_B + L_{AR}J_R \qquad (2)$$

where L_{AB} and L_{AR} are again coupling coefficients. Although quite general, (2) is uniformative as to both the nature of the coefficients and the molecular mechanisms involved. Nonetheless several fundamental ideas can be derived by consideration of simplified versions of (2).

i) Assume $\Delta\phi = 0$, $L_{AB} = 0$ for all B, $L_{AR} = 0$; hence:

$$J_A = L_{AA}\Delta\mu_A = \frac{D\Delta C_A}{\ell} \quad (3)$$

The chemical potential gradient in this case is due solely to the concentration (activity) gradient of A hence $\Delta\mu_A$ may be replaced by $\Delta C_A = A_{in} - A_{out}$.* In this case L_{AA} is readily identified as the permeability coefficient of Fick's law, in turn related to the diffusion coefficent D and the diffusion length, ℓ. This case is called **passive diffusion** in which J_A depends solely on the concentration gradient of A.

ii) Assume $\Delta\phi = 0$, $L_{AR} = 0$, $L_{AB} \neq 0$ for one B; hence:

$$J_A = L_{AA}\Delta\mu_A + L_{AB}J_B \quad (4)$$

The flux in A has a composite dependence, in part dependent upon its own concentration gradient and in part on the flux of B. This general case is described as the **coupled diffusion** of A and B. It may also be a passive diffusion example depending on the nature of J_B.

iii) Assume $\Delta\phi = 0$, $L_{AB} = 0$ for all B, $L_{AR} \neq 0$; hence:

$$J_A = L_{AA}\Delta\mu_A + L_{AR}J_R \quad (5)$$

The flux in A has again a composite dependence, in part passive, but in part dependent upon the reaction flux J_R. This is the general case of **active transport** where J_A is coupled to a reaction.
In favourable cases, the free energy change of the reaction drives A against its concentration gradient but this is not an essential relationship.

1.3 Facilitated Diffusion

The general cases discussed above make no assumptions about the molecular details of the translocation of A and B across the membrane. In a biological context it is axiomatic that specific transporters assist this process and provide the framework for both active and passive transport.[13,14] In the presence of such agents the transport is **facilitated** (hence facilitated passive diffusion etc.). Several modes of action can be envisaged for the transporter.

---

* The designations in/out are arbitrary. In general $[A]_{in} > [A]_{out}$. The square brackets denoting concentrations are deleted throughout hence A_{in} means both the species and its concentration.

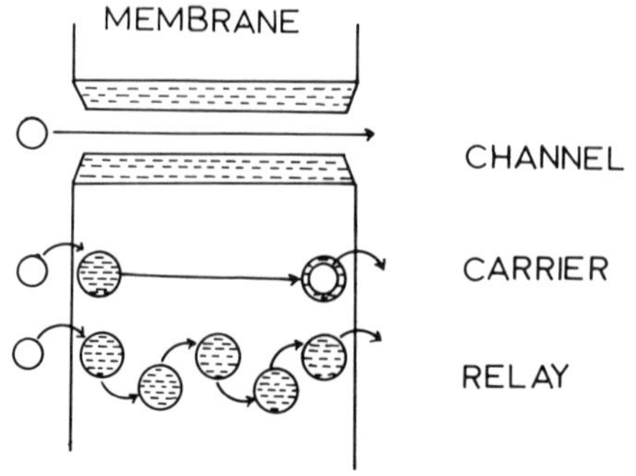

Figure 1: Schematic transport. ○ transported substrate species, carrier or binding site, channel forming compounds.

The transporter could form a channel, spanning the membrane, which provides a suitable internal environment for the diffusion of the substrate across the membrane. Alternatively the transporter could function as a carrier and transport would occur via diffusion of the substrate with the carrier. A third option is a relay in which the substrate hops between discrete binding sites within the transporter. A trivial analogy is presented by a river; one can cross by bridge or tunnel (channel), by boat (carrier) or on stepping stones (relay). Of course one could wade or swim but this is non-facilitated transport!

In comparison with non-facilitated transport, the transporter results in an increase in the permeation of the substrate. This is intuitively appreciated by consideration of the energetics of ion translocation across a non polar liquid membrane. Within the membrane, the ion has lost a great deal of solvation energy due to transfer from water to the low dielectric medium of the membrane. This loss can be considerable, up to 70 kcal/mole for a monovalent ion of 1Å radius in the middle of a lipid bilayer.[8] This loss of solvation energy is the energy barrier to translocation and because of its magnitude, results in slow permeation. The transporter functions to diminish the unfavorable energy cost of the translocation by providing, in the case of ions, a polar environment which compensates for the lost hydration energy. In essence the transporter functions as a **catalyst of translocation** by providing a lower energy pathway.

Incorporation of facilitated diffusion into (2) is straightforward but uninformative. It provides additional terms but the molecular mechanisms and the nature of the coefficients are still unknown. It will be more efficient to discuss the possibilities using schematic mechanisms.[14] For simplicity assume a carrier mechanism in which the carrier (L) and its complexes (LA, LB, LAB) remain membrane bound and where passive diffusion of A, B is negligible. As above, we assume that formation of the complex with the carrier facilitates diffusion.

Figure 2 illustrates several cases. Begin with Figure 2a; at the left interface A and B are coextracted by L to form the membrane complex LAB. This complex traverses the membrane by diffusion to the right interface where the reverse extraction occurs. Back diffusion of L

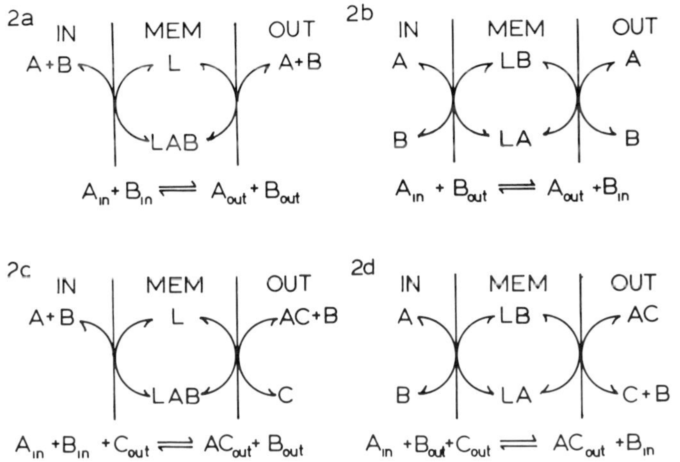

Figure 2: Schematic transport cycles: a) symport gradient pumping; b) antiport gradient pumping; c) symport reaction pumping; d) antiport reaction pumping.

completes the cycle. The substrate A and B travel in the same direction and their fluxes are coupled. This case is referred to as **cotransport or symport coupling.** In this case as well, the equilibrium noted defines an equilibrium constant. Since the two aqueous phases (IN, OUT) are identical, this constant will be unity or:

$$\frac{A_{out}}{A_{in}} = \frac{B_{in}}{B_{out}} \quad (6)$$

at equilibrium. Clearly the system will respond to maintain the relationship (6). For example, if a gradient in B is imposed ($B_{out} > B_{in}$) then A will be moved $A_{out} \rightarrow A_{in}$ until (6) is satisfied. This can result in the accumulation of A against its concentration gradient hence the designation **gradient pumping.** The scheme illustrated in 2b is closely similar with the principal difference that A and B move in opposite senses **(antiport coupling** or **coupled counter transport).** Again an imposed gradient in one species can drive the other against its concentration gradient (gradient pumping).

The remaining two cycles (2c/2d) are similar with respect to the composition of the IN and MEM phases; 2c is a symport and 2d an antiport in direct parallel with 2a/2b. The principal difference lies in the coupled reaction of A with C in the OUT phase to produce the covalent species AC. These cases are **active transport** cases and can result in **reaction pumping.** For example, addition of AC to system 2c at equilibrium will result in moving some B from OUT to IN. The transport of B is coupled the reaction AC $\rightarrow$ C + A. If $B_{in} > B_{out}$, then reaction pumping has occurred.

An analysis of the steady state and equilibrium positions of the systems of Figure 2 has been developed[14] using an approach from non-equilibrium thermodynamics. More generally there have been three main thrusts to the analysis of membrane behavior[3,12]: i) use of flux equations (Fick's law, Nernst-Planck flux equation)[3,15], ii) use of Eyring rate theory[16,17] and iii) use of laws of irreversible thermodynamics [14,17,18]. It has been established that all three approaches lead to identical or very similar results[3,17] and the formalism chosen will be primarily based upon the needs and predispositions of the experimentalist involved. The desire to link specific molecular properties of the transporter to the observed transport behavior tends to favor a choice of i) above. It will be this approach which is utilized below.

2. PRINCIPLES OF CARRIER TRANSPORT

Many of the developments in artificial membrane transport have explicitly or implicitly assumed a mechanism of facilitated diffusion using carriers. Various mathematical descriptions of carrier mediated transport mechanisms have appeared over the past 25 years[19-32] with experimental verifications in a biological,[22,23] engineering[24-27] and chemical context[28-32] spread over the same period. The field is **not** new, but despite the flurry of activity there remain a range of implications which are rarely considered. The goal of this section is to develop a basic framework within which to discuss the experimental examples of subsequent sections. This section also provides essential criteria for the design and execution of membrane experiments using carriers.

2.1 Assumptions[3,32]

a1) There are no gradients of pressure, temperature or electrical potential across the membrane. The sole driving force is the chemical potential gradient in the form of a concentration (activity) gradient.

a2) Thermodynamic equilibrium exists in all three phases and at the interfaces. This is equivalent to the assumption that interfacial reactions are fast with respect to diffusion.[3,19,25-27,32] The case of slow interfacial reaction is examined in sections 2.7 and 2.8.

a3) For simplicity assume the membrane is symmetrical. In the case of bulk liquid membranes (see 3.1) the three phases are stirred and the diffusion is limited to the unstirred boundary layers adjacent to the interfaces.[24-32]. The aqueous boundary layers can be included[29-31] or ignored[32] as their effect is usually negligible. Thus this symmetry assumption includes: equal aqueous volumes, equal interfacial areas and equal boundary layer thickness at the two interfaces. The assumptions are not essential, merely convenient.

a4) The carrier L and its complexes are entirely partitioned to the membrane phase while the substrates are entirely partitioned to the aqueous phases. Non facilitated transport is ignored.

a5) The diffusion coefficients of the carrier and its complexes are assumed to be equal.

This is not an essential set of assumptions as other approaches proceed from different ones[3,21,25] and yield identical results. Further, the assumptions a1 to a5 will be successively relaxed in subsequent sections as required for the experiment considered. These assumptions merely represent a convenient point of departure.

2.2 Single Substrate

The simplest process is given schematically in Figure 3; a substrate A is taken from the IN phase into the membrane to form the complex LA. the complex traverses the two boundary layers by diffusion to release A to the OUT phase. The process is driven by the concentration gradient in A; $A_{in} >> A_{out}$.

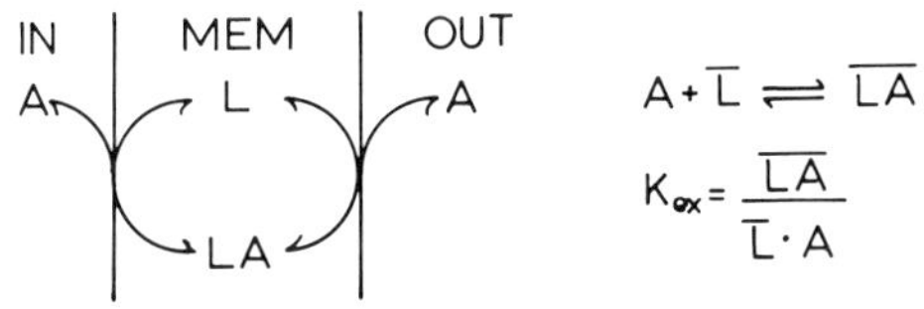

Figure 3: Schematic transport of a single substrate A by a carrier L. The overlined terms are concentrations in the membrane.

Since complex formation and dissociation is fast (equilibrium assumption) the concentration of various species at the interfaces are related through the overall extraction equilibrium constant, K_{ex}. The symmetry assumption requires that all concentration gradients be equal at the steady state; some examples are illustrated in Figure 4. The steady state flux of A is thus given by Fick's law (3) as:

$$J_A = \frac{D}{\ell}(LA_{i,in} - \overline{LA}) = \frac{D}{2\ell}(LA_{i,in} - LA_{i,out}) \tag{7}$$

where the i subscript indicates the interfacial concentration (aqueous boundary layers ignored).

The equilibrium assumption also provides a mass balance equation since the total concentration of carrier at every point must be equal (L_o). As a result, the concentration terms of (7) can be replaced by expressions involving K_{ex}, L_o and the observable bulk concentrations A_{in} and A_{out}. The final form of (7) is thus[32]

$$J_A = \frac{DL_o K_{ex}}{2\ell} \frac{A_{in} - A_{out}}{(1 + K_{ex}A_{in})(1 + K_{ex}A_{out})} \tag{8}$$

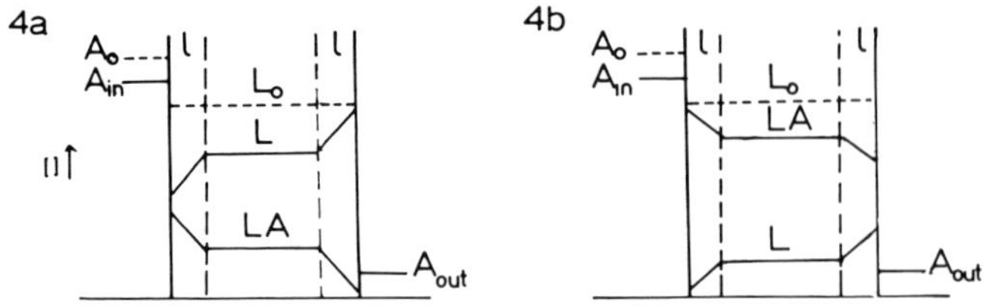

Figure 4: Steady state concentration profiles: a) moderately high carrier saturation at the IN interface, b) high carrier saturation at both interfaces. Redrawn from ref. 32 with permission of the American Chemical Society.

Consideration of some limiting conditions on (8) is instructive. Assume that initially $A_{out} = 0$ and that throughout the experiment $A_{in} >> A_{out}$.

i) When K_{ex} is small $K_{ex}A_{in} << 1$ (and $K_{ex}A_{out} << 1$) thus (8) becomes:

$$J_A \sim \frac{DL_o}{2\ell} K_{ex} A_{in} \tag{9}$$

This is a simple equation, first order in both L_o and A_{in}. Further, J is a linear function of K_{ex}, but the flux is low because K_{ex} is small.

ii) As K_{ex} increases the term $(1 + K_{ex}A_{in})$ cannot be ignored, thus (8) becomes:

$$J_A \sim \frac{DL_o}{2\ell} K_{ex} \frac{A_{in}}{(1 + K_{ex}A_{in})} \tag{10}$$

Equation (10) is identical in form to the Michealis-Menton equation derived from enzyme kinetic[32,33] and thus J_A behaves as a saturable process. As A_{in} increases, J_A eventually achieves a maximum at

$$J_{max} = \frac{DL_o}{2\ell} \tag{11}$$

At the maximal flux, the carrier is half saturated and the rate is independent of substrate concentration and K_{ex}. This case is discussed further in section 2.6.

iii) On further increase in K_{ex}, the $(1 + K_{ex}A_{out})$ term also cannot be ignored and (8) becomes:

$$J_A \sim \frac{DL_o}{2\ell} K_{ex} \frac{A_{in}}{(1 + K_{ex}A_{in})(1 + K_{ex}A_{out})} \tag{12}$$

The concentration profiles for this case are illustrated in Figure 4b. It is clear that the carrier is saturated at both interfaces and a further increase in K_{ex} will result in a **decrease** in J_A (already small). The Michealis-Menton approach is **invalid** under these circumstances.

The important conclusion from this simple examination is that the transport flux (or rate) is not **a simple function of either the concentration of A nor of its binding affinity with the carrier.** A more detailed view can be derived from an explicit solution of (8). This can be examined in terms of the parameter J/J_{max} as a function of the equilibrium extraction constant K_{ex}. Since A_{in} and A_{out} must change during the experiment, the form of the function will vary with the extent of transport (f). If A_{out} is initially zero, the concentration in the OUT phase can be expressed as a fraction of the initial concentration of A in the IN phase (A_o).[32] There are thus two input parameters of interest (L_o and A_o) as well as the various stages of the experiment (range of f; $0 \leqslant f \leqslant 0.5$). This function can be computed[32,34] and some examples are given in Figure 5.

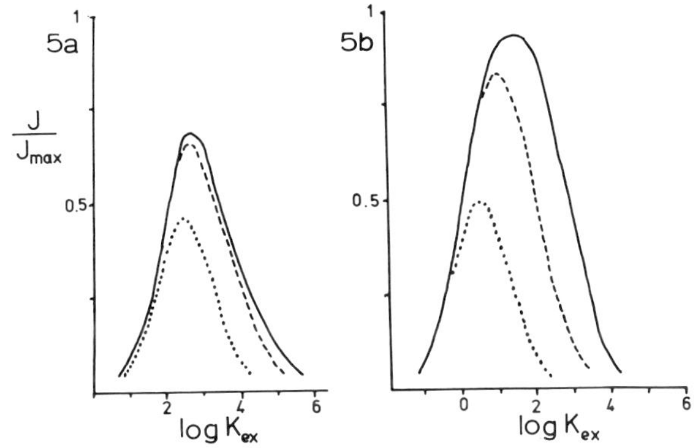

Figure 5: J/J_{max} as a function of K_{ex} for a single substrate.[34] a) $A_o = 0.01M$, $L_o = 0.001M$, b) $A_o = 1.0M$, $L_o = 0.0001M$. The function also depends on the extent of transport f; ——— f = .1%, - - - - - - - f = 1%, ··········· f = 10% of A_o transferred from IN to OUT.

The general form of the curves is as deduced above showing a maximum at some value of K_{ex}. Clearly the range of optimal K_{ex} is a function of the initial conditions of the experiment. Further, since a given experiment occurs with fixed A_o and L_o, the observed rate will be a vertical section through the curves (range of f). At $K_{ex} < K_{ex}(max)$, the rate will be apparently a constant but at $K_{ex} > K_{ex}(max)$, may diminish rapidly with extent of transport. In this latter instance the concept of an "initial rate" is invalid as is the use of Michealis-Menton kinetics.

Experimental confirmation of this was given by Wilbrandt 30 years ago.[23] Since electroneutrality must be preserved (a1, a2 above) both L and A must be neutral. Consequently this case is most often associated with facilitated diffusion of neutral molecules: glucose,[23] NO,[24] CO and O_2[35] and aromatics.[36]

2.3 Cotransport - Facilitated Diffusion of Ion Pairs

This is the case most commonly examined for the transport of cations with their associated anions.[5,6,10,25-27,29-31] The transport cycle is illustrated in Figure 6 where K_{ex} is defined. Note that K_{ex} is different from the previous case.

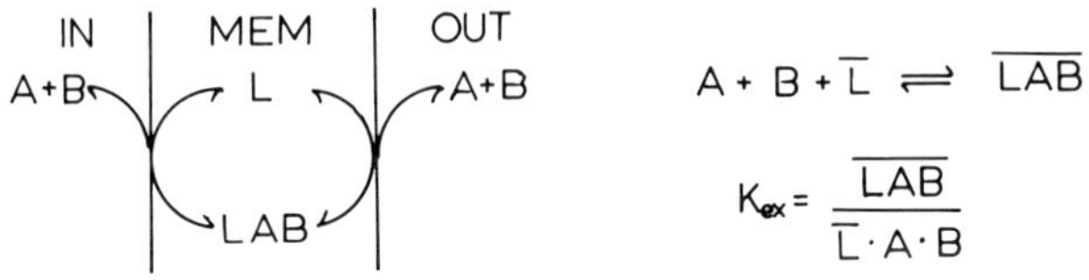

Figure 6: A schematic mechanism for cotransport

If we restrict ourselves to salts of monovalent cations and set $M^+ = X^- = A_o$ then the flux equation corresponding to (8) is:[32]

$$J_A = \frac{DL_o}{2\ell} K_{ex} \frac{A_{in}^2 - A_{out}^2}{(1 + K_{ex}A_{in}^2)(1 + K_{ex}A_{out}^2)} \tag{13}$$

The forms of (8) and (13) are very similar, differing primarily in the squared terms in A_{in} and A_{out}. This is in fact due to the product of the cation and anion concentration[25,26] and relates to the presence of both species in the slow step (diffusion) of the cycle.

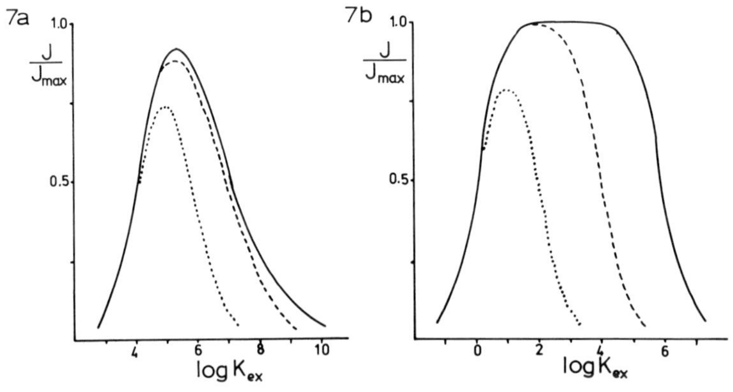

Figure 7: J/J_{max} as a function of K_{ex} for coupled cotransport.[32,37] Initial conditions and fractions of transport as in Figure 5.

As previously, J/J_{max} as a function of K_{ex} may be computed (Figure 7).[32,37] The curves closely resemble the previous case and the conclusions regarding the experimental behavior to be expected are similar. Note especially that for $L_o << A_o$ (7b) there is a very broad maximum where the "initial rate" is essentially invariant with K_{ex}. All carriers might then show identical transport fluxes for an experiment under these conditions. Normally L_o is in the range of millimolar and A_o is typically 5×10^{-2}M to 5×10^{-1}M. Under these conditions the range of K_{ex} for optimum transport ($J \sim J_{max}$) is $4 < \log K_{ex} < 6$. This is the origin of the oft quoted maxim of maximal rate at these values of extraction constant.[10,29-31] A discussion of the relevant experimental demonstrations of this case is deferred to section 4.

2.4 Countertransport

This condition typically occurs when the carrier bears a net charge, for example as a result of deprotonation[7,28] The electroneutrality condition requires the back transport to occur with a counterion to give rise to the situation depicted in Figure 8. The extraction constant is again redefined and represents the relative affinity of the carrier for the two substrates.

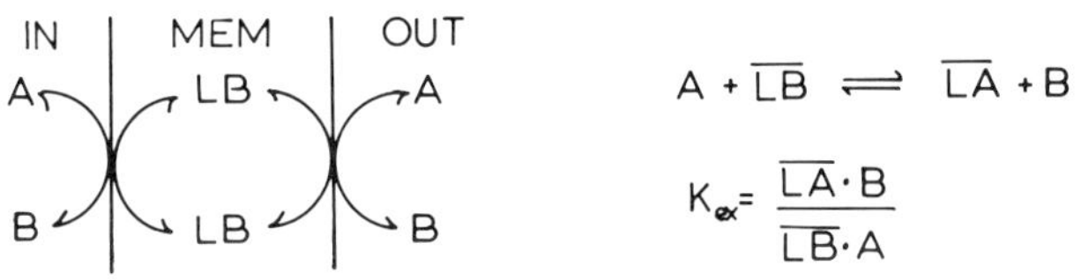

Figure 8: A schematic mechanism of countertransport.

Assume that initially only A occurs on the IN side and only B occurs on the OUT side and $A_{in} = B_{out} = A_o$ (and $A_{out} = B_{in} = 0$) at time = 0. As the transport proceeds $A_{in} = B_{out}$ and $A_{out} = B_{in}$ by the imposed stoichiometry of the coupling. Thus we can define $S_{in} = A_{in} = B_{out}$ and $S_{out} = A_{out} = B_{in}$ (hence $S_o = S_{in}$ at t = 0). Under these restraints (8) becomes:[32]

$$J = J_A = -J_B = \frac{DL_o K_{ex} S_o}{2\ell} \frac{S_{in} - S_{out}}{(S_{in} + K_{ex}S_{out})(S_{out} + K_{ex}S_{in})} \tag{14}$$

As in the previous cases J will increase with K_{ex} and S_o up to J_{max} whereupon J will decrease as carrier saturation occurs. The representation of J/J_{max} as a function of $\log K_{ex}$ shown in Figure 9[38] shows major differences with respect to previous cases in that the curves are symmetrically disposed around $\log K_{ex} = 0$, i.e., equal affinity of A and B for L. The maximal transport rate is thus totally independent of L_o, S_o, and f and occurs when ligand L has the **same affinity for both A and B.** The effect of extent of transport on "initial rate" now occurs for both $K_{ex} < K_{ex}(max)$ and $K_{ex} > K_{ex}(max)$. Again, the discussion of experimental examples is deferred until section 5.

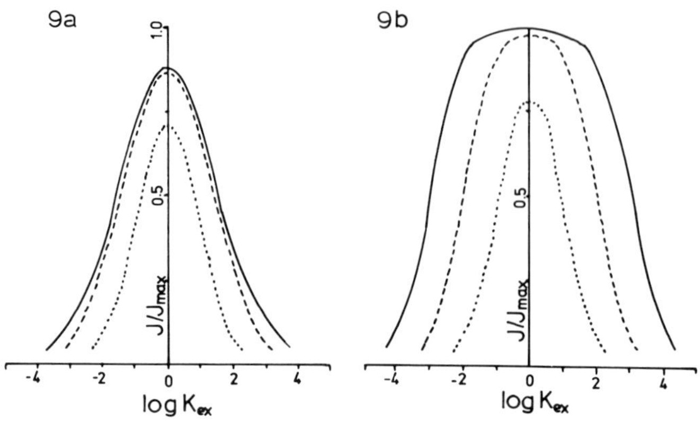

Figure 9: J/J_{max} as a function of K_{ex} for coupled countertransport.[32,38] The initial conditions and fractions of transport are the same as Figure 5.

2.5 Transport Selectivity - Substrate Competition

Inherent in the definition of membrane and the implicit goal of many transport experiments, is the idea of selective transport of one species from a mixture. Consider one case as illustrated in Figure 10 for the competitive transport of A and B coupled to the countertransport of C.

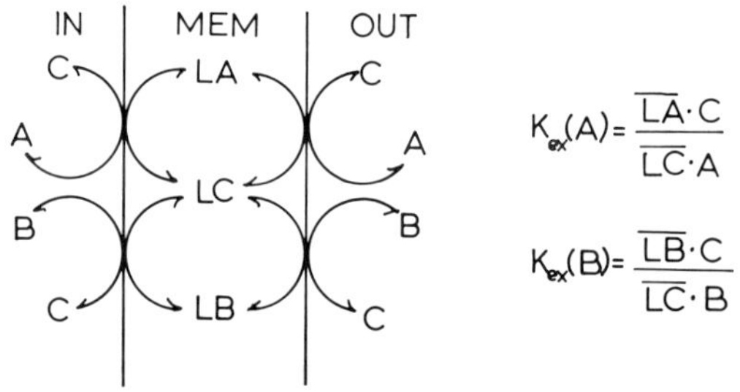

Figure 10: Competitive transport of A and B with countertransport of AC.

The general behavior of this system can be inferred qualitatively with reference to Figure 9 for a single cycle of countertransport. Assume that $K_{ex}(A) = 10\ K_{ex}(B)$. The thermodynamic extraction selectivity $K_{ex}(A)/K_{ex}(B) = 10$ indicates that a ten-fold selective separation of A in preference to B can in principle be achieved. Whether this in fact occurs during the experiment depends on the relative values of $K_{ex}(A)$, $K_{ex}(B)$ and $K_{ex}(max) = 1$. If $K_{ex}(B) < K_{ex}(A) \lesssim 1$, both species lie on the left hand side of the curves of Figure 9, $J_A > J_B$ and A will be initially selectively transported at near the expected value of 10A to 1B. This ratio will diminish as transport proceeds (a vertical slice through the curves) and this effect will depend on the initial conditions of concentration. In general however the transport can closely mirror the thermodynamic extraction selectivity for modest extents of transport. If on the other hand $1 < K_{ex}(B) < K_{ex}(A)$ then both species lie on the righthand side of the curves of Figure 9. As before, A is preferentially extracted but its rate is inherently slow compared to B. In this case the thermodynamic extraction selectivity is achieved only briefly and rapidly falls to modest levels as transport proceeds.

The argument above is qualitatively correct; computation of the explicit result is more difficult as the coupled fluxes of A, B and C must be calculated by numerical simulation of the concentration/time profiles of the species involved.[32] As is evident from the Figures above, the rate of transport can fall off quite quickly for some sets of initial conditions. Thus the commonly used selectivity criterion J_A/J_B is experimentally difficult to obtain. A more useful criterion is the separation actually achieved; if $A_{out} = B_{out} = 0$ initially then the selectivity can be defined as A_{out}/B_{out}. This will vary with the extent of transport (time) as illustrated in Figure 11, and of course will vary with the values of $K_{ex}(A)$ and $K_{ex}(B)$ as discussed above.

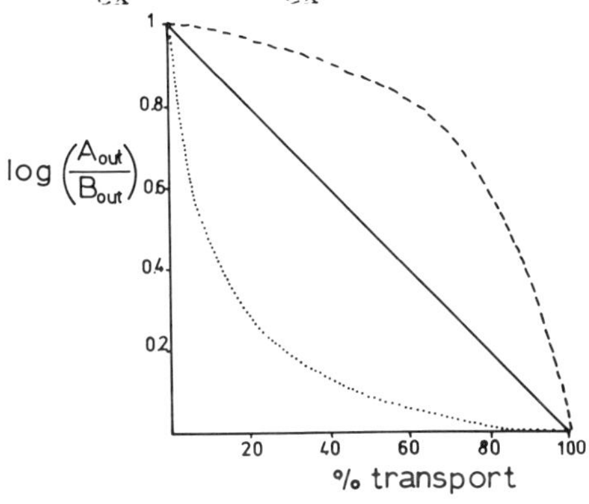

Figure 11: Variation in transport selectivity ($\log (A_{out}/B_{out})$) as a function of % transport for competetive transport of A and B with a counterflux of C (see Figure 10). Initial conditions: $C_{out} = 2A_{in} = 2B_{in}$ and $C_{in} = A_{out} = B_{out} = 0$. ——— $K_{ex}(A) = 10$, $K_{ex}(B) = 1$; ----- $K_{ex}(A) = 100$, $K_{ex}(B) = 10$; .... $K_{ex}(A) = 1$, $K_{ex}(B) = 0.1$. Redrawn from ref. 32 with permission of the American Chemical Society.

Figure 11 confirms the qualitative conclusions above, namely that when both extraction constants lie on the left of $K_{ex}(max)$ the extraction selectivity is sustained, while on the right, rapid loss of selectivity occurs. The membrane is **always** "selective for A" but the separation efficiency is much better in one case than in the other. The separation efficiency (for $K_{ex}(A) > K_{ex}(B)$) will be $1 \leq A_{out}/B_{out} \leq K_{ex}(A)/K_{ex}(B)$ in all symmetrical cases. In order to maximize the separation efficiency the K_{ex} for the best extracted substrate should lie equal to $K_{ex}(max)$ which in turn will be a function of **both** the carrier and the **experimental conditions.** This general result holds for cotransport cases as well but does not hold if the membrane is asymmetric (K_{ex} different at the two interfaces).[32] Experimental examples are discussed in sections 4 and 5.

2.6 Michealis-Menton Kinetics

Transport kinetics are often treated in the same fashion as enzyme kinetics due to formal similarities between the two processes.[33] As is clear from the above analyses, if $K_{ex} < K_{ex}(max)$ the transport will behave as a saturable process. In the case of an enzyme, the substrate binds to the enzyme to form a complex which in a slow step goes on to products. This is formally the same as a substate and carrier forming a complex which in a slow diffusional step results in translocation. The view of the carrier as a catalyst of translocation has been discussed above (1.3). For either process then, the rate (flux) can be related to the maximal rate (flux) through the Michealis-Menton equation:

$$\frac{V}{V_{max}} = \frac{S}{K_m + S} \text{ (enzyme) or } \frac{J}{J_{max}} = \frac{A}{K_m + A} \text{ (transport)} \qquad (15)$$

where S and A are the substrate concentrations, V and J are the rates of the two processes and K_m is the **disassociation** constant of the enzyme-substrate complex (or the carrier-substrate complex). Thus $K_m = 1/K_{ex}$ as defined above since K_{ex} is effectively the **association** constant of the carrier-substrate complex. With this relationship, (15) is identical to (10) with the definition of J_{max} given in (11) and thus (15) will hold when the appoximations giving (10) and (11) are valid i.e., $A_{in} >> A_{out}$ and $K_{ex} < K_{ex}(max)$.

Having recognized the formal relationship, (15) can be exploited using the methods of enzyme kinetics to yield the parameters of interest J_{max} and K_m (= $1/K_{ex}$). Rearrangement of (15) gives two equivalent linear forms; the Lineweaver-Burke (double reciprocal form) (16) and the Eadie-Hofstee form (17):[33]

$$1/J = \frac{K_m}{J_{max}} (1/A) + 1/J_{max} \qquad (16)$$

$$J = -K_m (J/A) + J_{max} \qquad (17)$$

Experimental values of J at various concentrations of A can be treated by either (16) or (17) to give K_m and J_{max} graphically. In the double reciprocal form $1/J_{max}$ appears as an intercept and K_m can be evaluated from the slope (K_m/J_{max}) or the intercept at $1/J = 0$ ($-1/K_m$). In the Eadie-Hofstee form, J_{max} again appears at the intercept and $-K_m$ can be derived from the slope. Both methods are widely used. Note as well that J_{max} contains the initial concentration of L_o An increase in L_o will increase the apparent J_{max} but will leave the affinity of the carrier for A unaltered. From (16) this will give a common intercept at $1/A = 0$ with the slope increasing as L_o, while from (17) the intercept will increase as L_o with a constant slope.

The selectivity of a carrier for two substrates can be examined as "competitive inhibition" of the transport of one substrate by another. An example would be the case illustrated in Figure 10 in which $K_{ex}(A)$ and $K_{ex}(B)$ are both less than $K_{ex}(max)$. Binding of one substrate e.g., B will diminish transport of A by decreasing the amount of L, and hence the amount of LA at equilibrium. The actual affinity of L for A is unaltered but the apparent affinity is diminished by an amount $B/K_m(B)$. Thus

$$K_m(A) \text{ apparent} = K_m(A)\ (1 - B/K_m(B)) \qquad (18)$$

At high concentrations of A, the carrier can still be shifted virtually entirely to the LA form, hence J_{max} is unaltered. In the double reciprocal form this gives a family of lines with a common intercept at $1/A = 0$ and with increasing slope with an increase in B. The Eadie-Hofstee form also given a common intercept for a family of lines of increasing slope with B. Numerous other competitive situations can be envisaged and treated similarly. Usually the parameters $K_m(A)$ and $K_m(B)$ will be accessible from single ion experiments and the validity of (18) can be directly determined. The advantage of these Michealis-Menton methods (when applicable) is that they give a rapid view of two parameters of interest namely K_{ex} and J_{max}. The former, in many cases, is not known from direct two phase extractions and the latter contains information on the diffusion coefficient D and the thickness of the diffusion layer ℓ (in principle).

2.7 Diffusion and Kinetic Regimes

To this point, we have regarded the interfacial region as a black box into which reactants disappeared, and from which translocating products appeared. This has provided a simple framework for analysis but has obscurred molecular detail. The purpose of this section is to focus on some general, but assumed, properties of the interface so that the nature of deviations from the above diffusional analysis can be more clearly recognized.

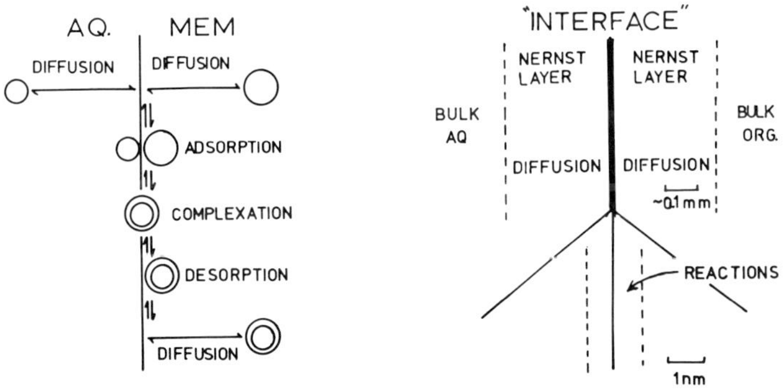

Figure 12: Two schematic views of a hypothetical planar interface.

Figure 12 (left) shows one general sequence of events for the functioning of the black box. Obviously others are possible but this is a reasonable minimum set. The cation (and anion) diffuse to the interface from the aqueous side and the empty carrier diffuses to the same interface from the organic side. Specific adsorption steps may occur[39] as well as the complexation and ion exchange reactions. Desorption (if necessary) followed by diffusion across the boundary layer in the stirred (bulk) organic phase then completes the extraction. As illustrated in Figure 12 (right), these processes occur on different scales. The aqueous and organic boundary layers where the diffusion steps occur are of the order of 100 μm thick whereas the interfacial reaction zone must be of molecular dimensions, perhaps 20-50 Å at most. At the molecular level the trivial concept of the planar interface vanishes but the scale of events is not altered.

This picture of the interface gives rise to three possible regimes governing mass transport:[40]

i) a "diffusion regime" in which chemical processes in the interfacial reaction zone are fast with respect to the time required to traverse the diffusion films (assumed above).
ii) a "kinetic regime" in which the chemical processes in the interfacial region are slow with respect to diffusion and;
iii) a "mixed regime" in which the two processes are of comparable rates.

One approach to determining the particular regime which is responsible for the transport behavior is to examine the rate of transport (flux) as a function of the stirring rate in the cell.[40] In general, at low stirring rates, the unstirred layers will be thick and diffusion will be slow. Under ideal hydrodynamic conditions the boundary layer thickness is a linear function of stirring rate hence the transport flux should increase in a linear fashion with stirring rate. Ideal hydrodynamics are rarely encountered (3.1) but there is nonetheless some general relationship between boundary layer thickness and stirring at low stirring rates. Any process which is controlled by the diffusion regime will therefore be accelerated by an increase in stirring and this regime can be unambiguously identified.[41] Should the process of interest be in the "kinetic regime" there will be no stirring effect since no amount of mixing on the macroscopic scale will alter events on the molecular scale (short of catastrophic collapse of the interfacial geometry). Thus in general a plot of flux as a function of stirring rate will initially increase in a "linear" fashion and may reach a plateau value at the onset of the kinetic regime. In the transition region the mixed regime is present.

2.8 Complexation Kinetics and Transport

Among the initial assumptions (2.1), one of the most problematic (to chemists!) is the assumption that all reactions are fast and equilibrium is achieved throughout the system. Other assumptions can be made concerning the interfacial exchange process: that it is rate limiting, that a mixed regime occurs, and so on.[20,21,32,42,43] In general, no matter what assumption is made, the general form of the resultant equation is similar to those developed above. The major differences occur at high carrier saturations; for a diffusional process J_{max} occurs for half saturated carrier while for a process controlled by the rate of complexation the maximal flux occurs at fully saturated carrier.[32]

The general influence of complexation kinetics in controlling transport can be estimated by examining the characteristic relaxation times associated with the two processes.[21] This will be the same as comparing the characteristic "relaxation distances" for the two processes.[42] The latter converts complex lifetimes into distances and the former compares the lifetime of the complex LA to the time required to traverse the total diffusion length.[21] It is this approach which is outlined below.

The complexation/decomplexation reaction for the simple case of section 2.2 is:

$$\overline{L} + A \underset{k_d}{\overset{k_f}{\rightleftharpoons}} \overline{LA} \qquad K_{ex} = k_f/k_d \tag{19}$$

where k_f and k_d are the rates complex formation and dissociation. The relaxation time for the complex (τ_{cplx}) is then given as:[21]

$$\tau_{cplx} = k_f\,\overline{L} + k_d \tag{20}$$

The corresponding relaxation time for diffusion (τ_{diff}) is given as:[21]

$$1/\tau_{diff} = 2D/\ell^2 \tag{21}$$

where D is the diffusion coefficent (of the complex) and ℓ is the boundary layer thickness.

The total flux, due to both diffusion and complexation rate limited processes is then given by (21):[21]

$$J_A = J_A^e \left(1 + \frac{\tau_{cplx}}{\tau_{diff}}\right) \tag{22}$$

where J_A^e is the flux determined from the assumption of rapid equilibrium. For the case of (19) this is J_A as given by (8). Other transport cycles will give different derived forms for both J_A^e and τ_{cplx} but these will be directly analogous to the case outlined above. The important conclusion is clear: if $\tau_{cplx} << \tau_{diff}$ then $J_A \sim J_A^e$ as provided by the analysis of sections 2.2 - 2.5. For $\ell \sim 100$ µm and $D \sim 10^{-5}$ cm^2 sec, τ_{diff} is of the order of 5 seconds. In order to achieve τ_{chem} of this magnitude the complex must be both stable and relatively kinetically inert. For the crown ethers and cryptands discussed below, this is rarely achieved hence transport limited by the kinetics of (19) is essentially never observed. Specific cases are discussed in sections 4 and 5.

3. EXPERIMENTAL ARTIFICIAL MEMBRANE SYSTEMS

3.1 Liquid Membranes

The most common type of artificial membrane consists of an immiscible organic liquid interposed between two aqueous phases. This is achieved using three main types of systems: i) bulk liquid membranes, ii) supported liquid membranes and iii) emulsion membranes. Many of the "polymer" membranes discussed in section 3.3 are closely similar

Bulk liquid membranes can adopt several geometries as illustrated in Figure 13. The earliest form was the "Schulman bridge"[44] (Fig. 13A) consisting of a cylindrical cell with a well at the center divided by a plane barrier. The membrane consists of a dense organic liquid in the well reaching above the base of the barrier and thus separating the two aqueous phases. All three phases are stirred (usually slower than 300 rpm) by means of mechanical stirrers in the aqueous phases and a magnetic stirrer in the organic phase.[45-47]

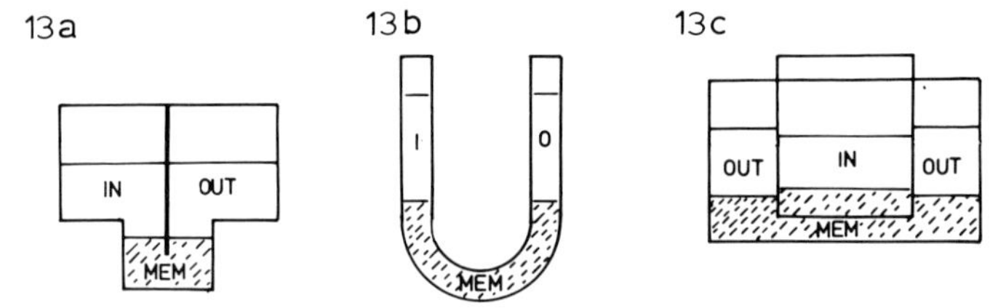

Figure 13: Bulk Liquid Membrane Cell Geometries: a, Schulman bridge; b, U-tube; c, concentric ring cell.

A related system is the U-tube cell (Fig. 13b) in which the organic phase sits in the base of the U separating aqueous phases in the arms. Stirring of the aqueous phases is mechanical and the organic phase is stirred magnetically. Rates to approx. 800 rpm can be achieved before the interfaces collapse.[41,48] A related geometry - a squared U - is typically stirred by mechanical stirrers which cross the interfaces. The maximum stirring rate for this geometry is about 250 rpm.[49] Yet another system uses concentric rings to define the three phases (Fig. 13c). The IN phase is typically in the inner ring with the OUT phase in the outer ring.[10,29-31] The organic phase is stirred (typically 120 rpm) but the aqueous phases are unstirred except by viscous drag. Note as well that the interfaces have different areas and the aqueous phase volumes are not the same.

All the systems described above involve a dense organic liquid for the membrane (chlorocarbons). Lighter liquids can be utilized either by density layering or by means of an H-tube. The former requires control of the densities of the three phases. The lowest (aqueous) phase is typically a concentrated sucrose solution, the membrane is a mixed solvent adjusted to have a density slightly greater than 1.0 g/ml and the other aqueous phase is pure water. The three layers sit in a test tube but the system is unstable to stirring.[4,50] The H-tube is just an upside down U in which the lightest phase is the membrane solvent which floats on the two aqueous phases in the feet of the H and fills the crossbar of the H. Stirring can be done mechanically through the interfaces or by magnetic stirrers in the feet and in the cross bar.[49,51]

None of these cells produces ideal stirring due to the off-center geometries. At best the stirring is nearly regular and at worst provokes catastrophic loss of the membrane barrier. Much of the rather large variation in experimental transport rates can be directly attributed to problems with reproducible stirring.[41] Attempts to calculate boundary layer thickness from hydrodynamic formulae are clearly unwarranted. The best that can be achieved is to use a closely matched sets of cells, a defined set of protocols and periodic routine checks. Even so reproducibility is rarely better than ± 10% and the off quoted value of ± 15% is quite realistic.

All these bulk liquid membranes are mechanically unstable and will not support even minor pressure gradients. An alternative type of liquid membrane - **supported liquid membrane** - improves the mechanical properties of the membrane without greatly altering the ease of fabrication and use. The membrane solvent in this case is imbibed into the pores of a porous support material and is held there by capilliary forces.[52] This support can be as simple as a piece of filter paper with octanol imbibed into it[28] or can be microporous teflon, polypropylene, polysulfone, etc.[52-54] The membrane is mounted as a flat sheet between two halves of a cell. Alternatively the porous support material can be in the form of hollow fibers.[52,55] The membrane, as above, is an organic solution imbibed into the porous fiber walls. The IN phase is typically pumped through the fiber lumens while the OUT phase circulates around the outside of the fibers.[52,55] Recycling[55] and cascade processes[56] can enhance overall performance and it is this type of system which holds the greatest promise for practical applications.[52] A related process, essentially a hollow fiber supported liquid - liquid extraction, is also of some practical interest.[57]

Supported liquid membranes are simple to make and use, and are more robust than bulk liquid membranes. The membrane phase is unstirred so the concentration gradient extends through the whole organic phase thickness.[54] The planar geometry is analytically simpler but hollow fiber geometries permit more efficient phase contact and consequently higher mass transport. One of the key problems for eventual industrial application is the loss of membrane solvent by entrainment into the aqueous phases.[52]

A third type of liquid membrane is also of some industrial interest due to the very high interfacial area which can be achieved. These are **"emulsion membranes"** first developed by Li[58,59] and Cussler[60,61]. Initial interest was in heavy metal extractions but other systems have been more recently studied.[62,63] The membrane is the organic phase of a water-in-oil-in-water ternary emulsion. The membrane is prepared as follows: the receiving phase (OUT) is vigourously dispersed in the membrane phase containing a surfactant to produce a water-in-oil emulsion. This emulsion is added to the IN phase (with less vigourous stirring) to produce a loose ternary emulsion. The innermost water droplets are coated by the organic phase hence are protected from the IN phase by the membrane. The membrane thickness is low and the droplet area is high, consequently the observed fluxes are high.[58-63] Product is recovered by an initial separation of the IN phase from the water-in-oil emulsion phase by settling, and subsequent destruction of the water-in-oil emulsion[64] to recover the OUT phase.

There are a number of features of emulsion membranes which restrict their general applicability. The "recipes" used to concoct the membranes are reasonably complex and the best membranes do not tolerate large changes in key variables such as solvent or surfactant concentration. This poses a problem since many common crown ether carriers are insoluble in the hydrocarbon solvent utilized.[62] The role of the surfactants is not at all clear: do they bind cations and anions? does this alter selectivity? None of this is too critical if the goal is to develop a practical separation system. These considerations are more critical if the intent is to study several different carriers.

Transport can be followed by analysis of aliquots withdrawn from the IN and/or OUT phases. From section 2 it is clearly desirable to work to a fixed extent of transport (e.g. 1%) but this implies some relatively close feed-back of analysis with sampling. Analysis of picrate anion by UV (as a result of cotransport with cations) is very common. Direct analysis of cations by atomic absorption and of both cations and anions by ion chromatography is essential if competitive selectivities are desired. Numerous other techniques (conductivity, polarography, potentiometry, pH-stat, radioactive tracers) have been used. No one has an inherent advantage although the techniques which perturb the phase volumes the least (in-situ monitoring, small aliquots) are to be preferred.

3.2 Bilayer membranes

From a biomimetic viewpoint the liquid membranes outlined above are very imperfect models of the membranes found in biological systems. These latter are very much thinner and are composed a complex mixture of amphiphiles formed into a bilayer structure.[1,11] Polar head groups, usually charged, face the aqueous solutions on both sides of a hydrocarbon core provided by the nonpolar tails of the amphiphiles. In a biological system, the membrane surfaces and core are littered with structural and functional proteins and carbohydrates leading to a complex molecular environment. Simpler systems can be prepared from purified membrane components or artificial amphiphiles either by formation of vesicles (lipsomes) or planar arrays (black lipid membrane - BLM - or Langmuir-Blodgett films). The latter type of system has been used for studies on natural ionophores.[8,17] In general the lifetimes of BLM are short and systematic studies with synthetic ionophores are few (but see isolated examples in refs. 8 and 17).

Naturally occurring phospholipids such as phosphatidyl choline, on dispersion in water, can form closed structures in which an internal aqueous compartment is separated from the bulk aqueous phase by a bilayer membrane. These are known as **vesicles** or **liposomes** and present an interesting membrane system for studies of transport. The nature of the vesicles depends on the structure of the amphiphile and on the method of preparation; single or multi walled structures of large (micron) or small (500 Å) size are known. Artificial bilayer forming compounds act similarly but provide a range of properties not available from natural compounds.[65,66] In addition, vesicle stability can be enhanced and permeability controlled by polymerization of the vesicles.[67,68]

This field is in the midst of explosive development and it is not possible to provide a full picture of all the ramifications. Rather, a few recent reports of transport by artificial ionophores can be examined to underscore the types of experimental considerations inherent in this field.

Carrier mediated transport by simple crown ethers has not yet been examined in a systematic fashion although occasional comparisons with natural ionophores have been reported.[4] Carboxylate containing crown ethers have been examined for the countertransport of Na^+ and H^+ across phospholipid membranes of large unilamellar vesicles (LUV's).[69] Countertransport occurs but at rates 2-3 orders of magnitude slower than related natural carboxylate ionophores such as nigericin.[69] Neutral crown ethers also transport Na^+ in the presence of an "uncoupler" - a weak acid for countertransport of H^+.[69] A more recent and extensive study of a cryptand carrier in the same vesicle system gives rates which are comparable to valinomycin.[70] The rates are of less interest than the experiments performed, which outline what can be achieved.

The vesicles are "large", ca 1μm diameter. This provides a sufficient internal volume to permit reasonable amounts of material to be transported. One can readily calculate that the small (500 Å) vesicles prepared in 0.01 M KCl contain only 10 K^+ ions inside. Small amounts of transport will thus produce very large concentration changes inside or very small changes externally. The LUV's contain larger reservoirs and provoke measurable concentration changes (over longer time periods). For the experiments reported,[69,70] proton efflux was measured by pH-stat and several hundred μL of mM base was added over 30-90 min. to maintain the set pH.

Another experimental consideration concerns the possibility that the carrier simply provokes vesicle breakdown. This can be monitored by an NMR technique using a paramagnetic probe in the external solution. The internal ^{31}P signals from the phosphate head groups are not affected by the paramagnetic probe over the period of the experiment so there has been little lysis.[70] A similar technique has been used to follow the (slow) transport of Pr^{+3} ions using natural ionophores[71] - the goal in this study was to use the probe to quantify transport. The general problem of monitoring is non-trivial and the pH-stat method[69] depends on a suitable co- or countertransport of protons. Use of UV or fluorescence probes is also feasible, and amenable to following faster processes.[4]

The principle factor which distinguishes carrier transport across vesicles from similar transport in bulk liquid membranes is the fact that the **transmembrane potential gradient cannot be ignored.** This can provoke transport via charged complexes utilizing the driving force of the potential gradient.[70] This factor is negligible for "electrically silent" countertransport[69] but intervenes in all cotransport systems. Experimentally, addition of an "uncoupler" which provides a proton pathway converts a cotransport to a countertransport system and effectively "short-circuits" the problem.[72]

The major additional feature available in bilayer membranes is the possibility of constructing channels for the transport of substrates. Naturally occurring channel forming compounds such as gramacidin[8,17] and amphotericin[73] have been studied in BLM and vesicle systems. Synthetic polypeptides which mimic gramacidin have also been prepared[74,75] suggesting that totally "abiological" channels could also be prepared. The principal difficulty in construction of channels of defined structure is the size of the object required, perhaps 3000-4000 g/mole to span a 40 Å bilayer.[76] This is in the range of small polymers and indeed a polymeric crown ether does induce vesicle permeability.[77,78] The transport rate is however, very modest. The actual proof of mechanism for the transport is similar here to that for biological transporters. In general, channels saturate at higher concentration, are less sensitive to the fluid state of the bilayer and have lower activation energies that carriers.[11,13] On the basis of activation parameters, the polymeric crown ether resembles a channel rather than a carrier.[78]

Another approach to the size of object suitable for channel formation is to attempt to mimic the antibiotics amphotericin and nystatin[73] which form pores by formation of aggregates containing 10-15 molecules. Experiments in this vein have been reported by Furhop using an amphiphile derived from monensin.[79,80] The "extended" length of this amphiphile can be no more than 24 Å hence a thin artificial membrane system has been used.[81] Vesicles containing the pore-forming compound loose Li^+ during purification but the process can be blocked by diammonium salts.[80] Whether this is evidence for a pore or simply a defect which is repaired in a mixed system awaits more detailed kinetics. Defects due to poor packing around the added "pore-former" would be expected to have an apparent kinetic order of 1-2 in the "pore-former". A cooperative interaction leading to aggregates such as amphotericin should have a much higher kinetic order (12-15 for amphotericin[73]).

Clearly there is considerable promise in the use of bilayer systems for transport studies. There are significant experimental problems and potentially formidable synthetic challenges. The simple systems examined to date provide a solid foundation for future work.

3.3 Polymer membranes

With the exception of supported liquid membranes, the systems of 3.1 and 3.2 may be described as **mechanically fragile.** They are sensitive to pressure and temperature gradients, often have complex preparation procedures (emulsions, vesicles) and may have limited lifetimes (vesicles, BLM). None of these features is particularly desirable with respect to eventual applications of membrane processes in analytic or separations problems.[82,83] For routine use the membrane should be robust, and in many areas an extended lifetime (months, years) is essential. The obvious response to these considerations is to prepare polymeric materials which exhibit suitable mechanical properties but which retain desirable transport properties as well. In addition, new properties could also appear which are not attainable in other systems. Several types of polymeric membranes can be discerned, ranging from hyperfiltration and reverse osmosis membranes to "membranes" for conventional filtration.[81,82] In this section we consider only those membranes containing or constructed from synthetic ionophores. As such the major types are: i) solvent/polymer membranes, ii) polymer/liquid crystal and polymer/bilayer composite membranes and iii) "fixed-site" transport membranes.

Solvent/polymer membranes form the basis of numerous types of "liquid membrane" electrodes.[3,12] In general they consist of a support polymer, usually polyvinylchloride (PVC) containing a large amount of plasticizer. A typical membrane of this type would contain 1 wt% ionophore, 33 wt% PVC and 66 wt% plasticizer.[84,85] Common plasticizers are orthonitrophenyl octyl ether, dioctyl phthalate, dibutyl sebacate, trioctyl phosphate etc.[84,85] The membrane is similar to a supported liquid membrane, but the diffusion resistance is higher resulting in slower bulk transport rates.[47]

In an ion selective electrode application, the transmembrane potential is the important feature and cannot be ignored, although it is common for theoretical analyses to consider the "zero current" condition.[3] A portion of mass transport is due to potential gradients and under conditions of imposed electrodialysis the transport may be accelerated.[3,86] It has been demonstrated that a portion of the mass transport occurs by the "relay mechanism" of Figure 1.[86] Further, the electromotive force developed depends on the differential mobility of cations and anions with the membrane.[84] The theoretical framework for transport and ion-selective electrode behavior requires a more detailed and rigorous approach than can be presented here.[3,13]

Conceptually related are **polymer composite membranes** containing **liquid crystal** forming materials[87] or **bilayer** forming materials.[88-90] The former utilize the structural properties of the polymer (PVC, polycarbonate) to form the membrane barrier with the liquid crystal component acting as the plasticizer.[87] The principle difference between these systems and the solvent/polymer membranes of ion selective electrodes is due to the **temperature regulation** which can be achieved. Below the liquid crystal/solid phase transition temperature (T_c) the membrane is essentially impermeable since it is like an amporphous solid. Above T_c, the liquid crystalline component is a fluid of low viscosity. As a consequence the permeation of neutral solutes (e.g. O_2) increases and the permeation is temperature controlled.[87] Incorporation of carriers leads to ternary composite membranes in which the ion permeation is temperature controlled.[87,91] The "on" process due to a temperature jump to above T_c is rapid (2 min.) but the "off" process due to cooling to below T_c takes somewhat longer (15-20 min.).[91]

Polymer/bilayer composite membranes are similar. These occur as two main types; bilayer forming compounds supported on porous nylon capsules[90] and bilayer/polyion complexes supported on grid materials.[88,89] The former consist of large (1 mm) diameter nylon capsules prepared by interfacial polymerization. Bilayer-forming amphiphiles are imbibed into this porous support to seal the capsule contents from the external solution.[90] The latter consist of a salt prepared from bilayer forming amphiphiles and a poly ion supported on rigid grid materials such as Pt.[88,89] The role of the bilayer in these membranes is similar to that played by the liquid crystal above, namely to provide a means of regulating the permeability in response to temperature and other external influences.[90] At the present level of development, carriers for ions have not been incorporated; the observed ion transport is nonfacilitated passive transport.[165]

In all systems to this point, the carrier is a small molecule, free to diffuse (within the limits of viscosity) within the membrane, but also susceptible to loss into the aqueous phases. In principle this might be avoided by providing a covalent link between the carrier and the polymer support of the membrane. This will restrain the transport to the "relay" mechanism of Figure 1. Membranes of this type are called **"fixed-site carrier" membranes.**[92] The majority of such membranes described to date involve simple ionizable groups hence give rise to simple countertransport[92-94] with protons. Membranes from crown ether containing polymers, either with carboxylates on the macrocycle[95] or on the polymer backbone[96] have also been described. These system also exhibit cation proton countertransport.

The requirements of the synthesis of ionophore containing polymers and the control of the ultimate physical properties are often contradictory and in general the polymers prepared make poor membranes.[95,96] This can be overcome by preparing the polymer film within a porous support,[95] by preparing and cross linking a polymer blend of "active" and "support" polymers[96] or by preparing a membrane from a blend of "active" and high molecular weight support polymers.[97] All these methods lead to membranes with activities comparable to supported liquid membranes of similar thickness.[95-97] In some cases the membranes are very robust and can be reused without degradation for prolonged periods (months).[97] The "relay" mechanism has been relatively little examined for lack of suitable membranes and for persistent problems of poor mass balance; the expected concentration gradients can be achieved but these collapse to initial values[94,95] or the expected gradient is only a fraction of the total predicted from the transport stoichiometry.[92,93,96,97] In systems where all ionic species concentrations can be followed, this can be shown to be a result of multiple coupled transport pathways.[97] The ultimate utility of membranes of this type will require better control of such complicating factors.

4. COTRANSPORT

4.1 Systems

From a biomimetic perspective, carriers for cotransport are related to the neutral ionophores valinomycin, nonactin etc.[4,98] The essential feature of cotransport systems for ions, with the assumption of electroneutrality, is that the carrier be neutral and that the transport occur via ion pairs. Typical examples of carriers of this type are given in Figure 14.

Foremost among the carriers in Figure 14 are the crown ethers such as 18-6, its homologs and derivatives.[10,25-27,29-31] Other heteroatoms (N,P,S) in place of oxygen, various chiral groups (binapthyl, spiro bifluorene) and lipophilic side chain derivatives are known[99] and many have been reported in transport experiments. Cryptands

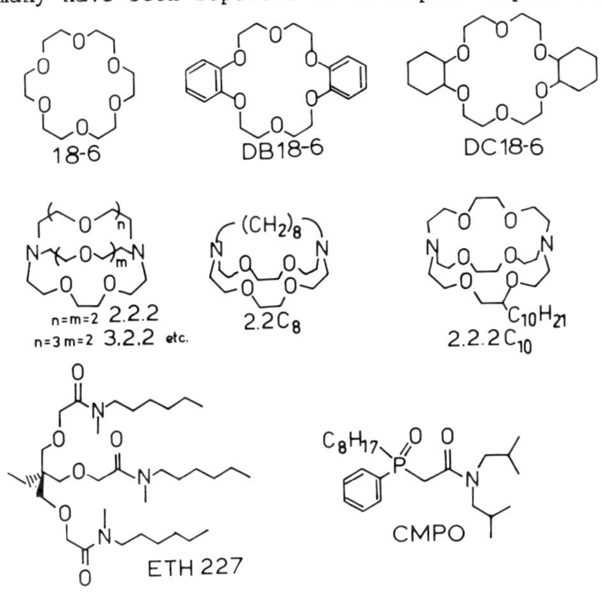

Figure 14: Typical carriers for cotransport of cations and anions.

such as 2.2.2, its homologs and derivatives fall into this class as well provided the aqueous phases are basic to avoid protonation.[6,46,47,70] Acyclic derivatives are possible as are neutral extraction reagents from liquid/liquid extraction systems; in Figure 14 an example of the former (ETH 227) is active in Na^+ ion-selective electrodes[100] and an example of the latter (CMPO) is reported for actinide transport.[101] Because of the emphasis on crown ethers and cryptands, the cations most commonly transported are alkali metal and alkaline earth cations.[10] Transport of related heavy metal cations such Ag^+, Pb^{+2}, Zn^{+2} etc. have also been reported.[10,102] Crown ethers also form complexes with ammonium salts hence transport of alkylamines, aminoacid esters[49] and phenylethylamines[102,103] is also possible. The carriers complex the cations but the cotransported anion may well be the species of interest. In this vein systems for the gradient pumping of picrate[104], aminoacid carboxylates[105] and complex anions ($AgBr_2^-$)[106] are of interest. There are a very large number of systems which fall into this class and the above is simply a brief account of the principal directions. The remaining sections examine a few examples in some detail with a view to comparing the theory developed in section 2.3 with experimental reality.

4.2 Theory and Experiment

Assuming that nonfacilitated transport can be neglected, the equation of interest is (13). In many cases the concentration on the OUT side is usually zero initially hence (13) becomes

$$J_A = \frac{DL_o}{2\ell} K_{ex} \frac{{A_{in}}^2}{(1 + K_{ex}{A_{in}}^2)} \tag{23}$$

If $K_{ex} << K_{ex}(max)$, (22) can be further simplified to give

$$J_A = \frac{DL_o}{2\ell} K_{ex} {A_{in}}^2 \tag{24}$$

Recall that the squared terms arises from a product of anion and cation concentrations, assumed to be equal initially.

Equations 23 and 24 have been verified under several sets of conditions and carriers:

i) at the steady state and for small amounts of transport, A_{in} is approximately constant. Thus A_{out} should increase as a linear function of time. This is clearly the case for a number of systems.[24-31]

ii) The flux should vary as a linear function of the initial carrier concentration. This has been assumed[10,29-31] as often as demonstrated[24-27,47,102] but can be regarded as well confirmed.

iii) The flux should vary with ${A_{in}}^2$ or more generally with the product $M^+_{in}\ X^-_{in}$. The former has been checked several times[25-27] including corrections for ionic activity.[30] The latter has been confirmed for mixtures of $KCl/LiCl/NiCl_2$ and the carrier DB18-6;[27] this system is sufficiently K^+ selective such that the fluxes of Li^+ and Ni^+ can be neglected which permits independent variation of K^+ and Cl^-.

Equation (23) or more generally (13) has a number of other features which provide a more testing analysis of mechanism. As noted in 2.2, J_{max} should be achieved with the carrier at half saturation. Fully saturated carriers should transport at $J_A << J_{max}$. In cases where % saturation at the steady are reported[47,102] the results are in general agreement with J_{max} at $L_o/2$. For example for the cell:[47] M^+ Picrate = 10^{-2} M‖2.2.2 in $CHCl_3$ 1.5 x 10^{-3} M‖H_2O, at the steady state gives rates (% saturation) of Na^+, 0.6 μmol h^{-1} (90%); K^+, 0.03 μmol h^{-1} (>>95%); Cs^+, 2.7 μmol h^{-1} (20%). Similar results are noted for other ligand/cation combinations.

More importantly J/J_{max} should vary with log K_{ex} as illustrated in Figure 7. Direct determination of K_{ex} is experimentally straightforward but lengthy and is subject to a variety of problems.[107-110] Consequently directly comparable data from two phase extraction experiments are seldom available for the three-phase liquid membrane system under study. Figure 14 presents one (perhaps the only) internally consistent block of data[6,47] The system utilized metal picrates (10^{-2}M) transported across a bulk $CHCl_3$ membrane by crown ethers, cryptands and valinomycin. The points are all experimentally determined pairs of fluxes and equilibrium extaction constants. The theoretical curve, calculated as for Figure 7,[37] used $S_o = 10^{-2}$ M, $L_o = 1.5 \times 10^{-3}$ M and a 1% fraction of transport. J_{max} was taken as 4 µmol h^{-1}.[47] All of these parameters are consistent with the conditions of the experiment. The agreement between theory and experiment is quite adequate despite the relatively small number of points and their associated uncertainties. Certainly the qualitative picture is strongly confirmed.

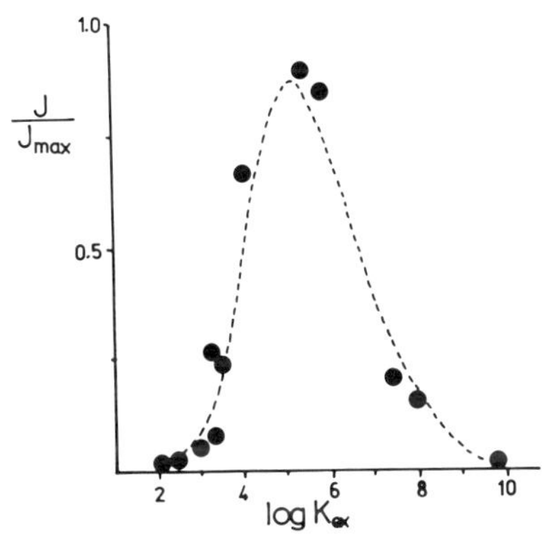

Figure 15: J/J_{max} as a function of log K_{ex} for the experimental system: MPic (10^{-2}M)|carrier in $CHCl_3$|H_2O. Data of ref. 47 used with permission of Dr. M. Kirch, theoretical curve calculated as described in ref. 34 and 37.

The other major block of directly comparable transport data has attempted to draw a relationship between J_A and the value of the stability constant for complexation in methanol (K_M).[10,29-31] This approach is based on the ready availability of values of K_M for a wide range of known and potential carriers[99] and upon the implicit assumption that K_{ex} is some constant function of K_M throughout the series of carriers examined. The model at its present level of complexity contains a number of cation dependent parameters.[10] Although it can correctly reproduce bell shaped curves like Figures 7 and 15, the nature of these additional parameters is not at all clear. A discussion of this system is deferred until section 4.5.

4.3 Factors Which Influence The Extraction Constant

The only key parameter which is not directly under experimental control is the value of the extraction constant K_{ex}. This term is a composite of several contributing processes, each one of which has a particular relationship between structure and final outcome. To examine the structural influences on K_{ex} it is therefore necessary to examine the structural influences on the contributing processes. Consider the following equilibria between water and an organic (membrane) solvent:

$$L + M^+ \rightleftharpoons LM^+ \qquad K_s \tag{25}$$

$$LM^+ \rightleftharpoons \overline{LM^+} \qquad K^+_{tr} \tag{26}$$

$$X^- \rightleftharpoons \overline{X^-} \qquad K^-_{tr} \tag{27}$$

$$\overline{LM^+} + \overline{X^-} \rightleftharpoons \overline{LMX} \qquad K_{ip} \tag{28}$$

where the overlined species are in the organic phase and the unsubscripted species are in water. Equation (25) defines the complexation stability constant in water (K_s), equations (26) and (27) define the single ion partitioning between the two phases (K^+_{tr} and K^-_{tr} respectively) and (28) defines electrostatic association (ion pairing) in the organic phase (K_{ip}). The sum of (25) to (28) gives the defining equilibrium for K_{ex} (Fig 6) for ion pairs in cotransport. Thus K_{ex} is simply the product of $K_s \times K^+_{tr} \times K^-_{tr} \times K_{ip}$. The contributing structural factors can then be discussed in terms of their influence on the elementary processes (25) to (28).

4.3.1 **Cation-ligand association.** This term is the best understood (in a quantitative fashion) of all the contributing equilibria. A large body of experimental data is available[99] and structural influences can be readily discerned. The principles outlined by Lehn[111] have been confirmed by a a dozen years of experimental activity. The principle lines are sketched below but many detailed reviews are available.[99,111-115]

(i) **Nature of the cation** - ionic radius, polarizability, hard-soft acid/base characteristics, preferred donor atoms. All these parameters influence the selection of an appropriate ligand with complementary properties.

(ii) **Relative cation/ligand topology** - complementary shape and size of the ligand with the bound ion is essential for optimum binding. This is the basis of the "hole size selectivity", macrocyclic effect, macrobicyclic effect etc.[99,111-115] A close "fit" of ion with ligand produces stable complexes.

(iii) **Donor atom type** - complementary donor sites matched with the preference of the bound ion favour complexation. Sites range from hard (O) to soft (S) and can include charged groups as well (see 5.2). Donor atom topology is important as well since the donor sites influence the ligand conformations.

(iv) **Ligand "thickness"** or hydrophobic bulk. For complexation in water, shielding of the cation by the ligand decreases long-range stabilization by the medium. Consequently "thicker", more bulky ligands destabilize complexes.[111] This effect is in opposition to the requirements of the transfer of LM^+ to the organic phase (26).

4.3.2 **Single ion partitioning between water and the organic phase.** Very little direct data on single ion transfer is available, and at the present time the only feasible approaches to analysis assume simple models of ionic solvation. Abraham and Liszi have described such a procedure for the calculation of the free energy of solvation of an ion in a solvent.[116] This procedure focusses on an electrostatic component (Born term) and on a neutral component (for displacement of solvent to place the ion in the solvent sphere).[116] For ions with known hydration energies, the free energy of transfer is simply the difference of the ion solvation in the organic phase and the hydration energy.[117] Unfortunately, the hydration energies of the complexes LM^+ are not known and must be calculated as well. In principle, all this is possible, but at present little has been published in this vein (but see ref.10).

Qualitatively, the critical parameter with respect to the ions transferred is their "radius". In general, as the radius increases, the solvation energy decreases. The effect is complex since the electrostatic and neutral terms act in opposition.[116,117] Consequently, the difference between two solvents, as required for (26) and (27) will be attenuated as the radius increases. In effect, the transfer from water to the organic solvent will become more and more favourable as the ion radius increases. The effective "radius" of LM^+ is problematic, especially for crown ether complexes,[111] but can generally be associated with the ligand as a whole, as opposed to just those parts which bind the ion.

The effect of K_{tr}^- (eqn 27) on the ultimate magnitude of K_{ex} is quite marked as has been demonstrated in both two phase extraction experiments[110,118] and in transport experiments[29]. A generally linear correlation of transport rate and anion hydration energy has been reported.[29] By the analysis here, this is only one part of one term in the series leading to K_{ex}. Some favourable cancellation has occurred for the series examined, but still the effect of K_{tr}^- is large enough to give rise to the correlation.[29]

The cation transfer component, K_{tr}^+ (eqn 26) is related to the concept of "lipophilicity" or "hydrophobicity" of the ligand.[2,7,10,111,119] Qualitatively, as the lipophilic character of the ligand increases, the complexes will be more favourably partitioned to the organic phase. The effect depends on the hydrophobic bulk ("radius" above) and has as its origin the disruption of water structure by the substrate.[119,120] An empirical method for the estimation of log K_{tr}^+ is the use of Hansch parameters for structural units in L.[111,119,121] The majority of the necessary parameters refer to partitioning to 1-octanol and in addition have been developed for solutes dissimilar from the materials of Figure 14.[121] As a result, the trend in these lipophilicities is usually reliable but the magnitude of derived partition coefficients is less reliable.[10,47] This method is useful for comparisons between homologs differing only by alkyl substitution especially if these can be calibrated with an experimental value.[48]

4.3.3 Ion Pairing in the Organic Phase. As in 4.3.2, direct experimental values are rarely available[107] but the situation can be examined with respect to methods for calculation of ion pairing free energies.[118,122] In general, electrostatic interaction to form the ion pair $\overline{LMX}$ will depend on the solvent dielectric constant and upon the sum of the radii of the cation and anion.[118,122] In addition, there may be additional structural contributions but these are generally ignored by "continuum" theories of ion pairing. This term is strongly favourable, especially in low dielectric solvents and the energies involved can be considerably larger than those involved in the other contributing equilibria (25-28). Unfortunately, at the present level of understanding, it is only the general trends which are clear: ion pairing increases as the solvent dielectric decreases and as the sum of the cation and anion radii decreases. This latter includes an effect for "localization" of anionic charge as well.

4.3.4 Trade-offs. There are numerous subtle trade-offs between a given structural factor and its composite influence on equations 25-28. This makes it difficult to ascribe any particular relationship between a structural factor and its influence on K_{ex}. Consider:

(i) **ligand thickness/hydrophobic bulk.** Increased bulk destabilizes LM^+ (lower K_s) and LMX in the organic phase (lower K_{ip}) but increases the partition coefficient (raise K_{tr}^+).
(ii) **anion size.** As this factor increases, the ion pairing component decreases (lower K_{ip}) but the partition coefficient for the anion increases (raise K_{tr}^-).
(iii) **solvent.** A decrease in the solvent dielectric constant will increase ion pairing (raise K_{ip}) but will decrease both K_{tr}^+ and K_{tr}^-.

Obviously, the net result of a change in any one of the parameters above will depend upon the system under consideration. It is fortunate indeed that any trends can be discerned in the raw data; this is undoubtedly due to the narrow range of physical properties of the ligands utilized.

4.4 Selectivity

The general principles of transport selectivity outlined for countertransport in section 2.5 are applicable in this cotransport case as well. To reiterate:

(i) The thermodynamic extraction selectivity $K_{ex}(A)/K_{ex}(B)$ can in no way be related to the values of J_A and J_B determined in separate experiments. The very common plots of flux as a function of cation radius measured from different experiments with a given carrier have a long tradition[24] (and many more recent references!). The "success" of these plots depends on all values if K_{ex} being less than K_{ex} (max).

(ii) In competitive experiments, the selectivity depends upon the relative values of $K_{ex}(A)$, $K_{ex}(B)$ and $K_{ex}(max)$. Since $K_{ex}(max)$ is a function of initial concentrations (Figure 6, section 2.3), transport selectivity will depend upon the experimental protocol. Variation in the initial ratio A_{in}/B_{in}, could alter the attainable selectivity considerably.

(iii) Transport selectivity is traditionally defined as the ratio of the fluxes J_A/J_B. This parameter is a function of the fraction of transport (2.3, 2.5) and in some cases ($K_{ex}(A)>K_{ex}(B)>K_{ex}(max)$) the initial rates diminish rapidly as does the selectivity. The parameter A_{out}/B_{out} is a more reliable indicator of selectivity.

When we wish to compare theory and experiment, we come rapidly to the conclusion that the majority of reports of "selective" transport refer only to single ion experiments. The **failure** of a plot of cation radius or log K_s with J_A is, however, more instructive. Reference 10 discusses examples for cryptand transport of alkali cations. In these cases[10,31] the log K_{ex} (M^+) values lie on either side of K_{ex} (max), hence J_M is not a linear function of log K_{ex} (or of log K_s (methanol)).

Competitive experiments on binary and ternary cation mixtures have been reported for a range of carriers for both bulk liquid[123-126] and emulsion membranes[62,63,127]. Insofar as the majority of carriers fulfil the condition $K_{ex}(A)$ and $K_{ex}(B)<K_{ex}(max)$ the selectivities (J_A/J_B) tend to follow the complexation selectivities in methanol[126]. (This assumes K_{ex} is some linear function of K_s(methanol) see 4.5). Note, however, that complexation selectivity can be quite marked while the transport selectivities are more modest. This is especially true of carriers where both $K_{ex}(A)$ and $K_{ex}(B)$ are remote from $K_{ex}(max)$.

Competitive experiments with emulsion membranes [62,63,127] reveal additional features. The selectivity of a Pb^{+2}/Sr^{+2} separation ($Pb^{+2}{}_{out}/Sr^{+2}{}_{out}$) by DC18-6 varies as a marked function of fraction of transport[62]. The system is highly Pb^{+2} selective but the concentration ratio in the OUT phase falls from a high of 76 at a small extent of transport to 1.8 near the end of the experiment. Qualitatively, this is the behaviour predicted by Figure 11.

A second interesting feature is the perturbation of transport selectivity due to membrane asymmetry. In an emulsion membrane, the internal volume is relatively small, thus a minor amount of transport produces a significant increase in concentration in the internal compartment. As a result, the concentration gradient collapses rapidly. In order to maintain the gradient, a complexing anion is added to the internal solution ($P_2O_7{}^{4-}$, OH^- etc.[62]). Without such complexing anions, transport rates are modest but reflect the thermodynamic selectivity of the carrier (if $K_{ex}(A)$ and $K_{ex}(B) < K_{ex}(max)$). Thus, the apparent Pb^{+2}/K^+ selectivity without a complexing anion is about 0.55, while with OH^- it is 450[62]. The latter estimate is low since a large fraction of transport has already occurred[62]. This too is in qualitative agreement with diffusion controlled mechanisms of section 2. Since the membrane is asymmetric, A_{out}(for Pb^{+2}) is held essentially at zero while B_{out} (for K^+) increases as in the simple analysis.

One of the clearest demonstrations of the kind of selectivity which can be achieved with crown ether containing membranes has been reported by Cussler[128,129] for the separation of KNO_3 and KI from brines containing Na^+, Ca^{+2} and Mg^{+2} as competing cations and Cl^-, SO_4^{-2} and CO_3^{-2} as competing anions. The method turns on a cooperative interaction of all the contributing terms in equations (25-28). As a result, in an emulsion membrane, a 2-5 minute contact produces 95% pure product in 12% yield. With suitable staging and recycling there is clearly the basis for a purification based on this system. One observation from this study is the apparent independence of flux on alkyl substitution on the carrier (R_2DB18-6, Fig.14)[129]. A similar effect was noted for the other carriers[31]. In both these cases, the effect is probably a composite of several opposing trends i.e., increase in K^+_{tr} compensated by a decrease in K_s and K_{ip}. In addition, the diffusion coefficient may fall off with increasing size of the carrier. If, in addition, the flux is near J_{max} for high substrate/low carrier concentration (Figure 7b), the observed flux will be rather insensitive to the precise value of K_{ex} and may well be constant.

4.5 Kinetic regime (?)

Are there cotransport systems which are not diffusion controlled? Undoubtedly - but there are no examples conclusively demonstrated to date.

Consider first the implications of equations (19) and (20) in the light of available data on complexation kinetics[99,130]. In order that τ_{cplx} become large, i.e. the complex lifetime become long, then all the terms k_f, k_d and L should be small. Crown ethers have high formation rates (10^6-10^8 $M^{-1}sec^{-1}$)[99] so it is unlikely that any examples of long lived complexes can be found. Small cryptands (2.1.1, 2.2.1) show much slower formation rates[130] and have stable complexes with small cations (Li^+, Ca^{+2}) hence "small" k_d. At a carrier concentration of 10^{-4}M, τ_{cplx} for Li^+ 2.1.1 is about 1.2 sec (using values for water)[130]. Using equation (20), and ℓ = 100 μm, D = $10^{-5}cm^2sec^{-1}$ gives τ_{diff} of 5 sec. Thus, under these conditions, a 20% decrease in flux might be noted according to equation (22). The experiment has not been done and represents about as far as one can go with these carriers. Other metal ions known to form inert complexes might exhibit decomplexation limited transport but clear experimental evidence is unavailable.

Turning now to the block of transport data alluded to at the end of section 4.2, we can now consider the model proposed by Lamb et al[30] for the transport of a range of cations with a range of carriers. This model differs from the system of 2.3 in two respects: i) it utilizes K_M, the stability constant for complex formation in methanol in place of K_{ex} and ii) it incorporates additional boundary layers and gradients for the diffusion of uncomplexed ion pairs into the membrane[10,30]. These additional layers are equivalent to the "relaxation" distances of

Friedlander and Keller[42] and, as such, represent a "kinetic" component to the process[21]. In such a case, τ_{cplx} will be given by an equation like (20), where ℓ is replaced by the "relaxation" distance. According to (22), the incorporation of the numerical values utilized to produce good "fits" of J_M to log K_M[10,30] produces a negligible (0.4%) change in the flux under diffusion control. This conclusion is in agreement with the analysis immediately above - crown ethers simply have too rapid complexation kinetics to show a "kinetic" regime due to complexation/decomplexation control of rate. It is likely that the "relaxation" distances incorporated in the model simply correct for minor imperfections in the assumption that K_{ex} is a linear function of K_M.

An explicit relationship between K_M and K_{ex} **can** be made by consideration, as in equations 25-28, of the constituent equilibria. In order to transform K_M in methanol into K_{ex} for water/organic solvent the following must occcur: M^+ must be transferred from methanol to water, LM^+ and L must be transferred from methanol to the organic phase, an anion X^- must be transferred to the organic phase and finally LM^+ and X^-must ion pair in the organic phase. A procedure for the evaluation of the free energies of each of these steps gives estimates of K_{ex}.[131] When these values are used, in place of K_M, the data of ref. 30 "fit" the model of equation (13) within the accumulated errors in the procedure[131]. As concluded above, the "relaxation distances" are simply correction factors to relate K_M and K_{ex}.

There are two other types of system of interest in this context of "kinetic" control. The first of these types are systems involving transmembrane redox gradients.[133-137] The general problem of achieving transmembrane electron transport and the coupling of redox processes to transport is of interest from both a biomimetic[134-136] and practical[133] viewpoint. Two systems are sketched in Figure 15.

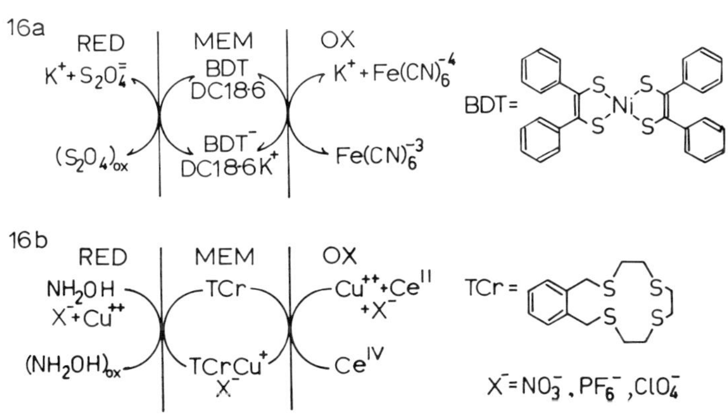

Figure 16: Cotransport of electrons and cations. System a, ref. 132; System b, ref. 133.

The first example (system a)[132], is an example of reaction pumping (Figure 2c) in which the redox reaction can be used to create a gradient in K^+.[132] The kinetics are complex and the details are incomplete. Nonetheless, the K^+ transport is essentially independent of the K^+ concentration but is a function of the dithionite concentration in the RED(IN) phase.[132] This is conceivably a case where the transport rate is limited by the rate of carrier reduction (itself either diffusion limited, or a result of the inherent kinetics). A possible slow interfacial process cannot be excluded but is not yet demonstrated.

The second example (system b) is apparently similar but is not a reaction pumping example: the gradient in the anion drives this process, and when absent the Cu^+ transport ceases. The role of the redox reaction is to provide a suitable extractable cation. Since Cu^+ does not build up in the RED (IN) phase[132], it is unclear (as above) whether the redox kinetics play a role. The rate is anion dependent however as would be predicted for a diffusion limited process.

A second type of "mixed" system is the use of photo responsive crown ethers, particularly the system of Figure 17.[5,138] The transport of K^+ with a suitable hydrophobic counterion (picrate$^-$, NDS$^-$, BNDS$^-$) occurs down the concentration gradient in K^+ and is thus a cotransport process with two interconverting carriers. The Z form has a markedly higher affinity for cations than the E form presumably due to a "bis crown" effect.[138] Thus irradiation produces a "better" extracting reagent. In addition, binding of metal ions to the Z form slows the thermal isomerization to the E form by up to a factor of ten.[138] Both forms transport K^+X^- and the rate in the dark follows the lipophilicity of the anion (Pic$^-$ > BNDS$^-$> NDS$^-$). Upon irradiation, the transport rates for Pic$^-$ and BNDS$^-$ **decrease** while that for NDS$^-$ **increases**. The effect depends on where the irradiation is directed; irradiation of only the IN interface slightly accelerates transport with Pic$^-$.

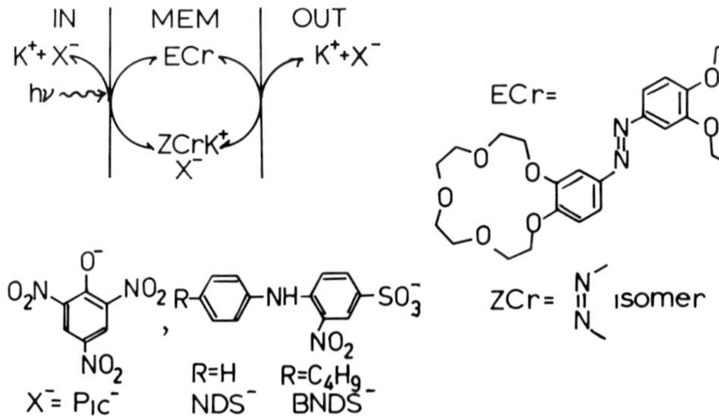

Figure 17: Photochemically mediated ion transport[138] by an azo bis crown ether.

The authors ascribe this set of behaviours to slow Z→E isomerization which slows the rate of ion release[5]. However, the results are in qualitative agreement with a diffusion model in which the E→Z photo isomerization shifts K_{ex} to greater than K_{ex}(max) for Pic^- and $BNDS^-$ butnot NDS^-. Localised irradiation would enhance the amount of K^+ in the membrane whatever the value of K_{ex}. It is true that thermal isomerization is slow, but the rate of cation loss to the aqueous phase need not be. The conclusion that transport is limited by ligand dynamics is not yet founded.

5. COUNTERTRANSPORT

5.1 Systems

From a biomimetic perspective these carriers are related to the carboxylate ionophores nigericin, monensin, lascalocid etc.[4,7,28,139] The essential feature is that the carrier forms a neutral complex in the membrane (ion pair, Zwitterion) for **both** directions of transport hence must itself be charged under some conditions. The common form for this to take, is to append ionizable groups on a neutral crown ether periphery. Examples in Figure 18 include carboxylic acids,[7,48,69,140-146] phenols[147] and other ionizable groups.[148,149]

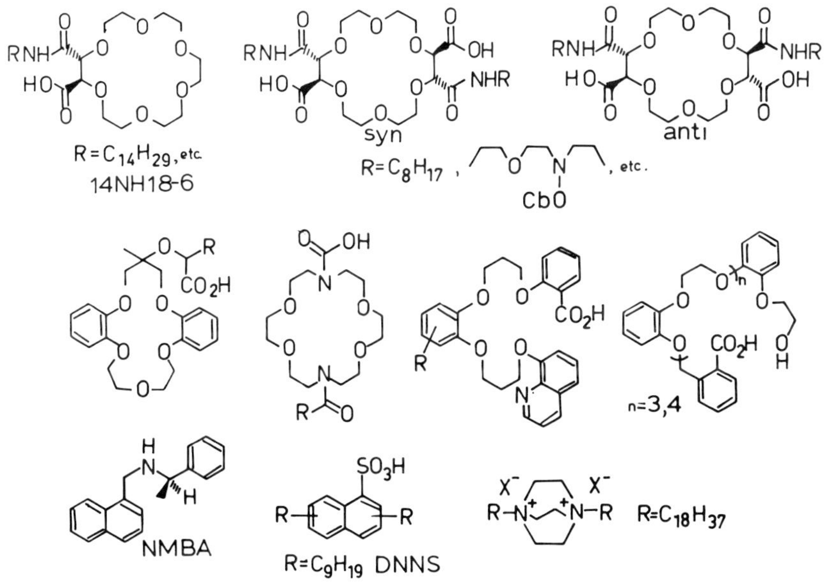

Figure 18: Carriers for countertransport

Acyclic relatives are also possible[7,51,150,151] in which the polyether fragments is presumed to encapsulate the metal ion in a helical arrangement. Many anion transport systems fall into this class such as simple ammonium salts for the transport of mandelate[152] and phosphates[6,153] as well as metal complexes for a range of anions.[105]

Metal ion transport with calixarenes also falls in this class.[154,155] There is a larger range of transport cycles open to countertransport than was the case with cotransport. Some examples are sketched in Figure 19. Cation-proton countertransport can occur with a range of stoichiometry. Anion/anion countertransport and cation/cation countertransport have both been demonstrated for simple carriers[45,152] but apparently have not been reported for macrocyclic carriers.

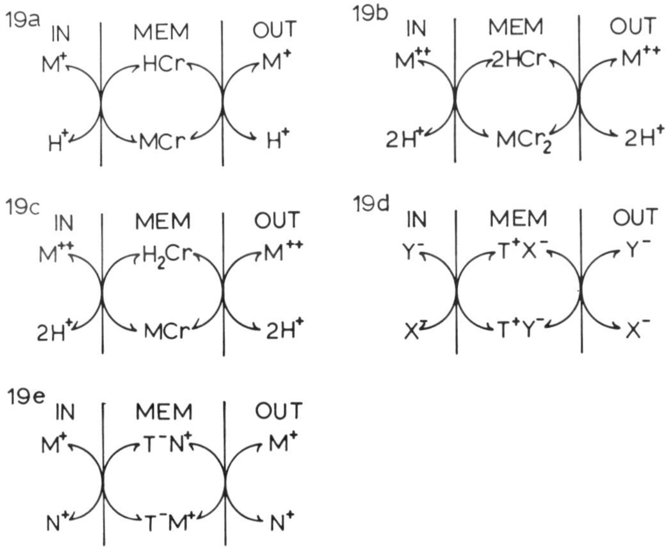

Figure 19: Schematic countertransport cycles

5.2 Theory and experiment

The applicable flux equation (14) and the curves of J/J_{max} as a function of log K_{ex} (Figure 9) have less relevance here than was the case in section 4.2. This is true for several reasons, both fundamental and theoretical. Firstly, there are few **reasonable** simplifications which can be made to equation (14).[28,32] This is especially true in cation-proton countertransport cases, where the goal is to accumulate metal ions using a proton gradient. In these cases the symmetry assumptions leading to (14) are not valid. Secondly, also in M^+/H^+ cases, the ionizable group usually has a pK_a fixed by the ligand structure. Recall that K_{ex} measures the relative affinity of the carrier for the two substrates. We can proceed to analyse the factors which enter into K_{ex} as previously, by examining the constituent equilibria (for a mono basic carrier HCr).

Equation 30 defines the aqueous solution acidity constant and equation 31 defines the aqueous solution stability constant for formation of the 1:1 complex. Equations 29 and 32 are partition equilibria. If we define them in the usual fashion, K_{tr} = org./aq., then 29 defines the reciprocal of H_{tr}^{HCr} and 32 defines, in the correct

$$\overline{HCr} \rightleftharpoons HCr \qquad 1/K_{tr}^{HCr} \tag{29}$$

$$HCr \rightleftharpoons H^+ + Cr^- \qquad K_a \tag{30}$$

$$Cr^- + M^+ \rightleftharpoons MCr \qquad K_s \tag{31}$$

$$MCr \rightleftharpoons \overline{MCr} \qquad K_{tr}^{MCr} \tag{32}$$

$$\overline{HCr} + M^+ \rightleftharpoons \overline{MCr} + H^+ \qquad K_{ex} \tag{33}$$

$$K_{ex} = K_a \times K_s \times K_{tr}^{MCr}/K_{tr}^{HCr} \tag{34}$$

sense, K_{tr}^{MCr}. The summation of (29) to (32) gives (33) which defines K_{ex} (as in Figure 8). For carboxylate crown ethers pK_a is usually greater than $\log K_s$.[140] Similarly, HCr is usually more lipophilic than MCr[141,142] hence the resultant value of K_{ex} by (34) is usually much less than one. In essence this means that the carriers of Figure 18 are all expected to lie on the same (low) side of K_{ex}(max) hence will not provide a range to construct a Figure like 9 from experimental values. In accord with this analysis is the observation that transport kinetics follow Michealis-Menton kinetics[48,142] for M^+/H^+ transport by carboxylate crown ethers.

Simple cations and anions which achieve anion/anion or cation/cation countertransport are not subject to the same restraints[45,146] The ammonium salt(-)NMBA exhibits chiro selective transport of (+) mandelate against a counterflux of anions such as Cl^-, formate$^-$, acetate$^-$ etc.[152] Consider only one mandelate isomer; K_{ex} is then defined by (35):

$$Man^- + \overline{NMBAX} \rightleftharpoons \overline{NMBA\ Man} + X^- \tag{35}$$

$$K_{ex} = \frac{\overline{NMBA\ Man}\ X^-}{\overline{NMBAX}\ Man}$$

Here K_{ex} varies principally with the lipophilicity of the counter ion X^- relative to the lipophilicity of mandelate. The actual values of K_{ex} are not known but J does vary with X^- and passes through a maximum at X^- = propionate. If X^- is too lipophilic (hexanoate) or too hydrophilic (Cl^-, SO_4^{-2}) then transport of mandelate is inhibited as would be predicted from Figure 9.

Other observations are in general agreement. For supported liquid membranes a diffusion model has been tested in some detail[28,156,157] for nH^+/M^{+n} countertransport where M^{+n} = Na^+, Cd^{+2}, Zn^{+2} and Eu^{+3}. Related work on Cu^{+2} transport by a hydroxyoxime carrier shows a range of behavior ascribed to a "mixed" regime of simultaneous rate control by chemical reactions and diffusion.[158]

Equations (29) to (32) closely resemble (25) to (28) which were used to analyse the constituent factors in K_{ex} for cotransport. A parallel analysis is applicable here as well, for cases involving crown ethers, cryptands and like materials. In cases of the simple carriers such a NMBA, the magnitude of the association constant in water will be small and it will be more productive to consider an alternative cycle (below). Neither cycle has any mechanistic significance; partitioning to water is quite small and kinetically negligible for the case of carboxylate crown ethers.[48] The decomposition into different constituent equilibria is solely an attempt to understand (and control) the factors which enter into K_{ex}.

Consider first ionizable crown ethers. The analysis here directly parallels the discussion of 4.3 with two important differences. Firstly, the partition equilibria involve "neutral" species although the complex MCr is certainly dipolar. The first order theory discussed above[116,117] would treat HCr and MCr as exactly the same i.e., large neutral molecules, and would thus predict $K_{tr}^{MCr}/K_{tr}^{HCr} = 1$. This is certainly not the case experimentally[141] even for a very lipophilic crown ether. Clearly, a more sophisticated model of solvation, or experimental values are required. The general factors which influence partitioning remain as discussed in 4.3.2.

The second factor of importance is the presence of the ionizable group and its influence on the bound metal ion. The actual acidity constant (30) is principally a consequence of the functional group chosen. Obviously this can be controlled by remote functionality, as in the introduction of nitro groups onto phenol crown ethers[147] and by more subtle factors such as proximal charges and geometries.[159,160] The stability constant for complexation depends on the factors discussed in 4.3.1 with the additional feature of electrostatic interactions between the ionized group and the bound ion. In general complex stability increases with the **proximity** of the negative charges to the cation[161] and with the degree of localization of the charges on both the anion and cation.[160] The effect is therefore rigorously electrostatic. In the limit, the cation and anion form a tight ion pair with only minor interaction with the crown ether portion of the carrier. More typically, a given crown ether carboxylate will exhibit a range of interactions for a series of cations.[160] This effect has been analysed by consideration of three contributing interations to the total complexation free energy:[162] i) a contribution due solely to the crown ether, as for example the binding of the cation to the neutral (protonated) form, ii) a contribution due solely to the anion, essentially an ion pairing term and iii) the **cooperative** interaction of the anion as an additional binding site leading to a crown ether complex **assisted** by the anion. The relative contribution of these three terms has been deduced for a range of carboxylate crown ethers[162] and the analysis is applicable to other ionizable carriers as well.[147] The findings[162] reflect the obvious: **cooperative interaction is enhanced when the geometry is optimal.**

The above discussion is applicable for carriers which might be expected to form complexes in the aqueous phase. Carriers such as NMBA form "complexes" only in the organic phase simply by ion-pairing. It becomes more useful to examine such cases in the light of an alternate cycle as shown below. Charges are omitted for generality.

This is a general form of K_{ex} as defined in Figure 8. Note that when $S_1 = M^+$, $S_2 = H^+$ and $L = Cr^-$ then (39) is identical to (34). As previously (36) and (37) are partition equilibria, subject to the

$$S_1 \rightleftharpoons \overline{S_1} \qquad K^{S_1}_{tr} \tag{36}$$

$$\overline{S_2} \rightleftharpoons S_2 \qquad 1/K^{S_2}_{tr} \tag{37}$$

$$\overline{LS_2} + \overline{S_1} \rightleftharpoons \overline{LS_1} + \overline{S_2} \qquad K \tag{38}$$

$$\overline{LS_2} + S_1 \rightleftharpoons LS_1 + S_2 \qquad K_{ex} \tag{39}$$

structural influences discussed in 4.3.2. The complexation reaction (38) is confined to the organic phase and has the characteristics of an ion exchange reaction. As such it is the sum of two ion pairing reactions and may be amenable to direct calculation for simple carriers such as NMBA and DNNS.[122] The real value of (36)-(38) is to show the influence of parameters, such as ligand and substrate radius, on the value of K_{ex} (39). It further underscores the role of lipophilicity on transport as discussed above with NMBA and mandelate.[152] It is likely as well that the transport of Cs^+ by calixareres falls within this group as well[154,155] as the selectivity follows the ΔG°_h of the cation.

5.3 Selectivity

As noted above, the majority of cases have $K_{ex} < K_{ex}(max)$ thus the full range of implications of 2.5 has not been examined in detail. Again, the case of NMBA as a carrier of mandelate is instructive. In this case the transport process distinguishes between the two enantiomers of mandelate by the formation of diastereomeric salts with the carrier (-)NMBA.[152]

$$\overline{(-)NMBAX} + (-)Man^- \rightleftharpoons \overline{(-)NMBA(-)Man} + X^- \qquad K_{ex}(-) \tag{40}$$

$$\overline{(-)NMBAX} + (+)Man^- \rightleftharpoons \overline{(-)NMBA(+)Man} + X^- \qquad K_{ex}(+) \tag{41}$$

Since the salts are diastereomeric they should have different physical properties e.g., $K_{ex}(-) \neq K_{ex}(+)$. In the event, (+) mandelate is preferentially extracted for all antiport anions X^-, **but the mandelate enantiomer released to the OUT phase depends on the antiport anion.**[152] Further, the selectivity (at a fixed fraction of transport) goes through the same maximum at X^-= propionate as did the flux described above. This is a consequence of the values K_{ex} (±) depending upon the anion lipophilicity and to a lesser extent on the ionic composition of the IN and OUT phases. For X = Cl^-, the extraction selectivity on the IN (high ionic strength side) is $K_{ex}(+)/K_{ex}(-) \sim 1{:}1$ but on the OUT side (low ionic strength) the selectivity is $K_{ex}(+)/K_{ex}(-) \sim 1.4$. As a consequence the (-) isomer is released to the OUT phase! The general conclusions drawn in 2.5 refer to symmetrical cases. The situation described above is an **asymmetrical** case and results in a **transport selectivity** which is the **reverse of the thermodynamic extraction selectivity.**

The extent to which this is generally true is essentially unexplored. It is common for experiments involving proton gradients to have vastly different IN and OUT phases.[140-146] Differences between two phase competitive extractions and comparable competitive membrane experiments have been noted[144,145] which further depends on the type of membrane system.[145] It seems reasonably clear that membrane asymmetry induced by large pH gradients must play a role.

One of the more striking demonstrations of the control of selectivity refers to the pH regulation of K^+/Ca^{+2} selectivity by a dicarboxylic acid carrier.[143] This carrier, like others of its type[142,159] can form a range of complexes with mono and divalent cations containing 0, 1 or 2 bound protons.[160] The neutral complexes of interest, in addition to the diprotonated carrier H_2Cr, are HCrK and CrCa. The latter is favored at high pH.[160] Competitive transport can be examined in terms of the schematic mechanism of Figure 20.

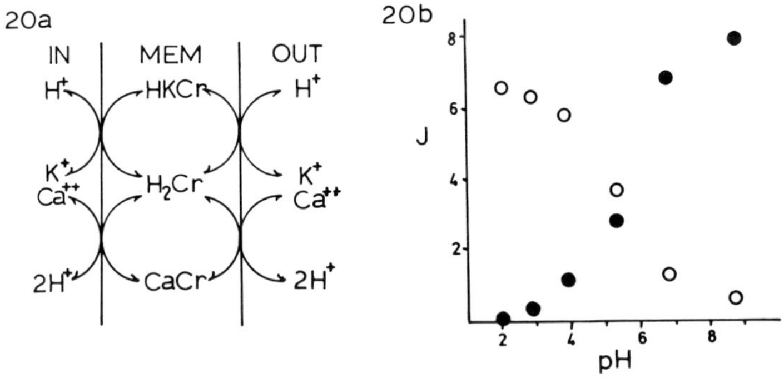

Figure 20: pH regulation of K^+/Ca^{+2} selectivity: a) schematic mechanism b) transport as a function of pH: ○ K^+ rate; ● Ca^{+2} rate in μmole h^{-1}, Redrawn from ref 143 with permission of Dr. J.M. Lehn.

At basic pH on the IN side, the ligand will be fully deprotonated and the lower cycle (CrCa 20a) will be favored. As the acidity of the IN side is increased, the mono protonated carrier becomes more favorable and the upper cycle (HCrK, 20a) will become important. The results of the pH influence on K^+/Ca^{+2} transport rates are sketched in 20b. The qualitative explanation above is confirmed and the effect is quite marked; the $Ca^{+2}K^+$ selectivity changes by a factor of 10^3, from 10 fold in favor of Ca^{+2} at pH8.6 to 160 fold in favor of K^+ at pH 2.0.[143] Again it is clear that **selectivity can be regulated by experimental variables.**

5.4 Kinetic regime

For the reasons outlined in 4.5, there seems to be little chance that the carriers of Figure 18 will have sufficiently slow complexation reactions to show any influence on diffusion controlled transport.

Nonetheless, there is a great deal of evidence to suggest that a "kinetic" regime can be established for a countertransport of alkali metal and alkaline earth cations by some carboxylate carriers[41,48,140-142] As discussed in 2.7, the diffusion regime is characterized by a (linear) relationship between stirring and flux, while in the kinetic regime, flux is independent of stirring. Figure 21 gives examples of this type of experiment for transport of K^+ and Ca^{+2} by the carriers 14NH18-6, anti(8NH)$_2$18-6 (Figure 18) and DB18-6 (Figure 14).

There is little doubt that above 300 rpm, two of the four processes achieve a plateau in flux as predicted for the onset of the kinetic regime. Figure 21a concerns transport under conditions where the loss of metal ion from the IN side of the cell, and the appearance of metal ions on the OUT side of the cell is linear with time. The lower curve of 21a provides additional proof that cotransport of KNO_3 by DB18-6 is a diffusion controlled process under the condition where the

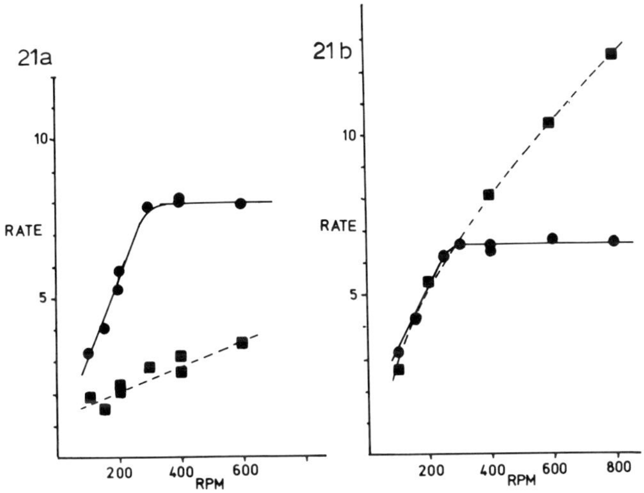

Figure 21: Transport rate as function of stirring of the three phases (U-tube). a) "zero order" regime 14NH18-6/K^+; DB18-6/KNO_3 b) "first order" regime 14NH18-6/K^+; anti(8NH)$_2$18-6/Ca^{+2}. Redrawn from ref. 41 with permission of Elsevier Science Publishers.

approximations of (24) are valid. In the same cell (a U-tube) the carrier 14NH18-6 initially shows similar behavior, but above 300 rpm achieves a plateau. In the plateau region the flux is given by an equation:[48,142]

$$J_K = k_o\overline{HCr} \tag{42}$$

Where k_o is a zero order rate constant (mol per time per area per concentration of carrier) and $\overline{HCr}$ is the concentration of 14NH18-6 in the membrane.

The case of Figure 21b is similar except that in these cases the concentration in the IN phase decreases by some first order process (exponential) while the coupled appearance in the OUT phase follows a similar first order process (hence "first-order regime").[48] The upper curve, for the carrier anti $(8NH)_2$18-6/Ca^{+2} is another example of a diffusion limited process. The rate law, deduced from initial rate studies is given by a set of consecutive, reversible first order reactions according to (43):[142]

$$Ca^{+2}_{in} \underset{k-1}{\overset{k_1}{\rightleftharpoons}} \overline{Ca^{+2}} \underset{k-2}{\overset{k_1}{\rightleftharpoons}} Ca^{+2}_{out} \qquad (43)$$

The apparent first order rate constants include diffusion coefficients, the ligand concentration, the boundary layer thickness etc.

In the same cell, and with a concentration/time profile closely similar, the carrier 14NH18-6 shows the onset of the kinetic regime about 300 rpm.[42] In the plateau region the flux is given by an equation:[48,141,142]

$$J_K = k_1 \frac{\overline{HCr}\ K^+}{H^+} \qquad (44)$$

where k_1 is a first order rate constant and the other species terms are concentrations. Equation 44 can be rearranged, using definitions for the acidity and stability constants of the carrier (equations 30 and 31) to a form showing that the flux is first order in the concentration of the carrier complex with K^+.[48,142]

For a given carrier, e.g. 14NH 18-6, the balance between the zero and first order regimes is controlled by the initial concentrations of K^+ and $\overline{HCr}$. The mechanism which unites these diverse behaviours focuses on the fact that the carriers are **surfactants**. Interfacial surface tension measurements show that the IN interface will be saturated at 10^{-5}M in carrier.[48,142] Acid-base and complexation equilibria at the interface are fast, for the reasons outlined in 4.5. Thus the interface is in equilibrium with the species in the aqueous phase. The flux is controlled by penetration of $\overline{HCr}$ from the bulk organic phase **into** the saturated interface (zero order regime), or under different concentration conditions by the desorption of the complex **from** the saturated interface (first order regime).[48,141,142] The flux is maximal, not when the bulk concentration in the membrane reaches $L_o/2$, but when the interface is **saturated with complex**. This will occur at a variety of concentrations[48,143] depending on **both** L_o and the initial metal ion concentrations.

This picture of transport dominated by kinetic processes at the interface shifts the emphasis from thermodynamic parameters such as K_{ex}. In fact, there are substantial differences in **selectivity** of competitive transport from those which would be predicted from thermodynamic selectivities.[163] The detailed analysis of the selectivity is complex but can be summarized as follows: the two phase (thermodynamic) selectivity of 14NH18-6 for K^+/Sr^{+2} is essentially 1, i.e. both cations are about equally partitioned to the organic phase.[163] Under "first order" regime conditions, however, the membrane is very strongly Sr^{+2} selective and the K^+ transport is greatly inhibited by Sr^{+2} until bulk concentration ratio K^+/Sr^{+2} is 10 or greater.[163] There are three factors which control the selectivity in this case. Firstly, there is an **electrostatic** contribution due to ionization of the carrier at the interface. This imparts a negative surface charge to the interface with enrichment of Sr^{+2} at the interface as a result of a double layer being created. This amounts to about a factor of 25 in favour of Sr^{+2}.[163] Secondly, there is a **thermodynamic** factor for ion binding at the interface; K^+ is bound somewhat more strongly than Sr^{+2} by a factor of about 4.[163] Finally, there is a **kinetic** factor; the complexes of Sr^{+2} leave the interface about 13 times faster than the K^+ complexes. This is again related to the **surfactant character** of the complexes. The 2:1 Sr^{+2} complex is globular[160] while the K^+ complex retains a polar head non-polar tail arrangement and is thus a better surfactant. If we take a kinetic view of surface activity, the tendency for a solute to occupy an interfacial position is the ratio of "on" and "off" rates. In the case discussed here, clearly the "off" rate for Sr^{+2} is faster than for K^+.[163] The net selectivity Sr^{+2}/K^+ is thus 25 x 13/4 ~ 80.

This example is of interest in that it provides a detailed look at one set of interfacial processes. To what extent these can be generalized is not clear but there are some anecdotal observations which suggest that absorption/desorption and consequent surface charge effects are important. The regulation of Ca^{+2}/K^+ selectivity was discussed above (Figure 21). The related dialkyl carrier anti$(8NH)_2$18-6 **failed** to show any hint of similar behaviour with $CHCl_3$ as solvent but reproduced the trends in Fig. 21 using the same solvent as Hriciga and Lehn[143] (10% 1-hexanol in $CHCl_3$).[164] One rationale for this behaviour rests with the observation that 1-hexanol is also surface active and the carrier thus does not saturate the interface in this solvent.[164] As a result, any surface charge due to ionization will be small, the electrostatic component of the selectivity will vanish, and H^+ and K^+ will be able to compete at the interface with Ca^{+2}.

The implication that loss of surfactant character can accelerate transport warrants consideration. Natural acyclic ionophores may achieve similar acceleration wherein cation binding alters the conformation to assist transfer into the membrane.[4] In the systems examined above, the onset of a diffusion "regime" for anti $(8NH)_2$18-6 suggests that this carrier leaves the interface very efficiently compared with other crown ether carboxylates; it has relatively weak surface activity.[164] The kinetic Sr^{+2}/K^+ selectivity noted above is due to the same effect. One is tempted to generalize that the best carriers have a **strong** surface activity with one substrate and a **weak** surface activity **after exchange of substrates.** The differences between **syn** and **anti** isomers may be related.[142] Acyclic ionophores[51,150,151] which presumably act by encapsulation might well exhibit a similar change of surfactant properties on ion binding.

6. SUMMARY

There is no doubt that the field of artificial membrane transport using synthetic ionophores has advanced remarkably in the past 15 years due primarily to the synthesis of new ionophores. Even though the theoretical framework substantially predated this activity, the merging of theory with transport experiment has often been sketchy. The purpose of this outline has been to examine key examples to illustrate the underlying principles and to suggest how **experimental variables dominate the results obtained.** To a very good approximation the assumption of a "diffusion" regime is often justified, is easily confirmed experimentally and provides a clear framework for exploitation of the inherent selectivity of a given ionophore. Thus for synthetic chemists who wish a "quick and nasty" experiment to examine the question of selectivity, the recipe is clear: a mixture containing all ions of interest in a standard experiment for each ligand of interest using a moderately stirred (100-200 rpm) cell and analysis of the mixture produced on the OUT side of the cell at a fixed, small extent of transport. Together with duplicates and controls, this modest set of experiments will place the results on an unambiguous footing from which clear conclusions about each ionophore's characteristics are readily obtained.

For those with more detailed interests in the transport process the demands are correspondingly higher. Is the system clearly in the "diffusion" regime or is it an example of the "kinetic" regime? What are representative values of K_{ex}? Where does the process fit with respect to the simple analyses of J as a function of K_{ex} as outlined in section 2? Does the selectivity directly follow the thermodynamic selectivity in a predictable fashion? Is the system asymmetric and does this alter selectivity? The results of these more extensive studies will no doubt uncover other manifestations of the basic principles. Hopefully, other confirmed cases of "kinetic" control will appear as well, to further probe events at the interface.

If the goal is to develop a specific separation, there are three distinct classes of interrelated considerations. Firstly there are the intrinsic properties of the ionophore. Its shape, size, ion binding properties, "lipophilicity" and "surfactant character" all influence the final outcome. Secondly, there are the choices of co- or counter-transported ions, their "lipophilicities" and ion-pairing properties. These provide further influences on K_{ex}. Finally, there are purely experimental considerations such as the initial concentrations, ionophore/substrate ratio and extent of transport. These define the range of K_{ex} for optimal transport. These three classes of consideration must be taken together in order to ensure that K_{ex} for the desired substrate is close to K_{ex}(max) and that the experiment is conducted in such a way as to fully exploit the available thermodynamic selectivity of the ionophore. Under the best conditions the method will achieve the goal of selective, efficient separation. On the other hand, subtle factors such as an unfortunate choice of a counter ion, could seriously impair the final separation. It is hoped that the principles sketched here will aid in identifying (and avoiding!) the pitfalls.

7. REFERENCES

1. P.L. Yeagle, **Acc. Chem. Res.,** 9, 321 (1978).
2. **Bioenergetics and Thermodynamics: Model Systems,** A.Braibraniti, ed., D. Reidel, Dordrecht, 1979; Contributions by J.J. Christensen, D.E. Fenton, A. Gliozzi, J. Jagur-Grodzinski, J.M. Lehn and G. Scibona.
3. **The Principles of Ion-Selective Electrodes and Membrane Transport.**, W.E. Morf, Elsevier, Amsterdam, 1981; M.B. McDonnell and P.M. Vadgama, **Ion Selec. Elect. Revs.** 11, 17 (1989).
4. G.R. Painter and B.C. Pressman, **Top Curr. Chem.**, 101, 83 (1982).
5. S. Shinkai and O. Manabe, **Top Curr. Chem.**, 121, 67 (1984).
6. J.M. Lehn in **Physical Chemistry of Transmembrane Ion Motions**, G. Spach, ed., Elsevier, Amsterdam, 1983, p. 181.
7. M. Okahara and Y. Nakatsuji, **Top Curr. Chem.**, 128. 37 (1985).
8. D.N. Urry, **Top Curr. Chem.**, 128. 175 (1985).
9. K. Cammann, **Top Curr. Chem.**, 128, 219 (1985).
10. D.W. McBride, R.M. Izatt, J.D. Lamb and J.J. Christensen, in **Inclusion Compounds Volume 3**, J.L. Atwood, J.E.D. Davies, and D.D. MacNicol, eds., Academic Press, London, 1984, p. 571; R.M. Izatt, G.A. Clark, J.S. Bradshaw, J.D. Lamb and J.J.Christensen, **Sep. Purif. Meth.**, 15, 21 (1986); R.M. Izatt, G.C. LindH, R.L. Breuning, J.S. Bradshaw, J.D. Lamb and J.J. Christensen, **Pure App'l. Chem.** 58, 1453 (1986).
11. **Membrane Mimetic Chemistry**, J.H. Fendler, Wiley, New York, 1982; H. Tsukube, **J. Coord. Chem**, 16, 101 (1987); T.M. Fyles, **Bioorg. Chem. Front.**, 1, 72 (1990).
12. **Transport Phenomena in Membranes**, N. Lakshminarayanaiah, Academic Press, New York, 1969.
13. **Transport Across Biological Membranes**, M. Hofer, Pitman, London, 1981.
14. J.D. Goddard, **J. Phys. Chem.**, 89, 1825 (1985); T.M. Fyles and S.P. Hansen, **Can. J. Chem.**, 66, 1445 (1990).
15. **Multi Component Diffision**, E.L. Cussler, Elsevier, Amsterdam, 1976.
16. B.J. Zwolinski, H. Eyring and C.E. Reese, **J. Phys. Colloid Chem.**, 53, 1426 (1949).
17. G. Szabo, G. Eisenman, R. Laprade, S.M. Ciani and S. Krasne in **Membranes: Volume 2, Lipid Bilayers and Antibiotics**, G. Eisenmann, ed., Dekker, New York, 1973, p. 181.
18. **Nonequilibrium Thermodynamics in Biophysics**, A. Katchalsky and D.F. Curran, Harvard U. Press, Cambridge, 1967.

19. T. Rosenberg and W. Wilbrandt, **Pharmacol. Rev.** 13, 109 (1961).
20. J.A. Jacquez, **Biochem. Biophys. Acta**, 79, 318 (1967).
21. R. Blumenthal and A. Katchalsky, **Biochem. Biophys. Acta**, 173, 357 (1969).
22. G.M. Shean and K. Sollner, **Ann. N.Y. Acad. Sci.**, 137. 759 (1966).
23. W. Wilbrandt, **J. Cell Comp. Physiol.**, 47, 137 (1956).
24. W.J. Ward, **A.I.Ch.E. Journal**, 16. 405 (1970).
25. E.L. Cussler,**A.I.Ch.E. Journal**, 17, 1300 (1971).
26. C.F. Reusch and E.L. Cussler, **A.I.Ch.E. Journal**, 19. 736 (1973).
27. F. Caracciolo, E.L. Cussler and D.F. Evans, **A.I.Ch.E. Journal**, 21, 160 (1975).
28. E.M. Choy, D.F. Evans and E.L. Cussler, **J. Am. Chem. Soc.**, 96, 7085 (1979).
29. J.D. Lamb, J.J. Christensen, S.R. Izatt, K. Bedke, M.S. Astin and R.M. Izatt, **J. Am. Chem. Soc.**, 102, 3399 (1980); R.M. Izatt, D.W. McBride Jr., P.R. Brown, J.D. Lamb and J.J. Christensen, **J. Membr. Sci.**, 28, 69 (1988).
30. Initial model: J.D. Lamb, J.J. Christensen, J.L. Oscarson, B.L. Nielsen, B.W. Asay and R.M. Izatt, **J. Am. Chem. Soc.**, 102, 6820 (1980); J.D. Lamb, R.M. Izatt, D.G. Garrick, J.S. Bradshaw and J.J. Christensen, **J. Membr. Sci.**, 9, 83 (1981); Revised model: R.M. Izatt, R.L. Breuning, M.L. Breuning, G.C. LindH and J.J. Christensen, **Anal. Chem.** 61 1140 (1989).
31. S. Yoshida and S. Hayano, **J. Membr. Sci**, 26, 99 (1986); S. Yoshida and S. Hayano, **J. Am. Chem. Soc.**, 108, 3903 (1986); S. Yoshida and T. Watanabe, **J. Coord. Chem.**, 18, 63 (1988); T.B. Stolwijk, E.J.R. Sudholter and D.N. Reinhoudt, **J. Am. Chem. Soc.**, 109, 7042 (1987); T.B. Stolwijk, E.J.R. Sudholter and D.N. Reinhoudt, **J. Am. Chem. Soc.**, 111 6321 (1989).
32. J.P. Behr, M. Kirch and J.M. Lehn, **J. Am. Chem. Soc.**, 107, 241 (1985).
33. **Elementary Kinetics of Membrane Carrier Transport**, K.D. Neale and T.G. Richards, Blackwell, Oxford, 1972.
34. Define $y = J/J_{max}$, $A_{in} = (1-f)\ A_o$ and $A_{out} = A_o + L\ y/2$.[32] Thus (8) becomes a quadratic in K_{ex}:
$a\ K_{ex}^2 + b\ K_{ex} + c = 0$
where $a = A_{in} A_{out}\ y$, $b = y\ A_{in} + y\ A_{out} - A_{in} + A_{out}$ and $c = y$. The roots are computed for a given set of L_o, A_o and f at intervals of 0.05 in y then plotted to produce the curves of Figure 5.

35. C.A. Koval, R.D. Noble, J.D. May, B. Lowe, Z.E. Reyes, B.R. Bateman, G.M. Horn and D.L. Read, **Inorg. Chem.**, 24, 1147 (1985); R.D. Noble, J.J. Pellegrino, E. Grosgogeat, D. Sperry and J.D. Way, **Sep. Sci. Tech.**, 23, 5195 (1988); H. Nishide, H. Kawakami, Y. Kurimura and E. Tsuchida, **J. Am. Chem. Soc.**, 111 7175 (1989).
36. F. Diederich and K. Dick, **J. Am. Chem. Soc.**, 106, 8024 (1984); A. Harada and S. Takahashi, **J.C.S. Chem. Comm.**, 527 (1987); C.A. Koval and T. Spontarelli, **J. Am. Chem. Soc.**, 110, 293 (1988).
37. As previously [34], the quadratic in K_{ex} is computed for

$a = y\ A_{in}^2\ A_{out}^2$, $b = y\ A_{in}^2 + y\ A_{out}^2 - A_{in}^2 + A_{out}^2$ and $c = y$.

38. As previously [34], the quadratic in K_{ex_2} is computed for $a = c = y\ A_{in}\ A_{out}$ and $b = y\ A_{in}^2 + y\ A_{out}^2 - A_o\ A_{in} + A_o\ A_{out}$.
39. P.R. Danesi, R. Chiarizia, M. Pizzichini and A. Saltelli, **J. Inorg. Nucl. Chem.**, 40, 1119 (1978).
40. P.R. Danesi and R. Chiarizia, **C.R.C. Crit. Rev. Anal. Chem.**, 10, 1 (1980).
41. T.M. Fyles, **J. Membr. Sci.**, 24, 229 (1985).
42. S.K. Friedlander and K.H. Keller, **Chem. Eng. Sci.**, 20, 121 (1965).
43. P.R. Danesi, E.P. Horowitz, G.F. Vandegrift and R. Chiarizia, **Sep. Sci. Tech.**, 16, 201 (1984).
44. H.L. Rosano, J.H. Schulman and J.B. Weisbuch, **Am. N.Y. Acad. Sci.**, 92, 457 (1961).
45. J.P. Behr and J.M. Lehn, **J. Am. Chem. Soc.**, 95, 6108 (1973).
46. M. Kirch and J.M. Lehn, **Angew. Chem. Int'l. Ed'n. Engl.**, 14, 555 (1975).
47. M. Kirch, These de Doctorat es Sciences, Université Louis Pasteur, Strasbourg, France, 1980.
48. T.M. Fyles, V.A. Malik-Diemer and D.M. Whitfield, **Can. J. Chem.** 59, 1734 (1981).
49. H. Tsukube, **J. Chem. Soc. Perkin Trans I**, 29 (1983).
50. B.C. Pressman and N.T. de Guzman, **Ann. N.Y. Acad. Sci.**, 264, 373 (1975).
51. K. Hiratani, K. Taguchi, H. Sugihara and K. Ito, **Bull. Chem. Soc. Jpn.**, 57, 1976 (1984).
52. P.R. Danesi, **Sep. Sci. Tech.**, 19, 857 (1984).
53. P.R. Danesi, R. Chiarizia and A. Castagnola, **J. Membr. Sci.**, 14, 161 (1983).
54. S. Djumovic, M.M. Kreevoy and T.S. Kerlak, **J. Phys. Chem.**, 89, 3151 (1985).
55. P.R. Danesi, **J. Membr. Sci.**, 20, 231, (1984).
56. P.R. Danesi and C. Cianetti, **J. Membr. Sci.**, 20, 201, 215 (1984).
57. B.M. Kim, **J. Membr. Sci.**, 21, 5 (1984).
58. N.N. Li, U.S. Patent 3,410,794 (1968); **Ind. Eng. Chem. Process Des. Dev.**, 10, 215 (1971).

59. R.P. Cahn and N.N. Li, **Sep. Sci. Tech.**, 9, 505 (1974).
60. K.H. Lee, D.F. Evans and E.L. Cussler, **A.I.Ch.E. Journal** 24, 860 (1978).
61. A. Hochhauser and E.L. Cussler, **A.I.Ch.E. Symp. Ser.**, 71, 136 (1975).
62. J.J. Christensen, S.P. Christensen, M.P. Biehl, S.A. Lowe, J.D. Lamb and R.M. Izatt, **Sep. Sci. Tech.**, 18, 363 (1983); R.M. Izatt, D.V. Dearden, D.N. McBride, J.L. Oscarson, J.D. Lamb and J.J. Christensen, **Sep. Sci. Tech.**, 18, 1113 (1983).
63. R.M. Izatt, J.D. Lamb and R.L. Bruening, **Sep. Sci. Tech.**, 23, 1645 (1988); R.M. Izatt, D.K. Roper, R.L. Bruening and J.D. Lamb, **J. Membr. Sci.**, 45, 73 (1989).
64. E.C. Hsu and N.N. Li, **Sep. Sci. Tech.**, 20, 115 (1985).
65. T. Kunitake, Y. Okahata, M. Shimomura, S. Yasunani and K. Takarabe, **J. Am Chem. Soc.**, 103, 5401 (1981).
66. T. Kunitake, N. Kimizuka, N. Higashi and N. Nakashima, **J. Am. Chem. Soc.**, 106, 1978 (1984).
67. R. Elbert, T. Folda and H. Ringsdorf, **J. Am. Chem. Soc.**, 106, 7687 (1984).
68. S.L. Regen, J.S. Shin and K. Yamaguchi, **J. Am. Chem. Soc.**, 106, 2446, (1984).
69. C. Thomas, C. Sauterey, M. Castaing, C.M. Gary-Bobo, J.M. Lehn and P. Plumeré, **Biochem. Biophys. Res. Comm.**, 116, 981 (1983).
70. M. Castaing, F. Morel and J.M. Lehn, **J. Membr. Biol.**,89, 251 (1986).
71. J. Grandjean and P. Laszlo, **J. Am. Chem. Soc.**, 106, 1472 (1984).
72. B. Cybulska, E. Borowski, Y. Prigent and C.M. Gary-Bobo, **J. Antibiotic**, 34, 884 (1981).
73. A. Finkelstein and R. Holz in **Membranes: Volume 2, Lipid Bilayers and Antibiotics**, G. Eisenman, ed., Dekker, New York, 1973, p. 377; Amphotericin mimic: T.M. Fyles, K.C. Kaye, T.D. James and D.W.M. Smiley, **Tetrahedron Letters**, 31, 1233 (1990).
74. R. DeSantis, A. Palleschi, M. Savino, A. Scipioni, B. Sesta and A. Verdini, **Biophys. Chem.**, 21, 211 (1985).
75. G. Spach, F. Hertz and Y. Trudelle in **Physical Chemistry of Transmembrane Ion Motions**, Elsevier, Amsterdam, 1983, p. 357.
76. J. P. Behr, J.M. Lehn, A.C. Dock and D. Moras, **Nature**, 295, 526 (1982); V.A. Carmichael, P.J. Dutton, T.M. Fyles, T.D. James, J. Swan and M. Zojaji, **J. Am. Chem. Soc.** 111, 767 (1989).
77. J.G. Neevel and R.M. Nolte, **Tet. Lett.**, 25, 2263 (1984).
78. U.F. Kragten, M.F.M. Roks and R.M. Nolte, **J.C.S. Chem. Comm.**, 1275 (1985); T.M. Fyles, T.D. James and K.C. Kaye, **Can. J. Chem.**, 68, 976 (1990).

79. J.H. Furhop and U. Liman, **J. Am. Chem. Soc.**, 106, 4643 (1984).
80. J.H. Furhop, U. Liman and H.H. David, **Angew. Chem. Int'l Ed'n. Engl.**, 24, 339 (1985).
81. J.H. Furhop, K. Ellermann, H.H. David and J. Mathieu, **Angew. Chem. Int'l. Ed'n. Engl.**, 21, 440 (1982).
82. H.K. Lonsdale, **J. Membr. Sci.**, 10, 81 (1982).
83. W. Pusch and A. Walch, **Angew. Chem. Int'l Ed'n. Engl.**, 21, 660 (1982).
84. A.P. Thoma, A. Viviani-Nauer, S. Arvantis, W.E. Morf and W. Simon, **Anal. Chem.**, 49, 1567 (1977).
85. G.J. Moody, R.B. Oke and J.D.R. Thomas, **Analyst (London)**, 95, 910 (1970).
86. H.K. Wipf, A. Olivier and W.Simon, **Helv. Chim. Acta**, 53, 1605 (1970).
87. T. Kajiyama, S. Washiza, A. Kumano, I. Terada, M. Takayanagi and S. Shinkai, **J. Appl. Poly. Sci., Appl. Poly. Symp.**, 41, 327 (1985).
88. Y. Okahata, G. Enna, K. Tagichi and T. Seki, **J. Am. Chem. Soc.**, 107, 5300 (1985).
89. Y. Okahata, T. Taguchi and T. Seki, **J.C.S. Chem. Comm.**, 1122 (1985).
90. Y. Okahata and T. Seki, **J. Am. Chem. Soc.**, 106, 8065 (1984).
91. S. Shinkai, S. Nakamura, S. Tachiki, O. Manabe and T. Kajiyama, **J.Am. Chem. Soc.**, 107, 3363 (1985).
92. N. Ogata, K. Sanui and H. Fujimara, **J. Appl. Poly. Sci.**, 26, 4149 (1981).
93. M. Yoshikawa, H. Ogata, K. Sanui and N. Ogata, **Poly. J.**, 15, 609 (1983).
94. M. Yoshikawa, Y. Imashiro, K. Sanui and N. Ogata, **J. Membr. Sci.**, 20, 189 (1984).
95. T.M. Fyles, C.A. McGavin and D.E. Thompson **J.C.S. Chem. Comm.**, 924 (1982).
96. K.Kimura, M. Sakamoto, M. Yoshinaga and T. Shono **J.C.S. Chem. Comm.** 978 (1983).
97. L.M. Dulyea, T.M. Fyles and G.D. Robertson, **J. Membr. Sci**, 34, 87 (1987).
98. **Membrane Active Complexones**, Yu A. Ovchinikov, V.T. Ivanov and A.M. Shkrob, Elsevier, Amsterdam, 1974.
99. R.M. Izatt, J.S. Bradshaw, S.A. Nielsen, J.D. Lamb, J.J. Christensen and D. Sen, **Chem. Rev.**, 85, 271 (1985).
100. R.A. Steiner, M. Oehme, D. Ammann and N. Simon, **Anal. Chem.**, 51, 351 (1979).
101. P.R. Danesi, E.P. Horowitz and P.G. Rickert, **J. Phys. Chem.**, 87, 4708 (1983).
102. E. Bacon, These de Doctorat es Sciences, Université Louis Pasteur, Strasbourg, France, 1984; E. Bacon and M. Kirch, **J. Membr. Sci.**, 32, 159 (1987); H. Tsukube, H. Adachi and S. Morosawa, **J.C.S. Perkin I**, 89 (1989).

103. E. Bacon, L. Jung and J.M. Lehn, **J. Chem. Res. (S)**, 136 (1980).
104. M. Sugiura and T. Shinbo, **Bull. Chem. Soc. Jpn.**, 52, 684 (1979); M. Sugawara, M. Omoto, H. Yoshida and Y. Umezawa, **Anal. Chem.**,60, 2301 (1988).
105. K. Maruyama, H. Tsukube and T. Araki, **J. Am. Chem. Soc.**, 104, 5197 (1982); J. Rebek Jr., B. Askew, D. Nemeth and K. Parris, **J. Am. Chem. Soc.**, 109, 2432 (1987); H. Kataoka and T. Katagi, **Tetrahedron**, 43, 4519 (1987); E. Kotufuta, K. Sumi and W.-C. Wu, **Chem. Lett.**, 637 (1989).
106. R.M. Izatt, G.A. Clark and J.J. Christensen, **J. Membr. Sci.**, 24, 1 (1985).
107. H.K. Frensdorff, **J. Am. Chem. Soc.**, 93, 4684 (1971).
108. B. Dietrich These de Doctorat d'Etat, Université Louis Pasteur, Strasbourg, France 1973.
109. P. Gramain and Y. Frére, **Nouv. J. Chim.**, 3, 53 (1979).
110. Y. Takeda, **Top. Curr. Chem.**, 121, 1 (1984).
111. J.M. Lehn, **Struct. Bonding**, 16, 1 (1973).
112. D.J. Cram and J.M. Cram, **Acc. Chem. Res.**, 11, 8 (1978).
113. J.M. Lehn, **Acc. Chem. Res.**, 11, 49 (1978).
114. B. Dietrich in **Inclusion Compounds Volume 2**, J.L. Atwood, J.E.Davies and D.D. MacNicol eds., Academic Press, London, 1984.
115. **Ionophores and Their Structures**, M. Dobler, J. Wiley and Sons, New York, 1981.
116. M.H. Abraham and J. Liszi, **J.C.S. Farad I**, 74, 1604 (1978).
117. M.H. Abraham and J. Liszi, **J. Inorg. Nucl. Chem.**, 43, 143 (1981).
118. A. Brandstrom, **Adv. Phys. Org. Chem.**, C. 15, 267 (1977).
119. **The Hydrophobic Effect.**, C. Tanford, J. Wiley & Sons, New York, 1973.
120. M.H. Abraham, **J. Am. Chem. Soc.**., 101, 5477 (1979).
121. C. Hansch, **Acc. Chem. Res.**, 7. 232 (1969).
122. R.M. Fuoss, **J. Am. Chem. Soc.**, 80, 5059 (1958).
123. J.D. Lamb, R.M. Izatt, R.A. Robertson and J.J. Christensen **J. Am. Chem. Soc.**, 102, 2452 (1980).
124. R.M. Izatt, D.V. Dearden, P.R. Brown, J.S. Bradshaw, J.D. Lamb and J.J. Christensen, **J. Am. Chem. Soc.**, 105, 1785 (1983).
125. J.D. Lamb, P.R. Brown, J.J. Christensen, J.S. Bradshaw, D.G. Garrick and R.M. Izatt, **J. Membr. Sci.**, 13, 89 (1983).
126. R.M. Izatt, D.N. McBride, J.J. Christensen, J.S. Bradshaw and G.A. Clark, **J. Membr. Sci.**, 22, 31 (1985).
127. R.M. Izatt, M.P. Biehl, J.D. Lamb and J.J. Christensen, **Sep. Sci. Tech.**, 17, 1351 (1982).
128. Z. Qi, T. Sasakura and E.L. Cussler, **Sep. Sci. Tech.**, 19, 963 (1984).

129. Z. Qi and E.L. Cussler, **J. Membr. Sci.**, 19, 259 (1984).
130. B.G. Cox, J. Garcias-Rosas and H. Schneider, **J. Am. Chem. Soc.**, 103, 1054 (1981).
131. T.M. Fyles, **Can. J. Chem.**, 65, 884 (1987).
132. J.J. Grimaldi and J.M. Lehn, **J. Am. Chem. Soc.**., 101, 1333 (1979).
133. A. Ohki, T. Takeda, M. Takagi and K. Ueno, **J. Membr. Sci.**, 15, 231 (1983).
134. S.S. Anderson, I.G. Lyle and R. Patterson, **Nature**, 259, 147 (1976); J.J. Grimaldi, J.M. Lehn and S. Boileau, **Nature**, 265, 229 (1977); T. Shinbo, K. Kurihara, Y. Kobatake and N. Kano, **Nature**, 270, 277 (1977).
135. E. Ozeki, S. Kimura and Y Imanishi, **J.C.S. Chem. Soc.**, 1353 (1988); L.E. Echegoyen, H.K. Yoo, V.J. Gatto, G.W. Gokel and L. Echegoyen, **J. Am. Chem. Soc.**, 111, 2440 (1989).
136. G. De Santis, M. Di Casa, M. Mariani, B. Seghi and L. Fabbrizzi, **J. Am. Chem. Soc.**, 111 2422 (1989).
137. S. Shinkai, K. Inuzuka, O. Miyazaki and O. Manabe, **J.Am. Chem. Soc.**, 107, 3950 (1985).
138. S. Shinkai, T. Nakaji, T. Ogawa, K. Shigamatsu and O. Manabe, **J. Am. Chem. Soc.**, 103, 111 (1981).
139. J. Bolte, C. Demuynck, G. Jeminet, J. Juillard and C. Tissier, **Can. J. Chem.**, 60, 981 (1982).
140. L.A. Frederick, T.M. Fyles, N.P. Gurprasad and D.M. Whitfield, **Can. J. Chem.**, 59. 1724 (1981).
141. T.M. Fyles, V.A. Malik-Diemer, C.A. McGavin and D.M. Whitfield, **Can. J. Chem.**, 60. 2259 (1982).
142. L.M. Dulyea, T.M. Fyles and D.M. Whitfield, **Can. J. Chem.**., 62,498 (1984).
143. A. Hriciga and J.M. Lehn, **Proc. Nat'l Acad. Sci. U.S.A.**.,80, 6426 (1983).
144. J. Strzelbicki and R.A. Bartsch, **J. Membr. Sci.**, 10, 35 (1982); R.A. Bartsch, N.A. Charewicz and S.I. Kay, **J. Membr. Sci.**, 17. 97 (1984).
145. Y. Kobuke, I. Tabushi, K. Oh and T. Aoki, **J. Org. Chem.**, 53, 5953 (1988).
146. S. Shinkai, H. Kida, T. Sone and O. Manabe, **J.C.S. Chem. Comm.**, 125 (1982).
147. M. Takagi and K. Ueno, **Top. Curr. Chem.**, 121, 39 (1984); E. Kimura, C.A. Dalimonte, A. Yamashita and R. Machido, **J.C.S. Chem. Comm.**, 1041 (1985).
148. W. Walkowiak, P.R. Brown, J.P. Shuckla and R.A. Bartsch, **J. Membr. Sci.**, 32, 59 (1987); M. Sugiura, M. Kikkawa and S. Urita, **J. Membr. Sci.**, 42, 47 (1989).
149. R.M. Izatt, G.C. Lindh, G.A. Clark, J.S. Bradshaw, Y. Nakatsuji, J.D. Lamb and J.J. Christensen, **J.C.S. Chem. Comm.**, 1676 (1985); R.M. Izatt, G.C. LindH, J.S. Bradshaw, C.W. McDaniel and R.L. Bruening, **Sep. Sci. Tech.**, 23, 1813 (1988).

150. H. Kuboniwa, K. Yamaguchi, A. Hirao, S. Nakahama and N. Yamazaki, **Chem. Lett.**, 1937 (1982); H. Kuboniwa, S. Nagami, N. Yamaguchi, A. Hirao, S. Nakahama and N. Yamazaki, **J.C.S. Chem. Comm.**, 1468 (1985).
151. K. Hiratani, K. Taguchi, H. Sugihara and K. Ito, **J. Membr. Sci.**, 35, 91 (1987).
152. J.M. Lehn, A. Moradpour and J.P. Behr, **J. Am. Chem. Soc.**, 97, 2532 (1975).
153. I. Tabushi, Y. Kobuke and J.I. Imita, **J. Am. Chem. Soc.**, 103, 6152 (1981); E. Kokufuta and M. Nobusawa, **Chem. Lett.**, 425 (1988).
154. R.M. Izatt, J.D. Lamb, R.T. Hawkins, P.R. Brown, S.R. Izatt and J.J. Christensen, **J. Am. Chem. Soc.**, 105, 1782 (1983).
155. S.R. Izatt, R.T. Hawkins, J.J. Christensen and R.M. Izatt, **J. Am. Chem. Soc.**, 107, 63 (1985).
156. P.R. Danesi, E.P. Horowitz and P. Rickert, **Sep. Sci. Tech.**, 17, 1183 (1982).
157. P.R. Danesi, R. Chiarizia and A. Castagnola, **J. Membr. Sci.**, 14, 161 (1983).
158. P.R. Danesi, E.P. Horowitz, G.F. Vandegrift and R. Chiarizia, **Sep. Sci. Tech.**, 16, 201 (1981).
159. T.M. Fyles, C.A. McGavin and D.M. Whitfield, **J. Org. Chem.**, 49, 753 (1984).
160. T.M. Fyles and D.M. Whitfield, **Can. J. Chem.**, 62, 504 (1984).
161. J.M. Lehn, **Pure. Appl. Chem.**, 50, 87 (1978).
162. R.J. Adamic, E.M. Eyring, S. Petrucci and R.A. Bartsch, **J. Phys. Chem.**, 89, 3752 (1985).
163. T.M. Fyles, **J.C.S. Farad.I**, 82, 617 (1986).
164. D.M. Whitfield, thesis, University of Victoria, Victoria, Canada, 1984.
165. Y. Okahata, **Acc. Chem. Res.**, 19, 57 (1986).

POLYMER/(LIQUID CRYSTAL) COMPOSITE MEMBRANES

Tisato Kajiyama
Department of Applied Chemistry
Faculty of Engineering, Kyushu University
6-10-1 Hakozaki, Higashi-ku
Fukuoka 812, Japan

ABSTRACT. Polymer/(liquid crystal) composite membranes were cast from a solution of a mixture of polymeric and liquid crystalline materials. Also, the ultrathin membranes, about 20 nm thick, were formed by carefully spreading a single drop of casting solution on the water surface (water-cast method). The thickness and the aggregation state of the water-cast membrane can be controlled by the kind of solvent and the concentration of a solution. Aggregation states of a polymer/(liquid crystal) composite membrane were investigated on the basis of differential scanning calorimetry, X-ray, sorption isotherm, sorption-desorption studies and scanning electron microscopic observation. Liquid crystalline materials form a continuous phase in the three-dimensional spongy network of a matrix polymer when the liquid crystalline fraction is above 45 wt%. Therefore, a liquid crystalline phase can serve as a low-viscosity diffusing phase for permeants such as gases or metal ions. The novel polymer/(liquid crystal) composite membranes can be applied to oxygen enrichment, molecular filtration, facilitated or active transport of metal cations and complete thermocontrol of ion permeation.

1. Introduction

The permeation of gases and liquids through polymer films has been studied by investigating the chemical nature, aggregation state, and thermal molecular motion of matrix polymeric chains[1]. The permeation process of low molecular substances through a homogeneous or nonporous polymeric membrane can be explained by a solution-diffusion mechanism. The solution characteristic depends on the affinity of the membrane for the permeant molecule, serving as a controlling factor to improve permselectivity. The diffusion process is influenced by the thermal molecular motion of polymeric chains (the magnitude of free volume or the size of the penetrant molecules). Therefore, the molecular design of a permselective membrane is based on two factors, solubility and diffusivity. Biological membranes are composed of various types of phospholipids, cholesterols and functional proteins. Bimolecular membrane of lipids are in a liquid crystalline state, capable of reversible

T. Osa and J. L. Atwood (eds.), Inclusion Aspects of Membrane Chemistry, 111–155.

structural modifications, and permeation properties of biomembranes depend upon such reversible changes[2,3]. For application to practical membrane systems without losing this essential controlling function, liquid crystalline compounds would be useful instead of phospholipids because they are much more stable than phospholipids in aqueous systems and show a similar gel-liquid crystal phase transition phenomenon. A polymer/(liquid crystal) composite membrane is a blend, a thin film cast from a polymer solution with nematic liquid crystalline materials [4-8]. Characteristics of nematic liquid crystalline materials include low viscosity comparable to water, and orientation of molecules in the presence of an electric field. A polymer/(liquid crystal) composite membrane can control solubility and diffusivity of gas permeation with chemical structure and/or excellent fluidity of liquid crystalline material[9]. For a composite membrane composed of mechanically tough matrix polymer, an ultrathin and fairly wide-area membrane is expected, in spite of the fact that the characteristics of this system are analogous to a liquid membrane[10-14]. The purpose of this study is to investigate the structural properties or compatibilities of the composite membrane composed of polymer and liquid crystalline material, the thermal molecular motion-permeation relationship, and also, the permeation mechanism of the composite membrane to hydrocarbon gases, oxygen or cationic ions. Electric field and photoresponsive effects on permeation have also been studied.

2. Materials and Methods

Figure 1 shows the chemical structures of polycarbonate(PC) and poly(vinyl chloride)(PVC) as a matrix polymer and N-(4-ethoxybenzylidene)-4'-butylaniline(EBBA), 4-cyano-4'-pentylbiphenyl(CPB) and 4-cyano-4'-heptyloxybiphenyl(CHOB) as liquid crystalline materials for the composite membrane. T_{KN} and T_{NI} are the phase transition temperatures from crystal to nematic and from nematic to isotropic, respectively. PC or PVC was used as the matrix substance for the composite membrane because it does not exhibit any apparent thermal molecular motion or phase transition in the temperature range studied here. The polymer/(liquid crystal) composite membranes contain liquid crystal(LC) of 15, 30, 45, and 60 wt%. These composite membranes are designated the 85/15, 70/30, 55/45, and 40/60 composite membranes, respectively. The composite membranes were prepared by casting from a solution of polymer with LC at 293 K(below T_{KN}). Aging was performed at 333 K, well above T_{KN} for 2hr in a nitrogen atomsphere.

In order to study oxygen permselective characteristics, the polymer /LC/(fluorocarbon monomer:FC) ternary composite membrane was prepared from a solution mixture. Perfluorotributylamine(PFTA, $M_w = 671$) and Tris(1H, 1H, 5H-octafluoropentyl)phosphate(TPP) were chosen as the FC monomers, because they have both an excellent oxygen solubility (for example, 0.43cm^3 O_2/cm^3 PFTA at 310 K) and a high thermal stability in comparison with other FC monomers. PFTA or TPP was contained in micelles formed by the surfactant Pluronic L44® (L44 : polyoxyethylene/ polyoxypropylene ABA-type block copolymer, Asahi Denka Co., Ltd.). PFTA or TPP and Pluronic L44 were sonicated for 30 min. Then, the micellar

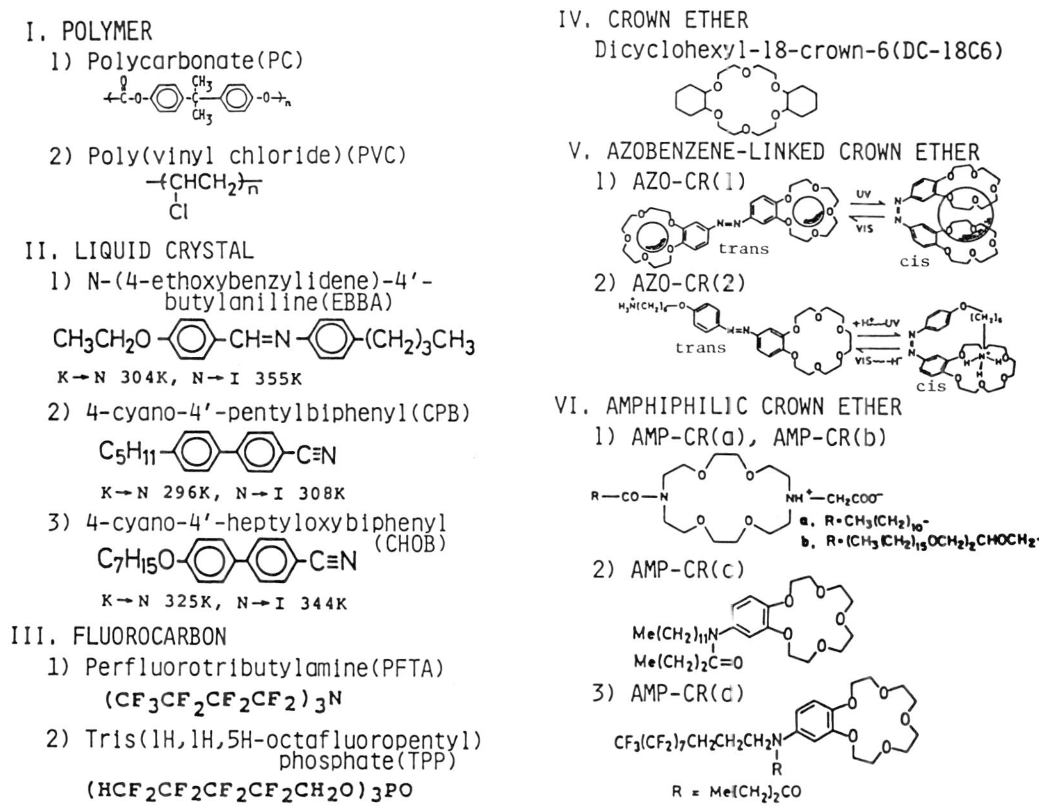

Figure 1. Chemical structures of polymers(I), liquid crystals(II), fluorocarbon monomers(III), crown ether(IV), azobenzene-linked crown ethers(V), and amphiphilic crown ethers(VI).

particles were dispersed in a tetrahydrofuran (THF) solution of a mixture of PVC and EBBA. The polymer/LC/(FC in L44 micelle) ternary composite membrane was prepared by means of casting from a solution of the mixture mentioned above. The weight ratio of each component was PVC/EBBA/FC/L44 = 40/60/7.2/7.2. The molar ratio of LC/PFTA was 20/1.

To investigate photoresponsive facilitated or active transport of K^+ through the composite film, the ternary composite thin film was prepared by the water-cast method: a solution of PVC, EBBA and crown ether bonded by an azobenzene group(AZO-CR) dissolved in a suitable solvent was poured onto the water surface at 283 K and allowed to spread. An ultrathin film was formed on the water surface after evaporation of the solvent. The weight ratio of PVC/EBBA/AZO-CR(1) and PVC/EBBA/AZO-CR(2) were 40/60/5.2 and 40/60/50, respectively. The solvent was a mixture of tetrahydrofuran(THF) and toluene in the case of PVC/EBBA/AZO-CR(1) and a mixed solvent of THF, toluene and chloroform in the case of PVC/EBBA/AZO-CR(2), respectively. These ultrathin composite membranes composed of the secondary system(polymer/LC) or the ternary system (polymer/LC/FC, AZO-CR or AMP-CR) were prepared changing the ratio of mixed solvents and the concentration of a solution in order to find the optimum condition for preparation of an ultrathin membrane. The 30∿35 layers of the water-cast membrane, of ∿50 nm thick, were built-up to prepare the thin film.

The composite membrane composed of PVC, EBBA and amphiphilic crown ether(AMP-CR) was prepared by evaporating solvent from a dichloroethane solution in order to study complete thermocontrol of ion permeation through the composite membrane.

X-ray diffraction, differential scanning calorimetry(DSC) and thermomechanical and density measurements were conducted, investigating the dispersing state of liquid crystalline molecules in the polymer matrix. The membrane and fracture surfaces of the composite membrane coated with gold were observed using Hitachi Scanning Electron Microscope(S-430). In order to observe the aggregation states of each component, EBBA in the composite membrane was extracted with ethanol at 333 K for 2hr. The permeability and diffusion coefficients of gases were measured by the volumetric method(high-vacuum method) originally developed by Barrer[15]. The relationship between liquid crystalline orientation and gas permeation was determined using the permeation cell with a pair of electrodes inserted and an electric field was applied during the gas permeation measurements.

3. Results and Discussion

3.1 Aggregation State of Polymer/LC Composite Membrane

In order to elucidate the aggregation or dispersing state of EBBA molecules in the composite membrane, thermal analyses are very useful. Figure 2(a) shows the differential scanning calorimetric(DSC) curves for PVC, EBBA, and PVC/EBBA composite membranes. Two endothermic peaks were observed with respect to EBBA, which correspond to the crystal-nematic (T_{KN} = 304 K) and nematic-isotropic(T_{NI} = 355 K) transition temperatures, respectively(curve 1 in Figure 2(a)). These transition processes were also observed at approximately similar temperature ranges for the PVC/EBBA composite membranes of which the EBBA fraction was above 45 wt% (curve 2 for the 40/60(PVC/EBBA) composite membrane). However, the PVC/EBBA composite membranes with the EBBA fraction less than 30 wt% do not exhibit such transition peaks on the DSC curve but this curve deviated from a flat base line at about 300 K (curve 3). This thermal behavior indicates that EBBA is molecularly dispersed at below 30 wt%. Therefore, in the case of the EBBA fraction at above 45 wt%, it is clear that EBBA exists in the both states of molecular dispersion and crystal domains. The glass transition temperature, T_g for PVC homopolymer was observed at about 355 K(curve 4), and it was depressed to about 330 K for the composite membranes(curves 2 and 3). The lowering of T_g indicates that PVC and EBBA are quite miscible, and EBBA in the composite membranes acts a plasticizer for PVC. The DSC thermogram of the PC/EBBA composite membrane exhibits a similar behavior to that of the PVC/EBBA one which has the same weight fraction of LC as shown in Figure 2(b).

This expected dispersion state of EBBA in the composite membrane was also confirmed by wide-angle X-ray diffraction measurements. The sharp Debye rings from EBBA crystallites were observed only in the case of the composite membranes containing EBBA of 45 and 60 wt%. The diffraction rings for the 85/15 composite membrane arise only from the crystalline phase of PC. Figure 3 shows the variations of the lattice

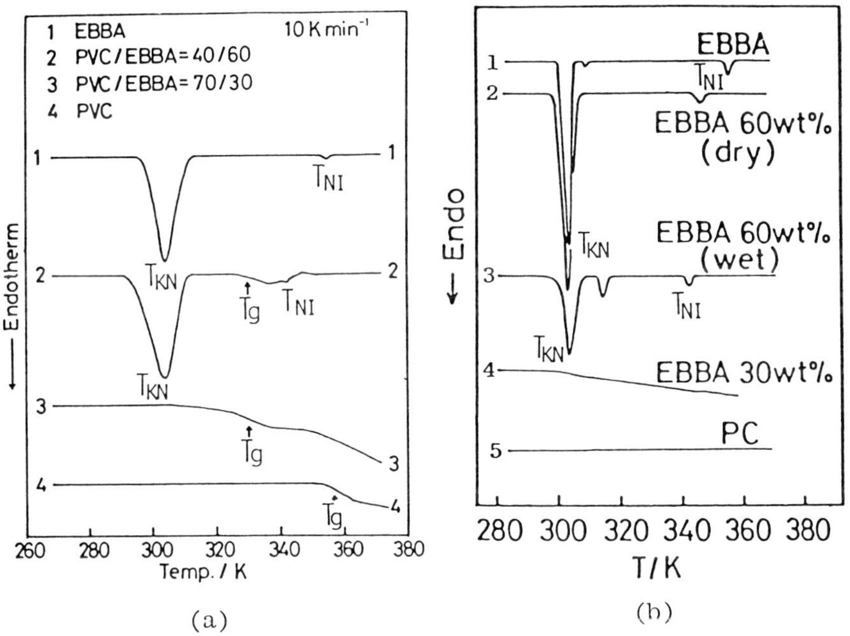

Figure 2. DSC curves of (a) PVC(4), EBBA(1), and PVC/EBBA(2)40/60; (3)70/30 composite membranes and (b) PC(5), EBBA and PC/EBBA (2),(3) 40/60, (4)70/30 composite membranes. T_g represents glass transition temperature of PVC.

constants of the a, b and c axes in the PC crystal with the weight fraction of EBBA. Though PC has a low degree of crystallinity of about 10%, as determined from density measurements, the lattice constants of the a, b and c axes can be evaluated from the 101, 020, 102, and 203 lattice spacings as a function of the weight fraction of EBBA[16]. These lattice constants of the a and b axes, correspoding to the interchain distance, increase to about 4 and 1%, respectively, by addition of EBBA of 30 wt%(33 vol%), but they remain constant in spite of further addition of EBBA above 30 wt%. On the other hand, the lattice dimension along the chain axis direction(c axis) is almost constant over the whole range of EBBA fraction.

Figure 4 shows the scanning electron microscopic(SEM) photographs of the fracture surface(upper portion in each photograph) and membrane one(lower portion)for the 40/60(PVC/EBBA) composite membrane after extracting EBBA with ethanol at 333 K for 2hr. AFS and SFS represent the air-facing surface and the substrate-facing one, respectively, when the membrane was cast. The sample presents an appearance of spongy networks of PVC material. Since about 95 wt% of total EBBA was extracted from the composite membrane with hot ethanol, it is evident that EBBA molecules form domains as an interpenetrating continuous phase among the 3-dimensional spongy networks of PVC matrix. Consequently, it seems reasonable to consider that the continuous EBBA phase takes a role of main diffusing region for gas or ion permeation. In addition, the

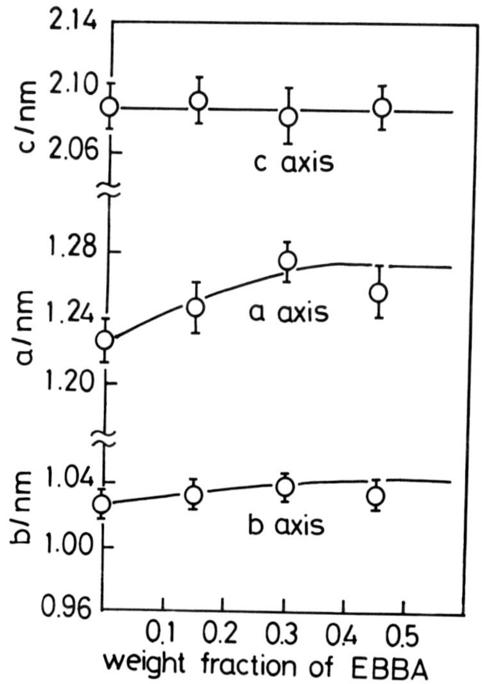

Figure 3. The variation of the lattice constants along a-, b- and c-axes for PC with weight colulme fraction of EBBA.

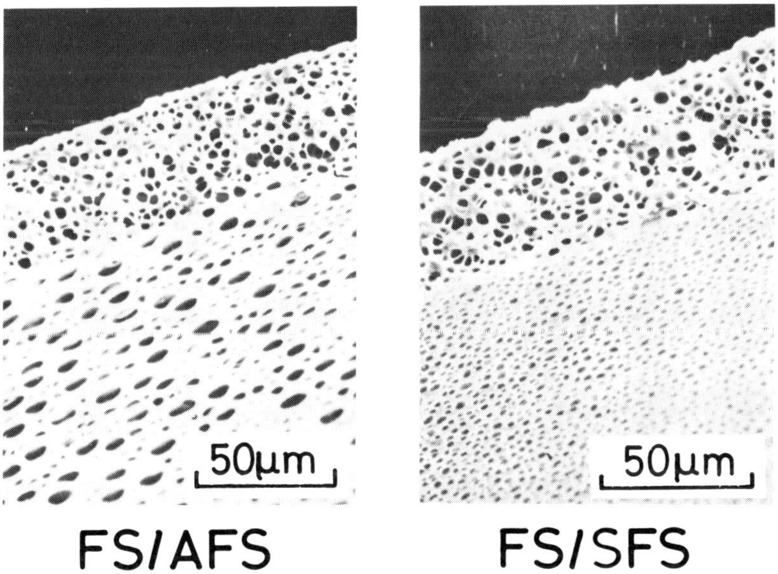

(a) (b)

Figure 4. Scanning electron micrographs of fracture surface(upper portion in each photograph) and membrane surface(lower portion) for 40/60(PVC/EBBA) composite membrane after extracting EBBA with ethanol at 333 K for 2h. (a) AFS and (b) SFS represent air-facing surface and substrate-facing surface, respectively.

concentration of EBBA on SFS was relatively a little higher than that on AFS due to higher hydrophobic character of LC. Thus, the gas permeation experiments in this study were consistently performed by use of the AFS side as a gas feed surface.

Furthermore, the formation process of the composite membrane was traced with a polarizing optical microscope(POM). Figure 5 shows the POM photographs for the 40/60(PVC/EBBA) composite membrane at the temperature of 293 K($<T_{KN}$) upon a solvent evaporation. In the case of an isotropic liquid state just after casting, a field of vision was dark [Figure 5(1)]. Liquid crystalline molecules were gradually clustering or aggregating with a solvent evaporation, resulting in an exclusion of PVC molecules as shown by a dark field in Figure 5(2). At a certain concentration, the crystallite-like domains of EBBA were produced and began to grow in PVC matrix[Figure 5(3)-(8)]. Simultaneously, an excluded PVC phase surrounded EBBA crystal domains as the networks in a similar manner to a coacervation effect. Consequently, distinct crystal domains of EBBA were formed in the composite membrane as a continuous phase and the characteristic interpenetrated structure composed of both components was completed in this process. Of course, the diameter of the EBBA cluster in Figure 5 is greater than that of holes in Figure 4. The EBBA clusters may be divided into smaller ones, though the surrounding PVC networks cannot be distinguished under POM.

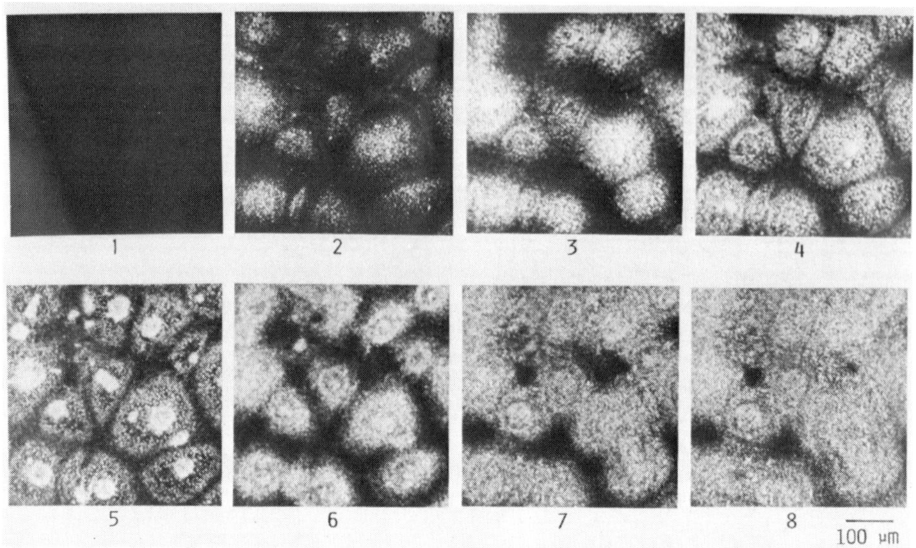

Figure 5. Polarizing optical micrographs under crossed nicols during a solvent evaporation for 40/60(PVC/EBBA) composite membrane at 293 K($<T_{KN}$).

3.2 Preparation of Polymer/LC Composite Ultrathin Film

Ultrathin membranes were prepared by careful spreading a single drop of a solution of the mixed phases on the water surface. After a considerable amount of solvent was evaporated, the ultrathin membrane was left floating on the water surface. A weight ratio of PVC/EBBA was 40/60. Figure 6 shows the transmission electron micrographs(TEM) of the one layer of composite membrane after extraction of EBBA with ethanol at 333 K. The aggregation state and the thickness of the ultrathin membrane were affected with casting conditions such as the kind of solvent, the concentration of a solution and the temperature of the water. The thickness of the water casting membrane can be easily controlled in the range of 20∿50nm depending on these variable preparation conditions. The PVC matrix is remarkably porous when the weight ratio of THF/toluene is 1/2 and also, the concentration of the solution ranges 8∿10 wt% as shown in an enlarged TEM photograph of Figure 7. Since the original thin membrane are nonporous before extracting EBBA, the porous portions surrounded by PVC fibrils must be filled with EBBA before extraction. Since EBBA exists as a continuous domain penetrating throughout the composite membrane, the EBBA phase can serve as a transporting or diffusing phase for permeates or carriers. The water-cast membrane prepared from a solution in the mixed solvent with the greater fraction of THF or toluene (e.g. THF/toluene = 1/1 or THF/toluene =1/3) exhibited an apparent phase-separated aggregate as shown in Figure 6. The mechanism of the phase separation has not been clarified yet. The ultrathin membranes for a photoresponsive active transport of K^+ were prepared

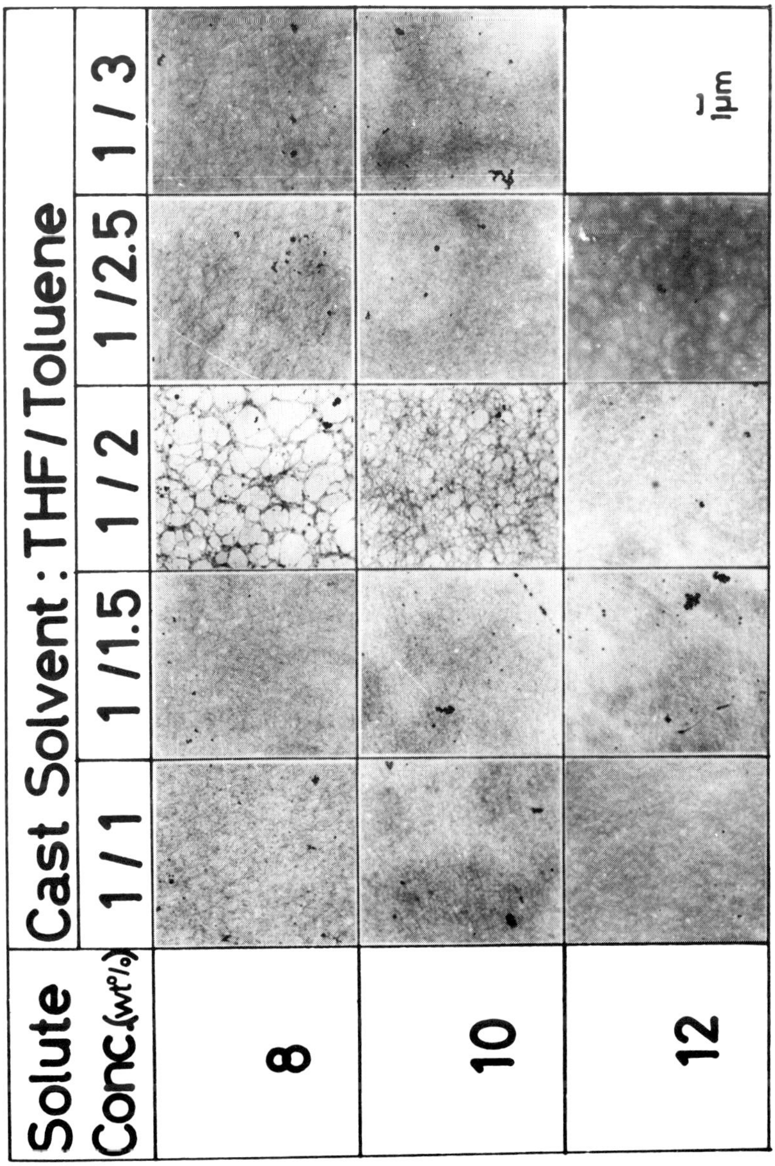

Figure 6. Transmission electron micrographs of PVC/EBBA water-cast membrane prepared under various conditions after extraction of EBBA with ethanol at 333 K.

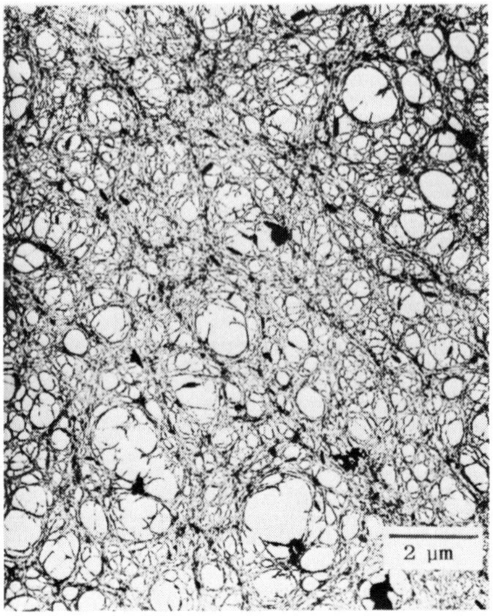

Figure 7. Enlarged transmission electron micrograph of PVC/EBBA water-cast membrane prepared from a 10 wt% solution of THF/toluene(1/2 wt/wt) after extraction of EBBA with ethanol at 333 K.

from a solution of PVC/LC/AZO-CR(10/15/1 in weight) in the mixed solvent of THF/toluene/chloroform(2/4/1 in weight). The aggregation state of the ultrathin membrane is very similar to that of the PVC/EBBA system prepared from a solution of PVC/EBBA in the mixed solvent of THE/toluene. The 30∿35 layers of the water-cast membrane with 20∿50 nm thick were built-up to form the thin film for permeation experiments of K^+.

3.3 Surface States-Permeation Mechanism Relationships of the 40/60 (Polymer/LC) Composite Membrane

Figures 8 shows the SEM photograph of the fracture(the upper portion) and surface(the lower portion) of the 40/60 composite membrane which was quenched with liquid nitrogen after annealing at 333 K($>T_{KN}$). It is clear that a large number of EBBA domains protrude from the membrane surface like a tongue. This suggests that EBBA domains reversibly stick out and draw back to the membrane surface below and above T_{KN}, respectively, since the volume expansion coefficient, β of EBBA is larger than that of PVC as shown in Figure 9. β is defined precisely in terms of the specific volume, v , that is, $\beta = (1/v)(dv/dT)$. T_g is the glass transition temperature of PVC. Over the entire temperature range studied here, β of PVC remained almost constant and of the order of $10^{-4}K^{-1}$. In contrast with PVC, β of EBBA showed a considerably drastic change with increasing temperature. At T_{KN} the magnitude of β exhibited the maximum of $6.31 \times 10^{-2}K^{-1}$, which was about 100 times as

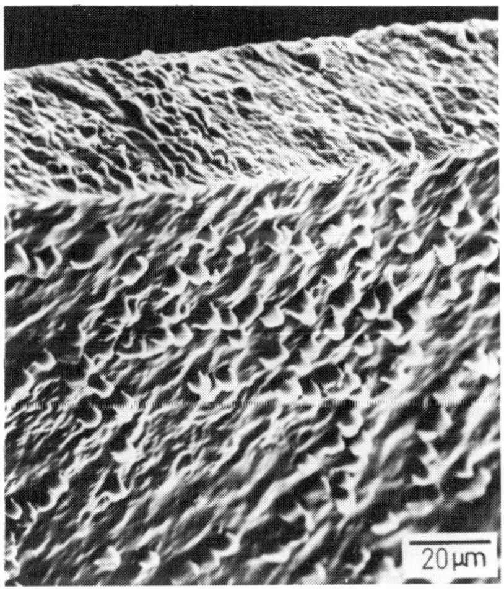

Figure 8. Scanning election micrograph of 60 wt% EBBA composite membrane quenched at a temperature in a liquid crystalline state.

large as that of PVC. Also the magnitude of β was of the order of $10^{-3}K^{-1}$ in the nematic and isotropic states of EBBA. Therefore, it is reasonable to consider that such a difference of β for PVC and EBBA may induce the composition change on the membrane surface depending on the temperature.

The chemical composition in the surface layer is determined by infrared(IR) spectroscopic analysis. IR spectra were obtained by the transmission and attenuated total reflection(ATR) methods, using a Perkin-Elmer 567 Grating Infrared Spectrometer. In the case of the ATR method, the depth of IR beam penetration, d_p, was calculated by

$$d_p = \frac{\lambda}{2\pi n_1(\sin^2\theta - n_{21}^2)^{1/2}}, \quad n_{21} = n_2/n_1 \quad (1)$$

where n_1, n_2, θ and λ are the refractive indices of the reflection plate (2.37 for KRS-5) and the sample(1.45 for the PC/EBBA composite membrane), the effective incident angle(45°) and the wavelength of the characteristic absorption band, respectively. n_2 was evaluated, assuming the additivity of the refractive indices of PC[16] and EBBA[17]. IR absorption bands due to the plane deformation of aromatic hydrocarbons in PC[18] (curve 1 in Figure 10) and EBBA[19] (curve 2) were observed at 1005 and 1040 cm^{-1} and designated as "a" and "b" in Figure 10, respectively. Curve 3 shows the IR spectrum for the 40/60 composite membrane. IR data down to a depth of 1.8 μm from the surface (calculated from eq.(1)) were obtained by the ATR-IR method. The weight fraction of EBBA in the surface region down to 1.8 μm was evaluated

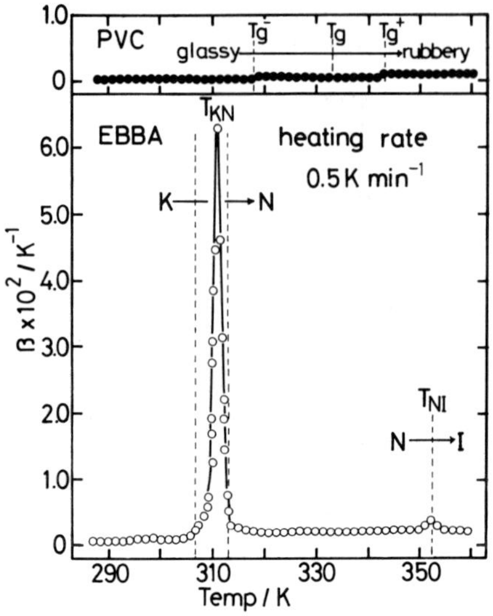

Figure 9. Change in volume expansion coefficients, β of PVC and EBBA as a function of temperture.

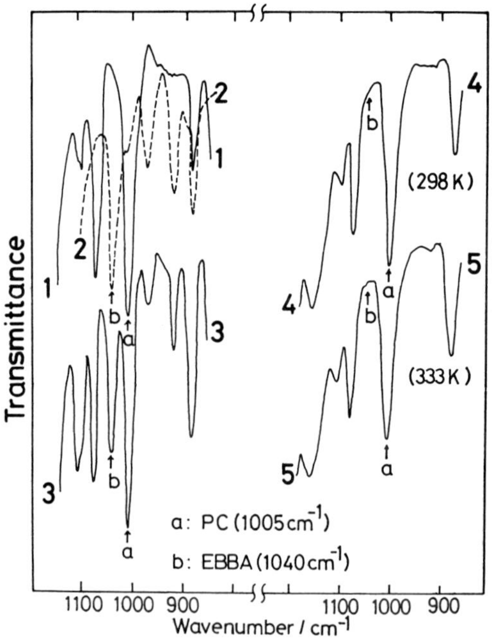

Figure 10. IR spectra of PC(1), EBBA(2), 40/60 composite membranes(3), and those laminated films (4 and 5).

using Sibilia's method[20]. The weight ratio of PC to EBBA, W_{PC}/W_{EBBA}, was calculated from

$$W \quad \frac{W_{PC}}{W_{EBBA}} = \frac{I_{PC}}{I_{EBBA}} \quad \frac{I^o_{EBBA}}{I^o_{PC}} \qquad (2)$$

where I_{PC}/I_{EBBA} and I^o_{PC}/I^o_{EBBA} are the ratios of the IR absorbance of the PC band(a) to that of the EBBA band(b) for the 40/60 and the 50/50 composite membranes, respectively. Thus, the weight ratio was normalized in terms of I^o_{PC}/I^o_{EBBA}. Since $W_{PC} + W_{EBBA} = 100$, the weight% of EBBA was calculated by

$$W_{EBBA} = \frac{100}{1+(I_{PC}/I_{EBBA})(I^o_{EBBA}/I^o_{PC})} \qquad (3)$$

When eq.(3) was used, W_{EBBA} of the 40/60 composite membrane was 65 70 wt%. This value indicates that the weight fraction of EBBA in the surface region is somewhat greater than the original bulk composition, and fairly close to that from the SEM observation of Figure 8.

The aggregation state of the components in the composite membrane

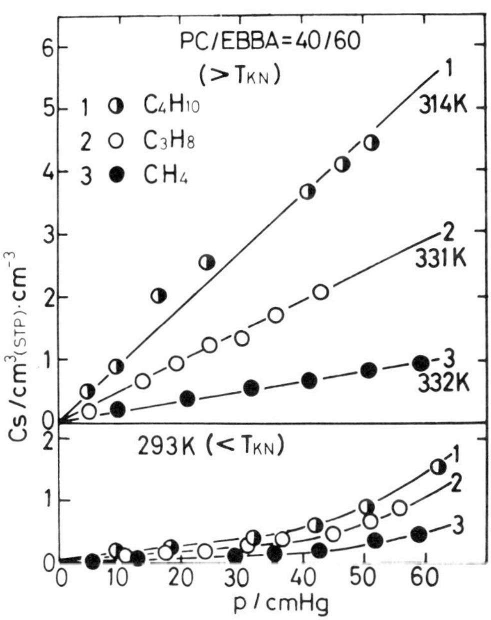

Figure 11. Sorption isotherms for C_4H_{10}, C_3H_8, and CH_4 in 40/60(PC/EBBA) composite membrane at 293(<T_{KN}) and above T_{KN}.

was investigated on the basis of sorption characteristics of gases such as sorption isotherms or sorption-desorption experiments.

Figure 11 shows the sorption isotherms of CH_4, C_3H_8 and C_4H_{10} to the 40/60 composite membrane. The measurements were carried out above and below T_{KN} of EBBA. The validity of Henry's law for the sorption isotherm of these hydrocarbon gases below T_{KN} is doubtful, but above T_{KN} it obeys the law. Calculated values of the solubility coefficients for CH_4, C_3H_8, and C_4H_{10} in the composite membrane are 1.69 x 10^{-2}, 4.62 x 10^{-2}, and 1.26 x $10^{-1} cm^3(STP)/cm^3cmHg$, respectively. Therefore, the solubility of C_4H_{10} is about 7.5 times greater than that of CH_4.

Figure 12 exhibits the sorption-desorption curves of CH_4 in the 40/60 composite membrane. Here, M(t) is the quantity of penetrant which has entered or left the membrane in time t, and M(∞) is the corresponding quantity at infinite time, that is, when equilibrium is reached between sorption and desorption. The linear relationship between M(t)/M(∞) and $t^{1/2}$ was recognized in the early stage of an experiment at 332 K (above T_{KN}), and this shape of the experimental curve suggests that this system follows Fickian sorption[21]. Therefore, in the temperature range above T_{KN}, it is reasonable to consider that a steady surface equilibrium is established immediately, and that the diffusion coefficient of CH_4 is a function of the concentration only. Similar behavior was observed for the polymer-organic vapor system in the rubbery state. In addition to the results of X-ray and DSC studies mentioned above and also, density measurement, these sorption or desorption experiments apparently indicate that the 40/60(PVC/EBBA) composite membrane can be handled as a homogeneous medium when considering gas permeation in the temperature range above T_{KN}[5,6,11].

3.4 Mechanical Properties of Polymer/LC Composite Membrane

Figure 13 shows the stress-strain curves for a series of PVC/EBBA composite membranes at several weight fractions of EBBA (at 293 K<T_{KN}).

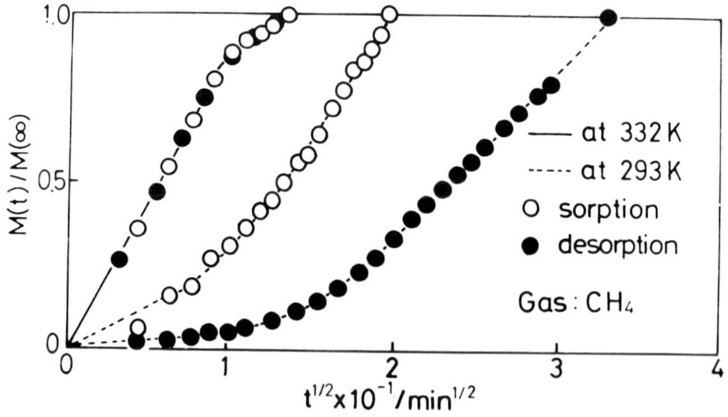

Figure 12. Sorption-desorption curves for CH_4 in the 40/60(PC/EBBA) composite membrane at 293(<T_{KN}) and 332 K(>T_{KN}).

The initial moduli,E for the PVC/EBBA composite membranes decreased with an increase in the weight fraction of EBBA, but the strain at break, ε_b increased up to 45 wt% of EBBA in comparison with that for PVC homopolymer(100/0). Also, the values of E for the composite membranes at above the EBBA fraction of 60 wt% increased again, which might be due to the contribution from the continuous crystal phase of EBBA. In particular, the 40/60(PVC/EBBA) composite membrane, for which the gas permeation characteristics were mainly investigated in this study, exhibited an excellent ductility, that is, ε_b = 413 %. In contrast with this, a series of the PC/EBBA composite membranes were fairly brittle[22], that is, ε_b = 35 % for the 40/60(PC/EBBA) composite membrane. In the case of practical applications as a permselective membrane, the ductile properties of membrane are required from a viewpoint of preparation of ultrathin films. Then, the mechanical property for the 40/60(PVC/EBBA) composite membrane proved to suffice for such a requirement. As mentioned above, the composite membrane is composed of the interpenetrating continuous phases of liquid crystalline molecules and polymer matrix. This kind of aggregation state should provide the excellent mechanical properties or stabilities to the composite membranes.

3.5 Aggregation State-Permeation Property Relationships of Polymer/LC Composite Membrane

The pervaporation system reported by Hung and Lin[23] was used for the measurement of the hydraulic permeability of water. The permeation cell was maintained in a water bath regulated by a temperature controller. Figure 14 shows the Arrhenius plot of the diffusive permeability coefficients,P, of water for the PC and the PC/EBBA composite membranes. The diffusive permeability coefficient was calculated from the hydraulic

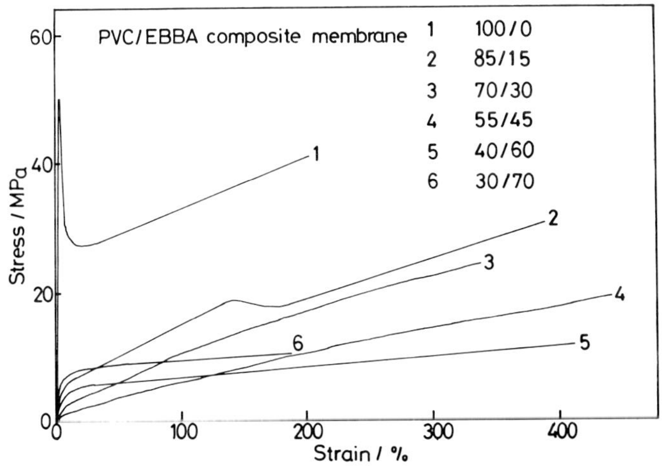

Figure 13. Stress-strain curves for a series of PVC/EBBA composite membranes at several weight fractions of EBBA(at 293 K, < T_{KN}):(1)100/0, (2)85/15, (3)70/30, (4)55/45, (5)40/60, (6)30/70.

permeability coefficient,K, determined in a temperature range of 273-333 K, using the equation of K = PV/RT, where V is the molar volume of water, R is the gas constant and T is absolute temperature, respectively [24]. The log P vs. 1/T plots for the 55/45 and 40/60 composite membranes reveal a distinct jump at about 304 K, corresponding to T_{KN}. The magnitude of P increases by about two and half times within an interval of a few degrees in the vicinity of the phase transition temperature. This striking increase of P may arise if the transition of EBBA from the crystalline phase to the nematic one induces activation of the thermal molecular motion of the membrane and hence, an increase of vacancy fraction around the domain boundary of EBBA and/or within the domain of EBBA. Indeed, these factors make the flux of water increased remarkably. This apparently indicates that the temperature dependence of P is closely related to the thermal molecular motion of the components composing the membrane.

The thermo-mechanical measurement was carried out in a homemade water bath after the permeation tests[4, 25], as shown in Figure 15. The sharp and broad tan δ peaks appear at about 300 K for the 40/60 and 55/45 composite membranes, respectively. This tan δ peak is associated with the onset of thermal molecular motion induced by the phase transition within the EBBA domain. From the results on DSC, X-ray and thermomechanical measurements, it is apparent that the tan δ curve starts to deviate from the flat base at around 280 K in the case of molecular dispersion of EBBA, and that it peaks at around 300 K when crystalline

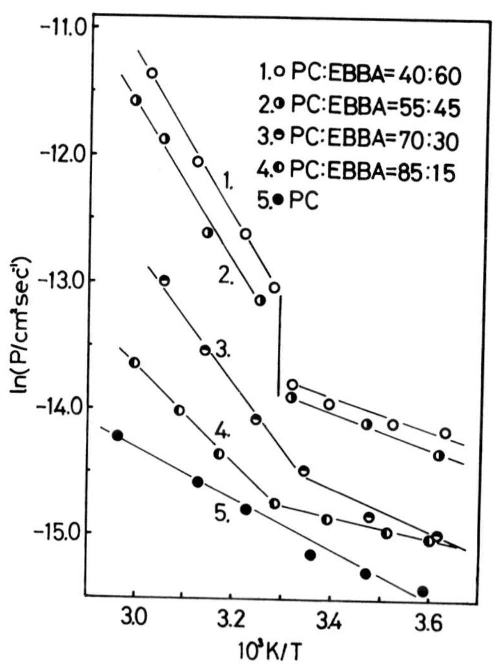

Figure 14. Arrhenius plot of diffusive permeability coefficient to water for PC and various PC/EBBA composite membranes.

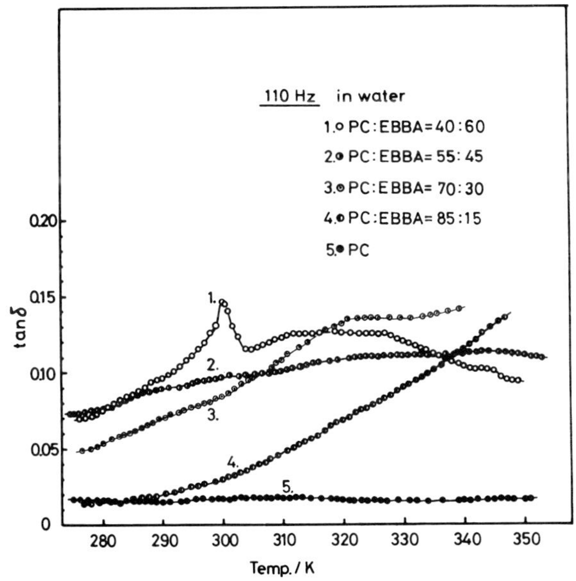

Figure 15. Temperature dependence of tan δ for PC and various PC/EBBA composite membranes.

domains of EBBA exist. When the tan δ peak was observed, the composite membrane revealed a distinct jump on the log P vs. 1/T plot. On the other hand, a slight increase of tan δ for the 85/15 and 70/30 composite films seems to contribute a distinct break of P at about 304 K. These results suggest that the magnitude of thermal molecular motion of the PC/EBBA blend film can control the degree of an increase of permeability in the temperature range of the crystal-nematic phase transition. The tan δ peak was not observed for the PC film, for which any phase transition or activation of thermal molecular motion was not detected by the DSC measurement. In this case, the PC film does not exhibit a discontinuous variation of P in the temperature range studied here.

It is concluded from the permeation and thermo-mechanical measurements that permeability of water through the PC/EBBA composite membrane can be controlled by quantitative or qualitative change of thermal molecular motion at the phase transition region such as the crystal-liquid crystal phase transition.

3.6 Gas Permeation Mechanism through Polymer/LC Composite Membrane

Figure 16 shows the Arrhenius plot of the permeability coefficients, P, to He, N_2 and various hydrocarbon gases for the 40/60(PC/EBBA) composite membrane. The distinct jump of P was observed in the vicinity of T_{KN} of EBBA. In particular, the magnitude of P to hydrocarbon gases increases by approximately 100∿500 times over a few degrees in the phase transition region shown by broken lines in Figure 16. In this region, permeabilities of He and N_2 gases also increase remarkably. As He gas is inert, its solubility property to the composite membrane may not change over the temperature range of the phase transition. Therefore, this discontinuous increase of P for inert gas may be caused by a remarkable increase of diffusion coefficient. This tendency may arise from the activation of thermal molecular motion of the membrane and/or an increase of the pore formation, the probability of which is enhanced due to a dynamic phase equilibrium within the EBBA domain in going from the crystal to the liquid crystalline states or vice versa. The dynamic phase equilibrium means coexistence of solid and liquid crystalline phases with respect to time and position during the process of phase transition. The magnitudes of an apparent activation energy of E_{app} for P of the 40/60 composite membranes are quoted in Table 1. In case of permeation of hydrocarbon gases, E_{app} values below T_{KN} of EBBA are about twice those above T_{KN}. This indicates that gas molecules can permeate more easily through the composite membrane in a state of mesomorphic phase of EBBA. With respect to permeation of various hydrocarbon gases, the magnitudes of P increase in the order i-C_4H_{10}, n-C_4H_{10}, C_3H_8 and CH_4 in the temperature range below T_{KN} of EBBA, and in the order CH_4, C_3H_8, i-C_4H_{10} and n-C_4H_{10} in the temperature range above the transition. The variations of P with temperature to CH_4, C_3H_8 and C_4H_{10} are nearly reversible across the phase transition range. In other words, below T_{KN} the magnitudes of P decrease with

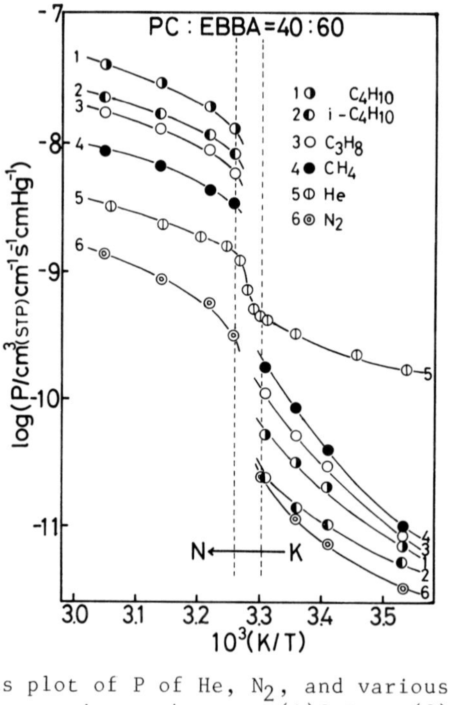

Figure 16. Arrhenius plot of P of He, N_2, and various hydrocarbon gases for 40/60(PC/EBBA) composite membrane. (1)C_4H_{10}, (2)i-C_4H_{10}, (3)C_3H_8, (4)CH_4, (5)He, (6)N_2.

Table 1. Apparent Arrhenius activation energy for P to various gases for 40/60(PC/EBBA) composite membrane.

Gases	E_{app}(kJ mol^{-1})	
	280-295 K	315-330 K
CH_4	64.2	32.3
C_3H_8	46.4	25.7
C_4H_{10}	51.5	23.5
i-C_4H_{10}	50.2	25.7
He	24.5	26.0
N_2	27.0	37.9

increasing number of carbon atoms or the occupied volume of gas, and this trend may suggest that the permeation process is predominantly governed by the diffusion process. On the other hand, above T_{KN} of EBBA, the magnitudes of P increase with increasing number of carbon atoms. Generally speaking, this tendency cannot be explained by the diffusion control mechanism, but it is reasonably expected that permeability is governed mainly by the solubility characteristis. Also, in case of permeation of CH_4, C_3H_8 and C_4H_{10} through the 55/45(PC/EBBA) composite membrane, the drastic jump on the log P vs. 1/T plot was observed in a similar manner to permeation of water or gas through the 40/60 composite membrane. The magnitudes of P to the 55/45 composite membrane increase in the order CH_4, C_3H_8 and C_4H_{10} above T_{KN} of EBBA, and this tendency is also opposite to that expected from diffusion control for gas permeation. It is very interesting that the permeation characteristics of hydrocarbon gases changes across the temperature range involving the crystal-nematic phase transition of EBBA. If the permeation characteristics of the homogeneous membrane are controlled by both solubility and diffusion behaviors, as well as validity of Fickian sorption and Henry's laws, the following relationship between the permeability,P, the diffusion,D, and the solubility,S, coefficients should hold

$$P = S\,D \qquad (4)$$

As described previously, the 40/60 composite membrane can be treated as a homogeneous medium and so, the permeability coefficient should be discussed separately in terms of the diffusion and the solubility coefficients, respectively, as written in eq.(4). Figure 17(a) presents the Arrhenius plot of the diffusion coefficients,D, to CH_4, C_3H_8, C_4H_{10} and i-C_4H_{10} for the 40/60(PVC/EBBA) composite membrane. The magnitude of D is calculated from the time lag, as described by eq.(5), where τ is the time lag, i.e. the intercept on the time axis of the experimental pressure-time plot, and ℓ is the thickness of the membrane.

$$D = \ell^2 / 6\tau \qquad (5)$$

The magnitudes of D beyond the phase transition region shown by the broken lines increase by approximately 9, 12 and 19 times for CH_4, C_3H_8 and C_4H_{10} or i-C_4H_{10}, respectively. The larger increase of D for C_4H_{10} and i-C_4H_{10} than for CH_4 or C_3H_8 may indicate an increase of free volume dimension, as well as that of free volume fraction in the crystal-nematic phase transition range of EBBA. The solubility coefficients, S, can be evaluated from the results shown in Figures 16 and 17(a) using eq.(4). Figure 17(b) shows the Arrhenius plot of the solubility coefficients for CH_4, C_3H_8, C_4H_{10} and i-C_4H_{10} in the 40/60 composite membrane. The magnitudes of S increase with increasing temperature up to the neighbourhood of T_{KN} exhibited by the broken line, and decrease gradually after maximum S. The negative and positive slopes of the Arrhenius plot correspond to positive and negative heats of solution for S, respectively, and this behavior on log S vs. 1/T means apparently the variation of the solubility mechanism of penetrant gases to the polymer/LC composite membrane across the phase transition range. The solubility mechanism must vary depending on the fractions of PC and EBBA in the surface layer because hydrocarbon gases dissolve into PC and EBBA to different extents. Beyond T_{KN}, however, the magnitudes of S increase by about 6 times for CH_4, 10 times for C_3H_8, and 20 times for C_4H_{10} and i-C_4H_{10}, and also, the absolute value of S increases in the order of CH_4,

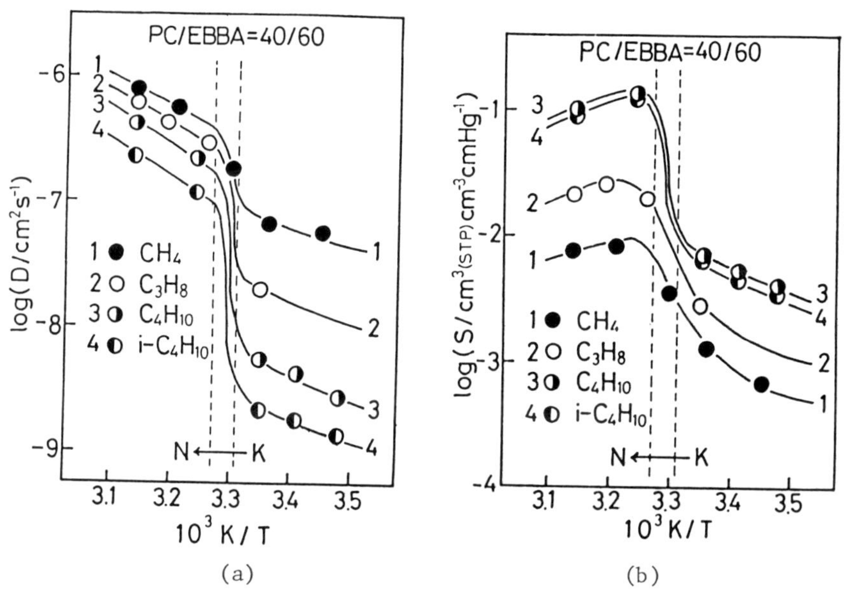

Figure 17. Arrhenius plots of (a)diffusion coefficient,D and (b) solubility coefficient,S for hydrocarbon gases in the 40/60(PC/EBBA) composite membrane.

C_3H_8, and C_4H_{10} or i-C_4H_{10} in the whole temperature range studied here. In as much as the thermal expansion coefficients of PC and EBBA are quite different, especially in the case of liquid crystalline state of EBBA, it is reasonable to consider that the weight fractions of PC and EBBA, or their mixture state in the surface layer of the membrane, vary with measuring temperature. Based on the scanning electron microscopic observation as shown in Figure 8, it is apparent that the EBBA domain alternately comes in and out of the membrane surface below and above T_{KN}, respectively, due to the larger volume expansion coefficient of EBBA than that of PC.

3.7 Oxygen Enrichment through Polymer/LC/Fluorocarbon Ternary Composite Membrane

Figure 18 shows a transmission electron micrograph(TEM) of an L44 micellar particle containing PFTA, which was stained with uranyl acetate. It is apparent that the PFTA/L44 mixture exists as a micellar particle and the mean diameter of it is about 0.4 μm. This micellar size is rather small, compared with the domain size of EBBA continuous phase in the composite membrane[6,26]. Although, judging from this TEM observation, we can consider the possibility that the PFTA/L44 micellar particles move through a liquid crystalline phase with very low viscosity above T_{KN}[27], we would like to focus on the enhancing effect of PFTA on oxygen solubility in the composite membrane surface. Based on X-ray photoelectron spectroscopy(XPS) study, it was confirmed that PFTA molecules were concentrated in the surface region of the ternary composite membrane due to a low surface free energy of FC material. It seems reasonable that the concentrated fluorocarbon layer on the composite membrane surface enhances solubility of oxygen rather than that of nitrogen as shown in Table 2. This indicates that oxygen-enriched air exist on the top surface of the composite membrane in prior to permeation of normal air.

Figure 19 shows the Arrhenius plot of permeability coefficients,P, for oxygen and nitrogen gases in the PVC/EBBA/(PFTA in L44 micelle) ternary composite membrane(PFTA-containing: curves 1 and 3) and the PVC/EBBA binary membrane(PFTA-free: curves 2 and 4). Broken lines represent the temperature regions of T_{KN} and T_{NI}. Over the whole temperature range studied here, values of P for both oxygen and nitrogen gases in the PFTA-containing composite membrane were greater than those in the PFTA-free one. In particular, an increase in oxygen permeability coefficient,P_{O_2}, was remarkable. An increase in P_{O_2} for the PFTA-containing composite membrane below T_{KN} indicates that the PFTA particles enhance the solubility of oxygen in the composite membrane. The degree of solubility of oxygen and nitrogen gas in FC monomer was evaluated by gas chromatography[28]. The solubility coefficient for oxygen gas in FC monomer is about 1.5 times that for nitrogen, as shown in Table 2, although FC monomer has considerable affinity for both gases. In addition, it has been reported that incorporation and release of gases in FC monomer are reversible, obeying Henry's law. Therefore, it is reasonable to consider that the additional increase in oxygen permselectivity in the PFTA-containing composite membrane may arise from

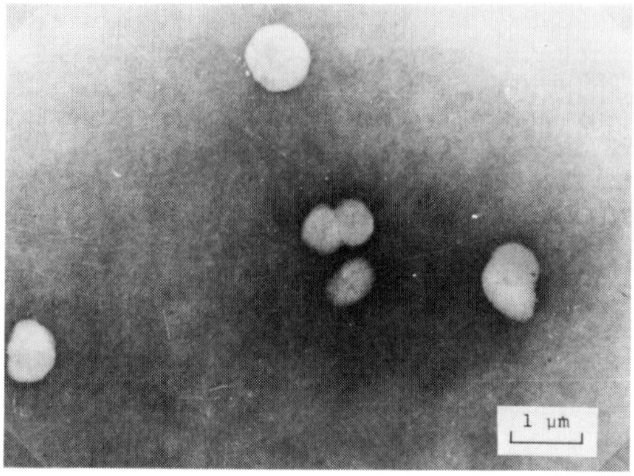

Figure 18. Transmission electron micrograph of a L44 micellar particle containing PFTA. The particle was stained with uranyl acetate.

Table 2. Solubilities of O_2 and N_2 in PFTA and L44 at 338 K

Compound	Temp.	S_{O_2} (cm^3(STP)/100 ml)	S_{N_2} (cm^3(STP)/100 ml)	S_{O_2}/S_{N_2}
PFTA	298 K	38.6	24.7	1.55
	338 K	35.4	22.8	1.56
TPP	298 K	18.9	13.3	1.42
	338 K	17.6	12.7	1.39
L44	338 K	8.6	5.6	1.53

the inherent solubility character of FC monomer.

Figure 20 shows the plots of P_{O_2}/P_{N_2} against P_{O_2} for the FC-contained(curves 1 and 2) and FC-free(curve 3) composite membranes, including the gas separation data for ordinary polymers. An increase in P_{O_2} corresponds to an increase of the measuring temperature. The broken lines reveal the border values for practical application as an oxygen enrichment membrane, that is $P_{O_2}>10^{-9}cm^3(STP)cm^{-1}s^{-1}cmHg^{-1}$ and $P_{O_2}/P_{N_2}>2.5$. The separation factor of the PFTA-contained composite membrane showed the maximum just above T_{KN}(P_{O_2}/P_{N_2} = 5.10). This value was about 2 times greater than that for the PFTA-free one. The temperature dependence of the solubility coefficient in the polymer/LC composite membrane reaches its maximum around T_{KN}. The order of magnitude of P for the PFTA-contained composite membrane was about 10^{-9} to $10^{-8}cm^3(STP)cm^{-1}s^{-1}cmHg^{-1}$ and the permeability coefficient ratio was about 3.5∿4.0 in a nematic or isotropic state of EBBA. Furthermore, above T_g of PVC in the composite membranes(shown by an arrow in Figure 20), the FC-contained composite membrane as well as the FC-free one have a unique effect on the P_{O_2}/P_{N_2}-P_{O_2} relationship; the magnitudes of

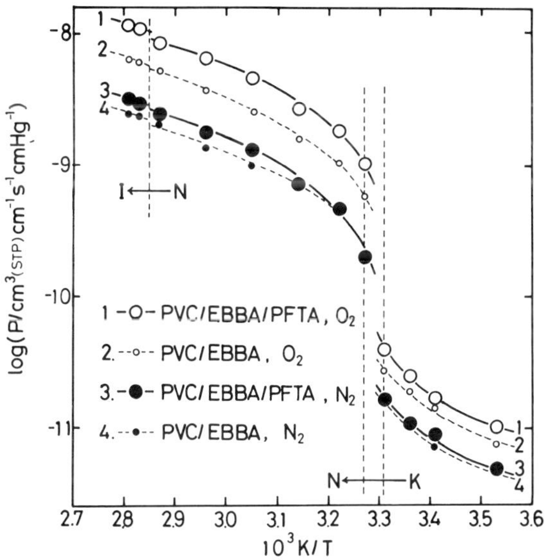

Figure 19. Arrhenius plots of permeability coefficients for oxygen and nitrogen gases in PFTA-containing (curves 1 and 3) and PFTA-free(curves 2 and 4) membranes.

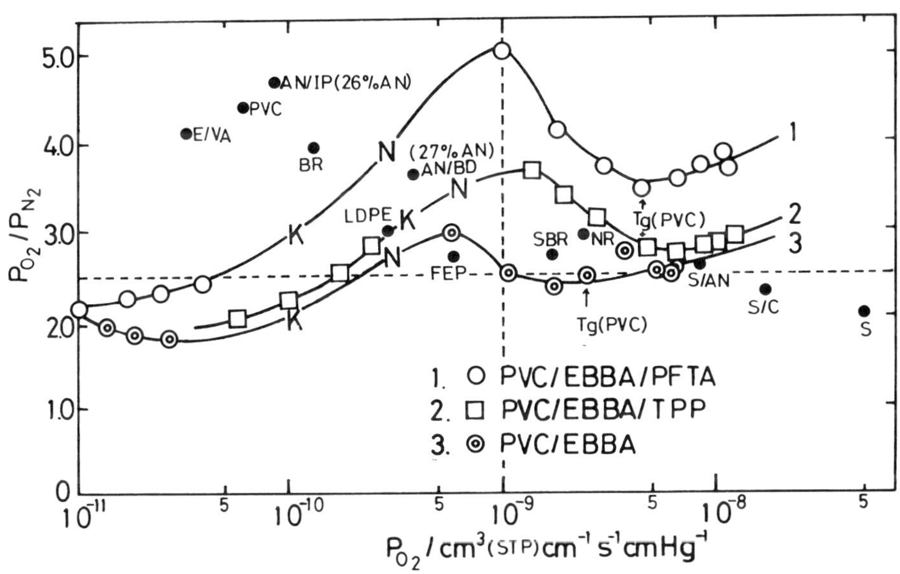

Figure 20. Variation of P_{O2}/P_{N2} with P_{O2} for (1) PVC/EBBA/PFTA, (2) PVC/EBBA/TPP and (3) PVC/EBBA composite membranes. Data for commonly used polymers are included.

P_{O2}/P_{N2} increased with increasing P_{O2}. This contrasts behavior of ordinary polymeric membranes for gas separation; P_{O2}/P_{N2} generally decreases with an increase in P_{O2}. Although the mechanism for this unique behavior has not yet been clarified, we speculate the reason for the FC-free composite membrane is that the frequency with which the polymeric chains make direct contact with oxygen and nitrogen gases on the upstream surface is considerably increased by the activated thermal molecular motion of matrix PVC chains above T_g. In this case, the much greater oxygen selectivity of PVC over liquid crystalline material may contribute to improving permeation characteristics above T_g[29]. Furthermore, in the case of the FC-contained composite membrane, this unique effect is deduced from properties of FC monomers. Although the solubility of both oxygen and nitrogen gases in FC monomers decreases with increasing temperature, a decrease to oxygen is less than nitrogen. Consequently, this effect may contribute to an increase of P_{O2}/P_{N2}, despite the continuous increase in P_{O2} above T_g of PVC.

There have been a few reports on successful results in attempts to improve the oxygen permselectivity by adding a material with good affinity for oxygen to the membrane[30]. The PVC/EBBA/FC ternary composite membrane can be an example of a characteristic permselective membrane for oxygen[31].

3.8 Molecular Filtration through Polymer/LC Composite Membrane

The polymer/LC composite membranes were cast from a 1,2-dichloroethane solution of PC and 4-cyano-4'-pentylbiphenyl(CPB) or 4-cyano-4'-heptyloxybiphenyl(CHOB) by using the homemade casting apparatus as shown in Figure 21. The PC/CPB and PVC/CHOB composite membranes contain CPB of 60 wt% and CHOB of 70 wt%, respectively, which have the positive dielectric anisotropy due to the direction of the dipole moment of a -C≡N group along with the molecular axis. When an electric field is applied during evaporation of solvent, CPB or CHOB molecules are oriented preferentially in the direction of an electric field. A pair of electrodes was set in the permeation cell as shown in Figure 22.

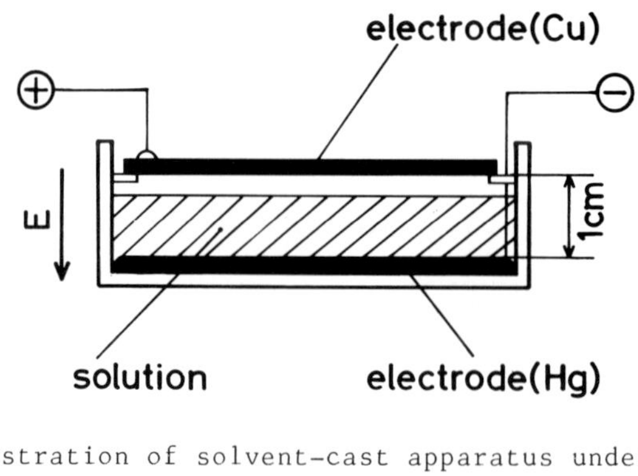

Figure 21. Illustration of solvent-cast apparatus under application of electric field.

Therefore, an electric field could be applied during measurements of gas permeation. The space between the electrodes was 5 mm.

Figure 23 is the wide-angle X-ray diffraction photograph taken when an electric field is applied to the 40/60(PC/CPB) composite membrane. The inner and the outer reflections correspond to the molecular length of CPB and the intermolecular distance, respectively. When an electric field was applied perpendicular to the membrane surface, it emphasizes

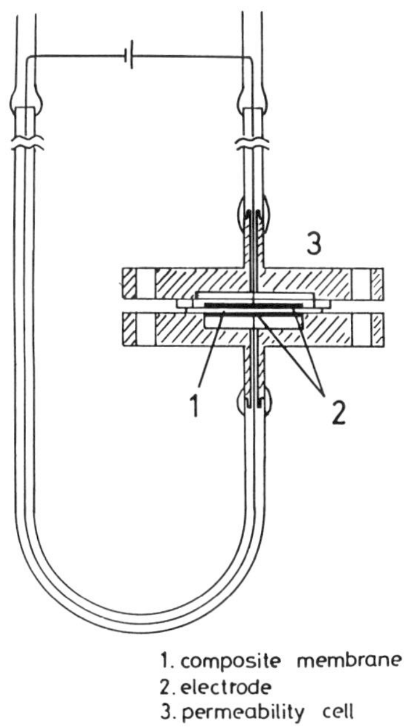

Figure 22. Permeation cell with a pair of electrodes inserted.

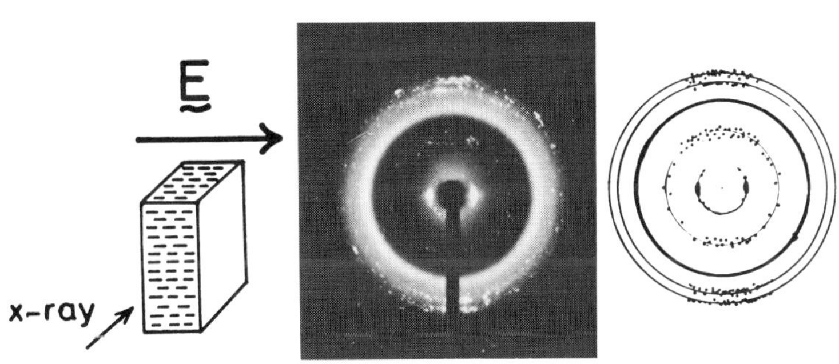

Figure 23. Wide-angle X-ray diffraction of 40/60(PC/CPB) composite membrane. The X-ray pattern was taken under the conditions of an applied electric field.

the normal alignment to the film surface. Since LC molecules aggregate as a continuous phase in the composite membrane, it may be possible to control diffusivity of gas molecules through the composite membrane under application of an electric field, on the bases of the relationship between the dimension of channel formed in the intermolecular regions and the sectional dimensions of permeating gases.

Figure 24 shows the time dependence of downstream pressure(P_1) to C_4H_{10} gas for the 40/60(PC/CPB) composite membrane in a liquid crystalline state of CPB when an applied voltage was varied at 298 K. The slopes of downstream pressure vs. time corresponding to the permeability coefficient,P became larger with an increase in the applied voltage.

Figure 25 shows the applied voltage dependence of P for C_4H_{10} and iso-C_4H_{10} gases. The diameter difference of the sections for C_4H_{10} and iso-C_4H_{10} is only 0.07 nm($d_{C_4H_{10}}$ = 0.49 nm and $d_{iso-C_4H_{10}}$= 0.56 nm). As an applied voltage was increased, the magnitudes of P for both C_4H_{10} and iso-C_4H_{10} gradually increased. The tendency on P with an applied voltage can be explained by the relative difference of diffusion path that gas molecules diffuse along a fairly straight path in the case of the oriented composite membrane and on the other hand, they do along a tortuous path in the case of the unoriented membrane. The separation ratio of $P_{C_4H_{10}}$ to $P_{iso-C_4H_{10}}$, α was 2 at zero volt(random orientation of LC molecules), and increased to 5 at 730 volt(oriented state). An increase of α with the magnitude of an applied voltage indicates that the distribution of channel dimension existing among LC molecules becomes narrower due to more progressed molecular orientation perpendicular to the membrane surface. This result indicates that the

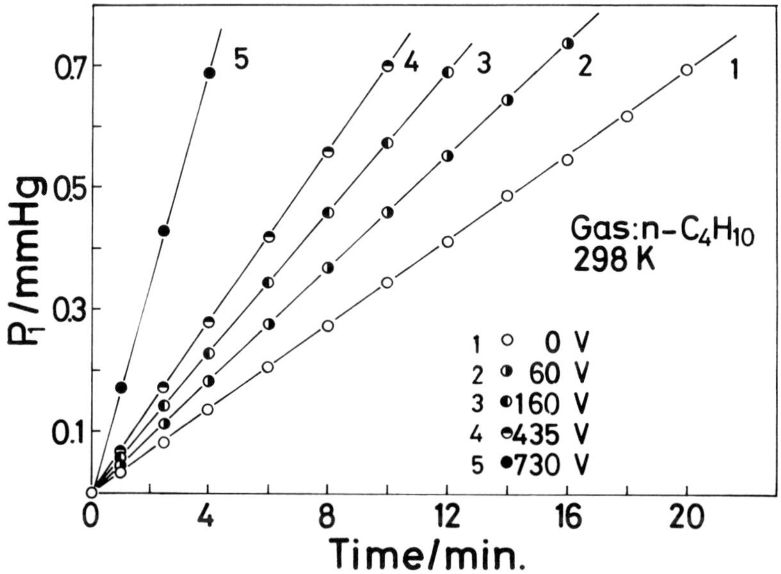

Figure 24. Time dependences of downstream pressure to n-C_4H_{10} and iso-C_4H_{10} gases for 40/60(PC/CPB) composite membrane as a function of voltage: (1)0V, (2)60V, (3)160V, (4)435V, (5)730V; 298 K.

greater degree of orientation of LC molecules induces more effective separation between C_4H_{10} and iso-C_4H_{10}, even though their diameter difference of molecular sections is smaller than one Å.

It has been reported that the magnitude of P of C_4H_{10} through the oriented membrane is larger than the unoriented[32]. Similar tendencies to C_4H_{10} and iso-C_4H_{10} gases were observed for C_5H_{12} and iso-C_5H_{12} gases. The difference in P may be due to the distances required for gas permeation between oriented and unoriented membranes as gas molecules in the oriented composite membrane diffuse along a fairly straight path, and in an unoriented membrane, they diffuse along a convoluted(irregular) path. Figure 26(a) and (b) are the Arrhenius plots of P of C_5H_{12} and iso-C_5H_{12} for both the oriented and unoriented composite membranes, and the ratio of P of C_5H_{12} to that of iso-C_5H_{12} through each membrane, respectively. The difference in P between C_5H_{12} and iso-C_5H_{12} for the oriented composite membrane is larger than the unoriented. In a liquid crystalline state of CHOB, the average intermolecular distance is 0.48 nm. Since this value is nearly equal to the effective diamiter of C_5H_{12}, (0.49 nm), while that of iso-C_5H_{12} is 0.56 nm, it is expected that C_5H_{12} molecules easily permeate along the intermolecular channels existing among CHOB molecules. On the other hand, iso-C_5H_{12} molecules may not permeate as easily as C_5H_{12} molecules diffuse in the region among CHOB molecules. As the fluctuation of the intermolecular distance of CHOB for the oriented composite membrane may be smaller than for the unoriented membrane, separation of pentane gas isomers through the oriented composite membrane may be more effective 33,34

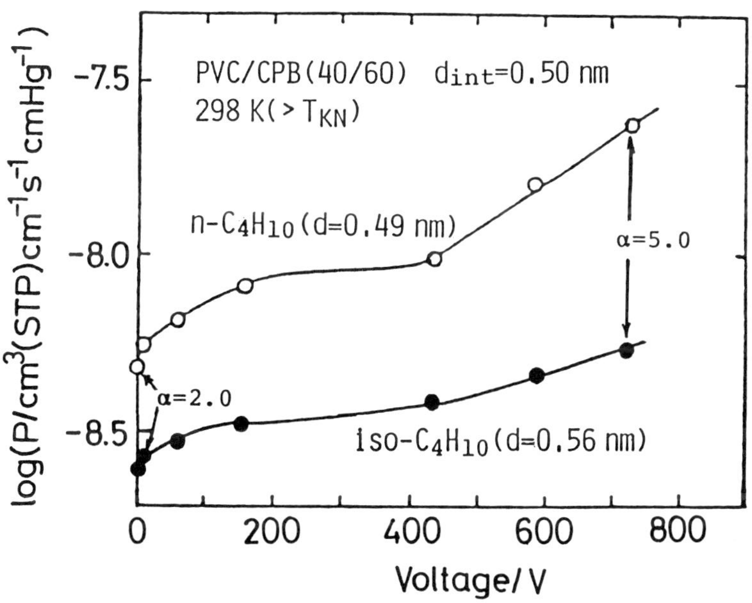

Figure 25. The variation of permeability coefficients for C_4H_{10} and iso-C_4H_{10} through the 40/60(PC/CPB) composite membrane with the magnitude of applied voltage.

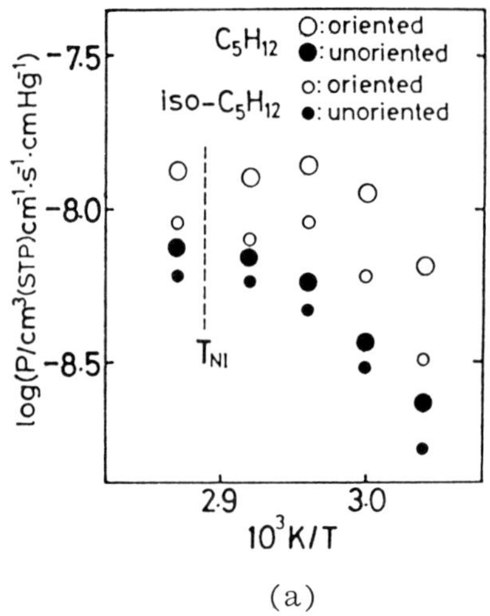

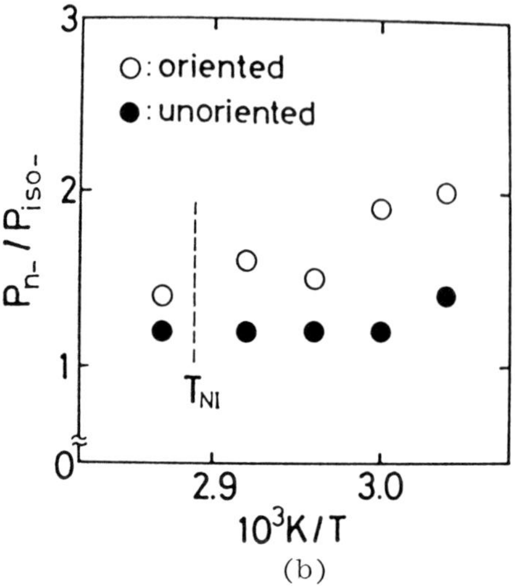

Figure 26. Arrhenius plots of (a) P and (b) separation factor to C_5H_{12} and iso-C_5H_{12} for oriented and unoriented 40/60(PVC/CHOB) composite membranes.

3.9 Facilitated Transport of Metal Cations through Polymer/LC/(Crown Ether) Ternary Composite Membrane

Permselectivity of alkali metal ions has been investigated by use of a polymer/LC/(crown ether) composite membrane as an engineering liquid membrane model. The polymer/LC/(crown ether) composite membrane was cast from a 1,2-dichloroethane solution of PC, EBBA, and dicyclohexyl-18-crown-6(DC-18C6). The composite membrane was composed of PC, EBBA, and DC-18C6 with a weight ratio of 40:60:3. In the EBBA/DC-18C6 and PC/EBBA/DC-18C6 composite membrane systems, the melting behavior of DC-18C6 disappears and, also, T_{KN} decreases slightly in comparison with that of EBBA. This thermal behavior indicates that DC-18C6 is fairly homogeneously dispersed in the EBBA phase and EBBA is also compatible to the media of PC. A liquid crystalline phase makes a role of the matrix for transporting a crown ether, which corresponds to the organic solvent phase of a liquid membrane.

The permeabilities of water and ions were measured by the osmosis method. The cells were filled with 1 mol/dm^3 aqueous solution of salt and distilled water separated by the membrane. Ionic flux was determined by measuring the concentration change evaluated by an atomic absorption spectrophotometry. Water permeability was calculated from the change in level height in the capillary. The diffusion coefficient of the composite membrane was evaluated by the sorption-desorption experiments.

Figure 27 shows the time dependence of concentration of transported metal ions(K^+, Li^+) across the PC/EBBA and the PC/EBBA/DC-18C6 composite membranes at 313 K, at which EBBA is in a liquid crystalline state. Ion concentration in the downstream side was normalized to the thickness of

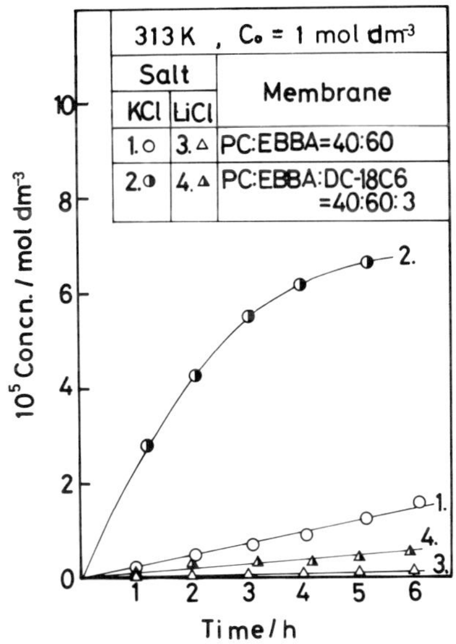

Figure 27. Permselectivity of metal ions by composite membrane; 313 K, C_o = 1 mol/dm3, salt = KCI or LiCl, 40/60(PC/EBBA)(curves(1) and (3)) and 40/60/3(PC/EBBA/DC-18C6)(curves(2) and (4)).

0.1 mm in each membrane. For the PC/EBBA/DC-18C6 composite membrane (case 2), K^+ is about 50 times more permeable in comparison with Li^+ (case 4). This result agrees with the fact that the ion diameter of K^+ (0.266 nm) better fits the cavity size of DC-18C6(0.26-0.32 nm) rather than Li^+(0.12 nm). Though the PC/EBBA composite membrane does not contain a selective cation carrier such as DC-18C6, it showed a slight selective transportation between K^+ and Li^+. This experimental result will be explained by the fact that K^+ is more hydrophobic because the ion diameter of K^+ is larger than that of Li^+. The effectiveness of transporting K^+ with DC-18C6 is apparently exhibited by the curves 1 and 2 as shown in Figure 27. Transportation of K^+ through the composite membrane increases about 10 times by addition of DC-18C6.

Figure 28 shows the sorption-desorption curves of K^+ for the PC/EBBA/DC-18C6 composite membrane at 313 K. These measuring points fall on a single curve of which shape shows the characteristics of Fickian sorption. Therefore, the value of the diffusion coefficient,D was evaluated from the initial slope of $M(t)/M(\infty)$ vs. $t^{1/2}/\ell$. Table 3 compares the flux and the diffusion coefficient of K^+ for the PC/EBBA/ DC-18C6 composite membrane with those for a liquid membrane. A liquid membrane was prepared by impregnating decanol(DeOH) containing DC-18C6 in an ultrafiltration membrane of polytetrafluorethylene. Carrier concentrations of EBBA in DeOH were prepared to be the same for the composite membrane permeations of K^+. At 313 K above T_{KN}, the flux of K^+ through the liquid membrane is about 3.5 times as much as the flux through the PC/EBBA/DC-18C6 composite membrane. The diffusion coeffi-

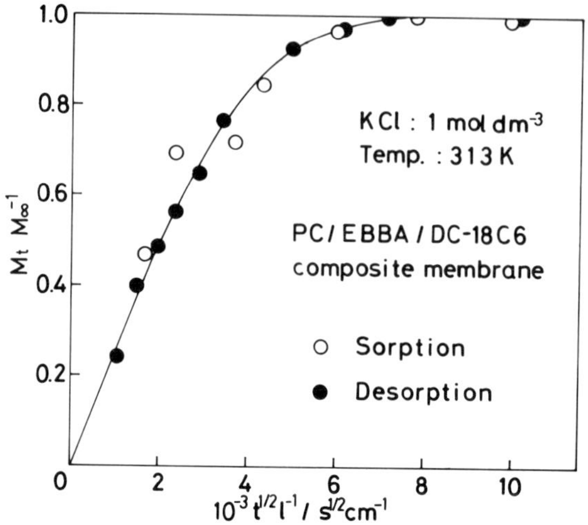

Figure 28. (○)Sorption-(●) desorption curves of K^+ in 40/60/3(PC/EBBA/DC-18C6) composite membrane. KCl;C_0 = 1 mol/dm^3, temperature =313 K

Table 3. Comparison of K^+ transport through composite membrane with liquid membrane.

	PC/EBBA/DC-18C6 composite membrane	MF[a]/DeOH/DC-18C6 liquid membrane
Carrier Conc.[b] (mol dm^{-3})	0.14	0.15
Flux[c] (mol cm^{-2} s^{-1})	6.3 x 10^{-11} 4.2 x 10^{-10}(343 K)	2.3 x 10^{-10}
Diffusion Coefficient (cm^2 s^{-1})	1.1 x 10^{-8} [d]	3.1 x 10^{-8}[e]

a) membrane filter(Fluoropore FP022)
b) porosity corrected
c) time = 0 ∿ 1 hr
d) evaluated from sorption-desorption curve
e) evaluated from time lag method

cient of K^+ through the liquid membrane is about 3.0 times as large as that through the composite membrane. These results suggest that the solubilities of K^+ in these membranes are approximately the same magnitude. Also, at 343 K in an isotropic state of EBBA, the flux of K^+ across those membranes is nearly equal or the K^+ flux for the composite membrane is slightly larger. The PC/EBBA composite membrane is composed of PC fibrillar networks and a EBBA phase interpenetrates as a three-dimensionally continuous one. This aggregation structure shows the characteristics of mechanical toughness or preserving effect of EBBA within the composite membrane. Therefore, over a temperature range of an isotropic liquid phase, EBBA may not easily effuse from the membrane into the water phase. In conclusion, metal ions can be selectively transported by use of the polymer/LC/(crown ether) composite membrane. Its permeability of metal ions is comparable to that of a liquid membrane when a liquid crystalline material is in an isotropic liquid state. Application of the polymer/LC/(crown ether) composite membrane is expected as a substitute for engineering liquid membranes[35]. Even, in the case of the composite membrane, crown ether compounds incorporate alkali metal ions into their cavities, forming complexes.

Permselectivity for alkali metal ions has been investigated using the polymer/LC/(photoresponsive crown ether) composite membrane. Shinkai et al. succeeded in synthesizing an azobenzene-bridged crown ether(AZO-CR)[36,37]. Azobenzene derivatives exhibit photoinduced reversible trans-cis isomerization[38]. The trans-AZO-CR(1) is transformed into the cis-isomer upon irradiation with ultraviolet(UV) light of wavelength 360nm, and the reverse takes place upon irradiation with visible(VIS) light of wavelength exceeding 460nm, or upon heating. The ion trapping ability of AZO-CR(1) for alkali metal ion is controlled by photoisomerization. Trans-AZO-CR(1) effectively extracts Na^+ 5.6 times as much as does cis-AZO-CR(1), whereas cis-AZO-CR(1) selectively extracts K^+ 47.5 times as much as that of trans-AZO-CR(1). Trans-AZO-CR (1) can form the sandwich-type 1:1 cation/AZO-CR complex, and extracts metal cations with larger diameters, such as K^+, than the cavity dimension of 15-crown-5. The ternary composite membrane was prepared by the water-cast method. Weight ratios of PVC/CPB/AZO-CR(1) and PVC/EBBA/AZO-CR(1) are 40/60/2.6 and 40/60/2.7, respectively. A molar ratio of liquid crystal and AZO-CR(1) is 52/1. The permeability to K^+ was evaluated using a normal osmosis method of the composite thin film built-up with

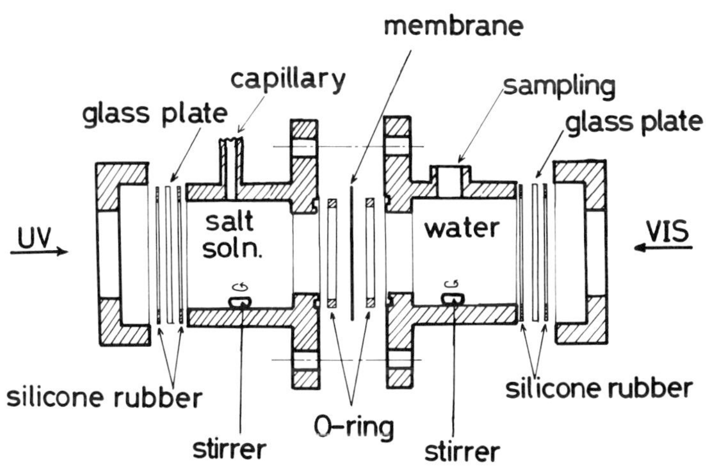

Figure 29. A permeation cell attached to the irradiation system of UV and VIS light for facilitated transport of K^+ through PVC/EBBA/AZO-CR(1) ternary composite film.

10∿30 layers of the water-cast membranes. The cell was filled with an aqueous solution of potassium p-toluenesulfonate of 0.01 mol dm^{-3} and distilled water separated by the thin film as shown in Figure 29. The concentration of permeated K^+ was evaluated by atomic absorption spectrophotometry. To investigate the photoresponsive permeation control, UV and VIS lights were irradiated perpendicular to the film surfaces from the salt solution and distilled water sides, respectively.
Figure 30 shows the irradiation time dependence of the UV absorption spectra for the ternary composite thin film at 300 K upon irradiation with UV and VIS light. For the UV absorption experiment, a liquid crystal of CPB was used because both PVC and CPB have no interfering absorption of UV in the wavelength range of 330-450 nm. The absorption maximum corresponding to trans-AZO-CR(1) was observed at 370 nm. The intensity of this absoption peak decreased with an increase in the irradiation time of UV light.
Also, its absorption peak returns to the original height upon irradiation with VIS light of the wavelength longer than 460nm. Therefore, Figure 30 indicates that the trans-cis photoisomerization of AZO-CR(1) occurs reversibly even in the ternary composite thin film in a manner similar to that in a dilute solution.

The time dependence of a relative increase of permeated K^+ in the permeation side through the PVC/EBBA/AZO-CR(1) composite thin film upon irradiation with UV and VIS light at 316 K is given in Figure 31. A relative increase of K^+ was calculated as the difference between permeated K^+ with and without light irradiation. At 316 K, EEBA is in a nematic state, and plays the role of a transporting or diffusing matrix for AZO-CR(1). It is apparent that UV and VIS light irradiation facilitates or activates permeation of K^+ through the PVC/EBBA/AZO-CR(1) ternary composite thin film. The permeation rate obtained in this study is about 10 times higher than that for the thick composite film of about

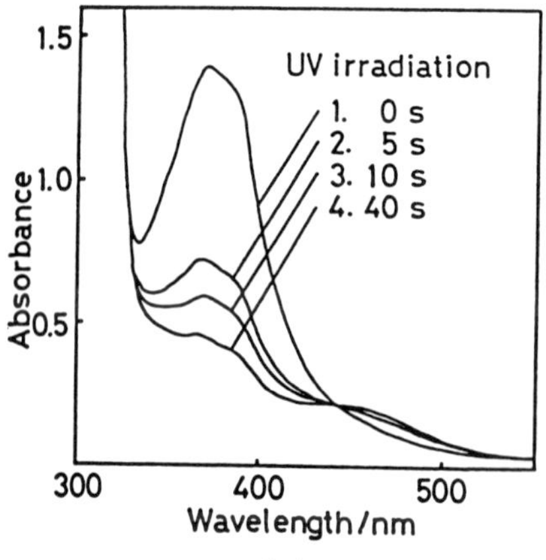

(a)

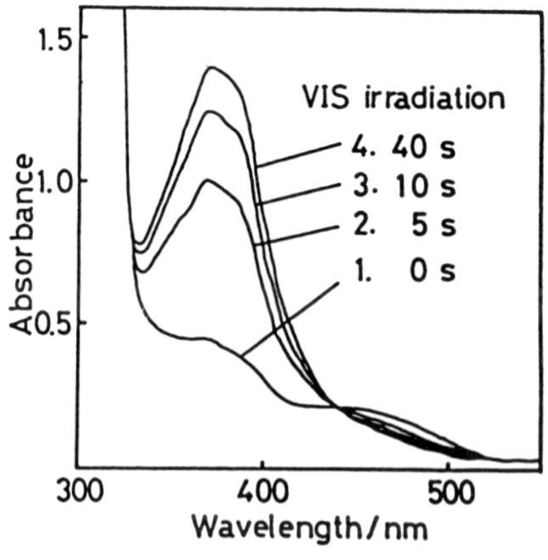

(b)

Figure 30. UV absorption spectra of PVC/CPB/AZO-CR ternary composite thin film upon irradiation with (a)UV and (b)VIS light.

50 μm[39,40]. A schematic representation of light-enhanced ion transport across the compsite thin film containing AZO-CR(1) ternary composite thin film is given in Figure 32. As the surface of the composite thin film makes contact with a salt solution, the cis-isomer easily forms the 1:1 K^+/AZO-CR(1) complex. This complex then diffuses to the opposite side of the water phase, due to the concentration gradient of cis-AZO-CR(1) in the composite thin film. At the surface of the thin film contacted to water, cis-AZO-CR(1), which traps K^+, transfers to the

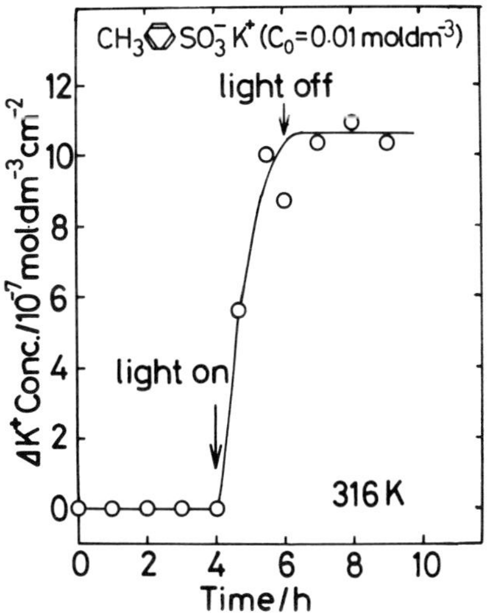

Figure 31. Photoisomerization effect on relative K^+ permeation through PVC/EBBA/AZO-CR(1) composite thin film.

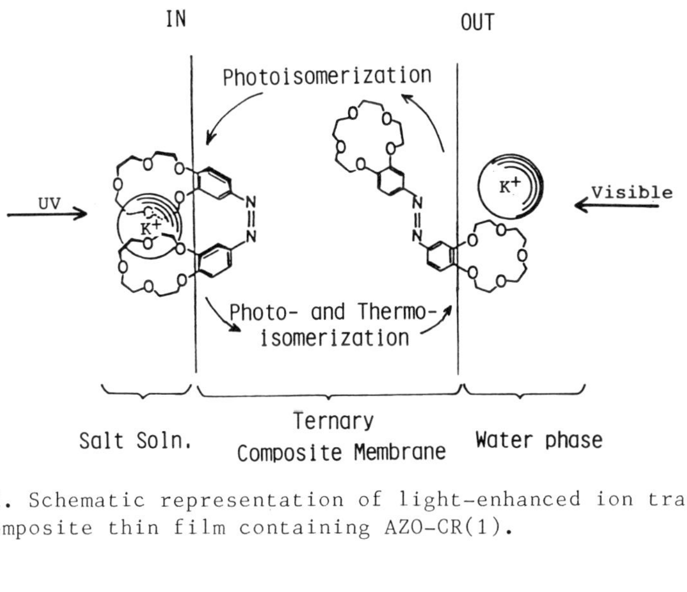

Figure 32. Schematic representation of light-enhanced ion transport across composite thin film containing AZO-CR(1).

trans-isomer upon irradiation with VIS light, so that trans-AZO-CR(1) releases K^+ into the water phase. Furthermore, trans-AZO-CR(1) returns to the salt solution phase side, due to the concentration gradient of the trans-isomer in the composite thin film. The affinity of AZO-CR(1) for K^+ must be strikingly varied on each side of the composite thin film due to the configurational change of AZO-CR(1) induced by irradiation with UV and VIS light. That is, the complex formation between K^+ and cis-AZO-CR(1) is more predominant around the contacting surface with a salt solution than that between K^+ and trans-AZO-CR(1) around the contacting surface with water. In the case of ordinary crown ether compounds, affinities with cations do not change on either side of the thin film. Therefore, this AZO-CR(1) system is advantageous for perm-selective facilitated transport of metal ions.

3.10 Photoresponsive Active Transport of Metal Cations through Polymer/LC/(Crown Ether) Ternary Composite Membrane

Photo-induced active transfer of K^+ for the polymer/LC/AZO-CR(2) ternary composite membrane is possible. This ternary composite thin film was prepared using the water-cast method. A solution of PVC, EBBA and AZO-CR(2) dissolved in a mixed solvent of tetrahydrofuran(THF), toluene and chloroform(solution concentration of 8 wt%) was dropped onto the water surface at 283 K. The weight ratios of PVC/EBBA/AZO-CR(2) and THF/toluene/chloroform are 40/60/4 and 2/4/1, respectively. 30∿50 layers of the water-cast membrane, approximately 50nm thick were built-up to form the thin film. The trans-cis photoisomerization of AZO-CR(2) in the composite thin film was confirmed by the UV absorption spectra upon irradiation with UV or VIS light. The permeation experiment was

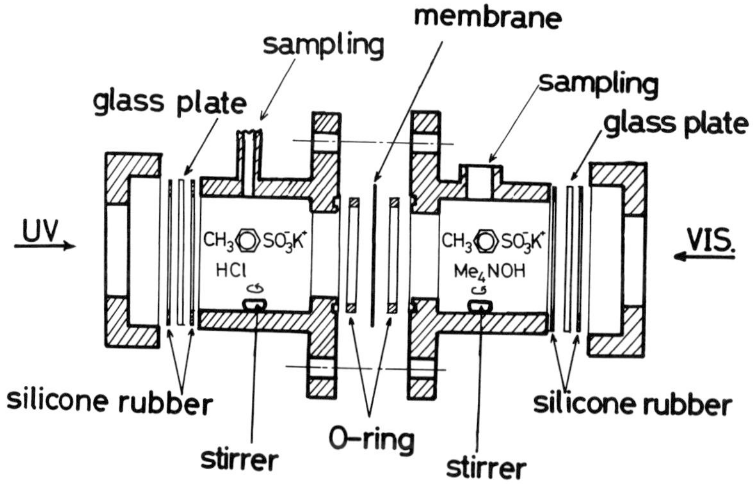

Figure 33. A permeation cell attached to the irradiation system of VIS and UV light for active transport of K^+ through PVC/EBBA/AZO-CR(2) ternary composite thin film.

carried out using a permeation cell attached to an irradiation system of VIS and UV light sources as shown in Figure 33. Basic and acidic aqueous phases with potassium p-toluenesulfonate of 1×10^{-4} M were separated by the ternary composite thin film. The VIS and UV light sources were arranged to irradiate perpendicular to the film surface from basic and acidic aqueous phases, respectively. The K^+ transport experiment was carried out at 313 K(a temperature above T_{KN} of EBBA). The concentration variation of K^+ was evaluated by atomic absorption spectrophotometry[41-43].

Figure 34 shows the time dependence of the amount of K^+ transported from the basic aqueous phase to the acidic phase upon photoirradiation and also, in the dark. Both the acidic and basic aqueous phases contained the same concentration of K^+ before permeation experiment started. Since there are two driving forces, photoirradiation and pH gradient(proton gradient), we have to evaluate the magnitudes of K^+ permeation upon photoirradiation and in the dark, separately, in order to discuss a photoresponsive active transport of K^+. It is clear that the permeation rate of K^+ upon photoirradiation is greater than the rate in the dark as shown in Figure 34. This indicates that an active transport of K^+ takes place from the basic aqueous phase to the acidic one through the ternary composite thin film upon photoirradiation. This is the first report to show a photo-induced active transport of metal cation through an organic thin film. Appearance of a slight peak on the amount of transported K^+ with permeating time is generally observed in the case of an active transport experiment in the batch system. This behavior may occur due to the balance between passive transport and an

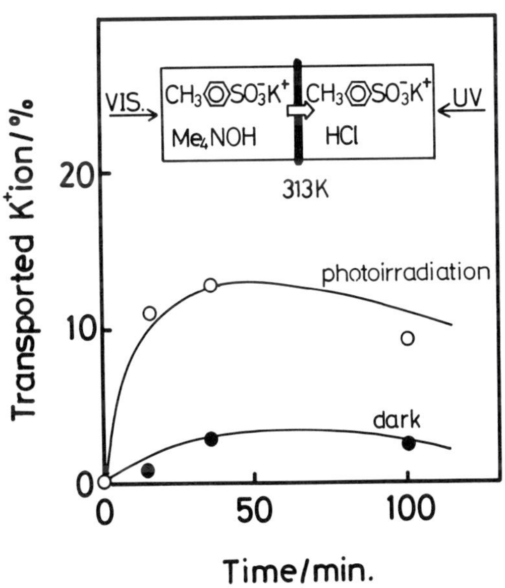

Figure 34. Time dependences of the relative amount transported K^+ into an acid aquous phase upon (1)photoirradiation and (2)in the dark.

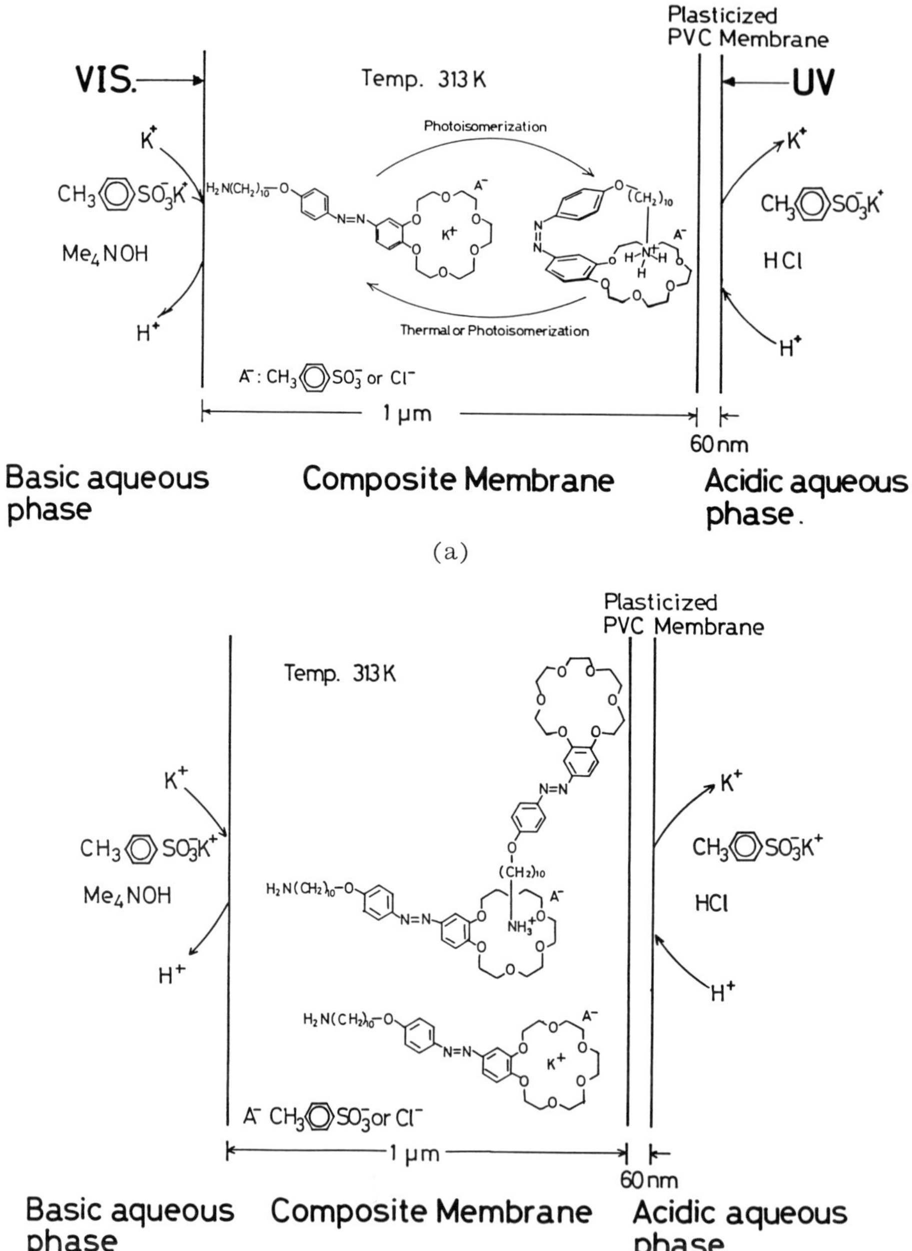

Figure 35. Permeation mechanism of K^+ through the ternary composite thin film upon (a)photoirradiation and (b)in the dark.

active one.

Figure 35 shows the schematic representation of the permeation mechanism of K^+ through the PVC/EBBA/AZO-CR(2) ternary composite thin film upon photoirradiation. At the surface of the composite thin film contacting the basic aqueous phase, trans-AZO-CR(2) exists preferentially upon irradiation with VIS light. Trans-AZO-CR(2) easily forms a complex with K^+ and trans-AZO-CR(2) entrapping K^+ diffuses to the opposite side of the film contacting the acid aqueous phase owing to the concentration gradient of the trans-isomer in the ternary composite thin film. At the surface contacting the acid aqueous phase, an amino tail group of AZO-CR(2) is protonated to form an ammonium group. When protonated trans-AZO-CR(2) is isomerized to the cis-isomer upon irradiation with UV light, the crown ether ring binds intramolecularly to the protonated ammonium group, so that cis-AZO-CR(2) effectively releases K^+ into the acidic aqueous phase. Furthermore, cis-AZO-CR(2) returns to the side of the basic aqueous phase owing to the concentration gradient of the cis-isomer in the ternary composite thin film and again, is isomerized to the trans-isomer upon irradiation with VIS light or heating. Since protonated cis-AZO-CR(2) becomes hydrophilic, the composite film surface contacting the acidic aqueous phase was coated with a plasticized thin PVC membrane 60nm thick in order to prevent leakage of protonated-AZO-CR(2) into the acidic aqueous phase. In the dark, active transport of K^+ is observed to the extent shown by the filled circles in Figure 34. When trans-AZO-CR(2) molecules entrapping K^+ form an intermolecular complex with in protonated trans AZO-CR(2) in the vicinity of the surface in contact with the acidic aqueous phase, an active transport of K^+ becomes possible even in the dark[39,40].

3.11 Thermoswitch Based on Ion Permeation through Polymer/LC/ (Amphiphilic Crown Ether) Composite Membrane

Preparation of amphiphilic crown ethers(AMP-CR) was reported in the papers described elsewhere[44,45]. The composite membranes for this purpose were prepared by evaporating solvent from a dichloromethane solution of PC, LC and an amphiphilic crown ether(AMP-CR) and a dispersion state of AMP-CR was investigated by DSC. The thickness of the membranes was 60 µm. The 40/60(PC/EBBA) binary composite membrane gave a DSC peak at 305 K, being comparable with the crystal-nematic phase transition of EBBA (T_{KN} = 304 K). The peak was not affected by the addition of AMP-CR(a) (2.9 mol% of EBBA) into the composite membrane. In contrast, the PC/AMP-CR(b) binary composite membrane(2.9 mol% of EBBA) gave a new peak, in addition to that of EBBA, at 313 K which is very close to a DSC peak of AMP-CR(b) in water(T_{KN}= 317 K). The DSC result supports that AMP-CR(a) is homogeneously dispersed in the composite membrane, whereas AMP-CR(b) exists as phase-separated aggregates not only in water but also in the composite membrane. Figure 36 shows the reciprocal absolute temperature dependence of the permeability coefficient for K^+, (P_{K^+}). Permeation of K^+ through the PC/EBBA/AMP-CR(a) composite membrane was observed both below and above T_{KN}, and a plot of P_{K^+} with a reciprocal absolute temperature exhibited a break point at around T_{KN}. From this Arrhenius plot, the magnitude of

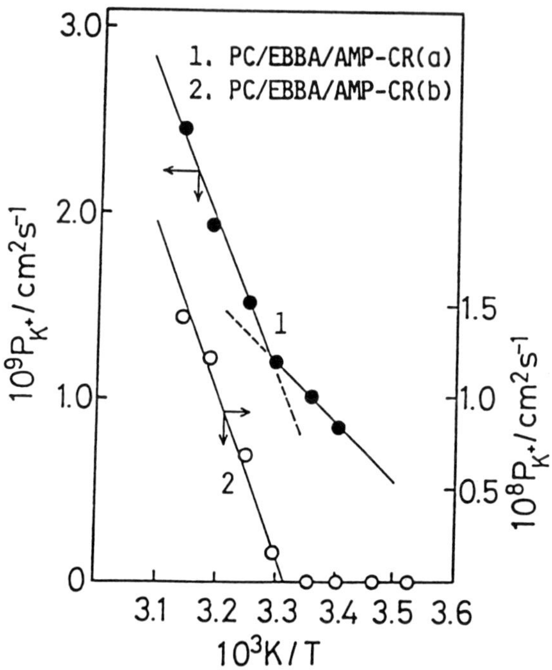

Figure 36. Arrhenius plot of P_{K^+} for (1)PC/EBBA/AMP-CR(a) and (2)PC/EBBA/AMP-CR(b) composite membranes.

activation energy for the K^+ transport was evaluated : E_{app} = 1.5 kJ mol^{-1} below T_{KN} and 2.1 kJ mol^{-1} above T_{KN}. Carrier-mediated K^+ permeation must be directly affected by the thermalmotion of LC molecules. In contrast, K^+ permeation through the PC/EBBA/AMP-CR(b) composite membrane was completely suppressed below T_{KN} and increased with increasing temperature above T_{KN} and the magnitude of E_{app} was 3.6 kJ mol^{-1}[46].

The permeation experiment mentioned above clearly demonstrates that K^+ permeation through a crown-containing composite membrane can be thermocontrolled and the rate is largely governed by the dispersion state(homogeneous or phase-separated) of the immobilized crown ethers as shown in Figure 37. Here, we demonstrate a new concept, namely, that the dispersion state can be readily changed by introducing a fluorocarbon chain into the crown ethers. As fluorocarbons are almost insoluble in hydrocarbon solvents, crown ethers covalently linked to a fluorocarbon chain would form phase-separated aggregates in the composite membrane phase. The thermocontrol of ion permeation through the ternary composite membranes composed of PC/EBBA/AMP-CR(c) and PC/EBBA/AMP-CR(d) was investigated. In order to assess the influence of the dispersion state on the ion permeability, we carried out thermocontrol of ion(K^+ and Cs^+) permeation in a U-tube immersed in a thermostated water bath. The membrane area was 3.46 cm^2 and IN and OUT aqueous phases were of 30.0 ml volume. The IN aqueous phase contained 0.30 M of alkali metal toluene-p-sulphonate and the concentration of alkali metal cations in

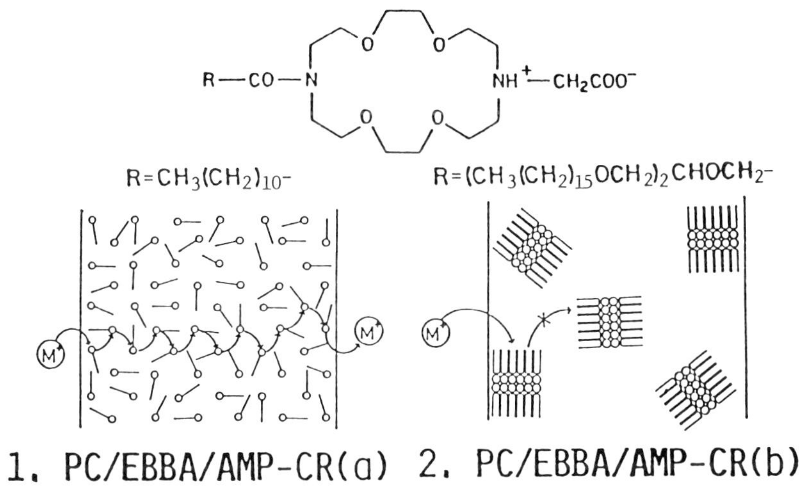

Figure 37. Schematic representations of (1)moleculary dispersed AMP-CR(a) and (2)phase-separated AMP-CR(b) in the composite membrane.

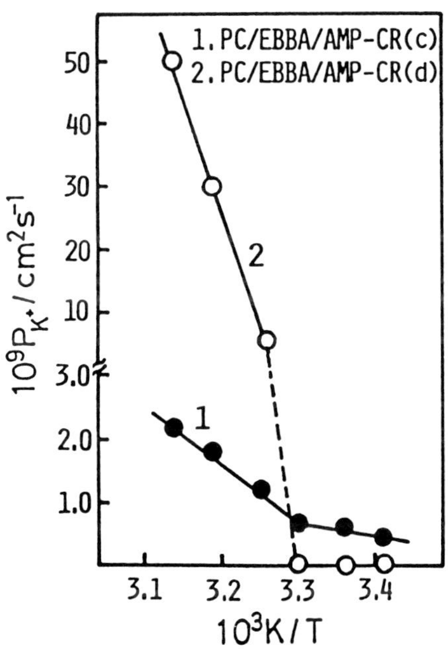

Figure 38. Arrhenius plot of P_{K^+} for (1)PC/EBBA/AMP-CR(c) and (2)PC/EBBA/AMP-CR(d) composite membranes.

the OUT aqueous phase was evaluated by atomic absorption spectroscopy. Figure 38 shows plots of the permeability coefficient for K^+(P_{K^+}) vs. reciprocal temperature[46]. Permeation of K^+ across the PC/EBBA/AMP-CR(c) composite membrane was observed below and above T_{KN} and the plot had a clear break point at around T_{KN}. This indicates that carrier-mediated K^+ permeation is directly affected by the molecular motion of the liquid-crystal phase. Interestingly, K^+ permeation across the PC/EBBA/AMP-CR(d) composite membrane was 'completely' suppressed below T_{KN} and greatly accelerated with increasing transport temperature above T_{KN}: the P_{K^+}($4.97 \times 10^{-8} cm^2 s^{-1}$) is greater by a factor of 22.7 than that for the PC/EBBA/AMP-CR(c) composite membrane at 318 K($2.19 \times 10^{-9} cm^2 s^{-1}$). This indicates that the fluorocarbon-containing AMP-CR(d) can carry K^+ much faster than the hydrocarbon-containing AMP-CR(c). This implies that the transport system is largely governed by diffussion of AMP-CR. K^+ permeation across the PC/EBBA/AMP-CR(d) composite membrane changed dramatically below and above T_{KN}. This novel phenomenon may be attributed to the phase-seperated dispersion state of AMP-CR(d)[47]. Since below T_{KN} the molecular motion of EBBA is frozen, in such a membrane system, ions are transported according to a site-to-site jump mechanism [50]. In fact, this jump mechanism can be operative when the carrier sites are dispersed homogeneously(i.e., PC/EBBA/AMP-CR(a)). However, when the carriers form phase-separated aggregates(i.e., PC/EBBA/AMP-CR(b)), aggregate-to-aggregate jump of ions would be almost impossible as shown in Figure 37. This may lead to the complete suppression of K^+ permeation below T_{KN}.

It can be concluded from the permeation experiment through the polymer/LC/AMP-CR composite membrane that ion permeation below T_{KN} is largely governed by a dispersion state of the carriers and the rate of K^+ permeation can be sensitively controlled as a thermoswitch.[48,49].

Acknowledgement

The author expresses his sincere gratitude to Dr. S. Shinkai of Nagasaki University and Drs. A. Takahara and H. Kikuchi of Kyushu University for their cooperative works and helpful discussions.

Nomenclature

AFS	air-facing surface
AMP-CR(a)	amphiphilic crown ether with monoalkyl chain
AMP-CR(b)	amphiphilic crown ether with dialkyl chains
AMP-CR(c)	amphiphilic crown ether(15C5) benzene-connected with dialkyl chains
AMP-CR(d)	amphiphilic crown ether(15C5) benzene-connected with fluorocarbon chain
ATR	attenuated total reflection
AZO-CR(1)	azobenzene-linked crown ether(15C5)
AZO-CR(2)	crown ether(18C6) azobenzene-connected with amino group
CHOB	4-cyano-4'-heptyloxybiphenyl
CPB	4-cyano-4'-pentylbiphenyl
D	diffusion coefficient

DC-18C6	dicyclohexyl-18-crown-6
DeOH	decanol
DSC	differential scanning calorimetry
d_p	depth of infrared beam penetration
E	initial modulus
E_{app}	apparent activation energy
EBBA	N-(4-ethoxybenzylidene)-4'-butylaniline
ε_b	strain at breake
FC	fluorocarbon
I_i	infrared absorbance of ith component (i=PC, EBBA) in composite membrane
I^o_i	infrared absorbance of ith pure component(i=PC, EBBA)
IR	infrared
ℓ	thickness of membrane
LC	liquid crystal
L44	polyoxyethylene-polyoxypropylene ABA type block copolymer(Pluronic L44)
M(t)	quantity of sorbed or desorbed penetrant in time t
M(∞)	quantity of sorbed or desorbed penetrant at infinite time
n_i	refractive index(i=1, 2)
P	permeability coefficient
PC	polycarbonate
PFTA	perfluorotributylamine
POM	polarizing optical microscope
PVC	poly(vinyl chloride)
S	solubility coefficient
SEM	scanning electron micrograph
SFS	substrate-facing surface
$\tan\delta$	dynamic loss tangent
T_g	glass transition temperature
TEM	transmission electron micrograph
THF	tetrahydrofuran
T_{KN}	crystal-nematic phase transition temeprature
T_{NI}	nematic-isotropic phase transition temperature
TPP	Tris(1H, 1H, 5H-octafluoropentyl)phosphate
UV	ultraviolet(light)
VIS	visible(light)
W_i	weight fraction of ith component(i=PC, EBBA)
λ	wavelength of characteristic absorption band
τ	time lag
θ	effective incident angle(45°)

References

1. R.E. Kesting, 'Synthetic Polymeric Membranes - A Structural Perspective', second ed., John Wiley & Sons, 1985.
2. D. Chapman, R.M. Williams, B.D. Ladbrooke, 'Physical Studies of Phospholipids, VI, Thermotropic and Lyotropic Mesomorphism of some 1,2-Diacyl-Phosphotidyl-Cholines(lecithins)', *Chem. Phys. Lipids,* **1**, 445 (1967).

3. M.C. Block, L.L.M. van Deenen, J. De Gier, 'Effect of the Gel to Liquid Crystalline Phase Transition on the Osmotic Behavior of Phosphatidylcholine Liposomes', *Biochim. Biophys. Acta,* **1**, 433 (1976).
4. T. Kajiyama, Y. Nagata, E. Maemura, M. Takayanagi, 'Molecular Motion-Permeability Relationships in Polycarbonate/Liquid Crystal (EBBA) Composite Membrane', *Chem. Lett.,* 679 (1979).
5. T. Kajiyama, Y. Nagata, S. Washizu, M. Takayanagi, 'Characterization and Gas Permeation of Polycarbonate/Liquid Crystal Composite Membrane', *J. Membrane Sci.,* **11**, 39 (1982).
6. S. Washizu, T. Kajiyama, M. Takayanagi, 'Function of Gas Permselectivity for Polymer/Liquid Crystal Composite Membrane and its Application to Oxygen Enrichment Membrane', *J. Chem. Soc. Jpn.,* 383 (1983).
7. S. Washizu, I. Terada, T. Kajiyama, M. Takayanagi, 'Gas Permeation through Polymer/Liquid Crystal Composite Membrane', *Polym. J.,* **16**, 16 (1984).
8. T. Kajiyama, S. Washizu, M. Takayanagi, 'Membrane Structure and Permeation Properties of Poly(Vinyl Chloride)/Liquid Crystal Composite Membrane', *J. Appl. Polym. Sci.,* **29**, 3955 (1984).
9. T. Kajiyama, S. Washizu, A. Kumano, I. Terada, M. Takayanagi, 'Permselective Characteristics of Polymer/Liquid Crystal Composite Membrane', *J. Appl. Polym. Sci., Appl. Polym. Symp.,* **41**, 327 (1985).
10. H. Kikuchi, A. Kumano, T. Kajiyama, M. Takayanagi, S. Shinkai, 'Characterization and Photo-Controlled Ion Permeation of Polymer/Liquid Crystal Composite Thin Film Containing Azobenzene-Bridged Crown Ether', *Rept. Progr. Polym. Phys, Jpn.,* **27**, 273 (1984).
11. T. Kajiyama, 'Membrane Structure and Permeation Property of Polymer/Liquid Crystal Composite', *Membrane,* **6**, 256 (1981).
12. S. Shinkai, K. Shimamoto, S. Nakamura, O. Manabe, T. Kajiyama, 'Thermocontrol of Electron Transport through Ternary Composite Membranes Composed of Polymer/Liquid Crystal/Electron Carriers', *J. Polym. Sci., Part C, Polym. Lett.,* **25**, 495 (1987).
13. T. Kajiyama, H. Kikuchi, A. Miyamoto, 'Functional properties of thin films of polymer/(liquid crystal) composites', *Frontiers of Molecular Science, IUPAC,* 505 (1989).
14. H. Kikuchi, J. Hattori, Y. Mori, T. Kajiyama, 'Enantioselective Transport of Amino Acid Salts through Polymer/(Liquid Crystal) Composite Thin Film Containing Chiral Crown Ether', *Kagaku Kogaku Ronbunshu,* **15**, 617 (1989).
15. R.M. Barrer and G. Skirrow, 'Transport and Equilibrium Phenomena in Gas-Elastomer System, I. Kinetic Phenomena', *J. Polym. Sci.,* **3**, 549 (1948).
16. Van A. Prietzshk, 'Die Kristallstruktur des Polycarbonats aus 4,4'-Dioxydiphenyl-2,2-Propan', *Kolloid Z. Z.,* **156**, 8 (1958).
17. H.S. Subramhanyan, J.S. Prasad, 'Refractive Indexes, Densities, Polarizabilities and Molecular Order in Nematic p-ethoxybenzylidine-p-n-ethylaniline', *Mol. Cryst. Liq. Cryst.,* **37**, 23 (1976).
18. C.A. Boye, 'Infrared Spectra of cis and trans isomers of Poly(1,4-cyclohexylenedimethylene) Terephthalate', *J. Polym. Sci.,* **55**, 263 (1961).

19. G. Vergoten, G. Fleury, N.R. Jones, A. Nadeau, 'Vibrational Spectroscopy of Liquid Crystals, IV. Infrared and Far Infrared Spectra of the MBBA and EBBA Liquid Crystals', *Mol. Cryst. Liq. Cryst.* **36**, 327 (1976).
20. J.P. Sibilia, 'The Surface Composition of Biconstituent Fibers', *J. Appl. Polym. Sci.*, **17**, 2911 (1973).
21. J. Crank, 'The Mathematics of Diffusion', Oxford Univ. Press, London 1956.
22. S. Washizu, I. Terada, T. Kajiyama, M. Takayanagi, 'Oxygen Enricher Effect by Means of Polymer/Liquid Crystal Composite Membrane with Oxygen Gas Carrier of Fluorocarbon Type Monomer', *Polym. Prepr. Jpn.*, **32**, 2973 (1983).
23. R.Y.M. Hung, V.J.C. Lin, 'Separation of Liquid Mixture by Using Polymer Membranes, I. Permeation of Binary Organic Liquid Mixtures through Polyethylene', *J. Appl. Polym. Sci.*, ***12***, 2615 (1968).
24. H. Yasuda, C.E. Lamaze, A. Peterlin, 'Diffusion and Hydraulic Permeabilities of Water in Water-swollen Polymer Membranes', *J. Polm. Sci.*, **9A**, 1117 (1971).
25. E. Maemura, Y. Nagata, T. Kajiyama, M. Takayanagi, 'Thermal Molecular Motion-Permeability Relationships of Thin Film of Poly(-methyl-L-glutamate), and Interpolymer of Poly(allyl-L-glutamate)/Silicone Rubber', *Rep. Progr. Polym. Phys. Jpn.*, **20**, 663 (1977).
26. T. Kajiyama, S. Washizu, Y. Ohomori, 'Oxygen Permselective Characteristics of Poly(Vinyl Chloride)/Liquid Crystal/Fluorocarbon Ternary Composite Membrane', *J. Membrane Sci.*, **24**, 73 (1985).
27. R.S. Porter, J.F. Johnson, 'Order and Flow of Liquid Crystals: The Nematic Mesophase', *J. Appl. Phys.*, **34**, 51 (1963).
28. Y. Ohmori, S. Washizu, I. Terada, H. Kikuchi, T. Kajiyama, 'Oxygen-Enricher Effect by Means of Polymer/Liquid Crystal Composite Membrane with Oxygen Gas Carrier of Fluorine Compound,' *Polym. Prepr. Jpn.*, **33**, 1679 (1984).
29. T. Nakagawa, H.B. Hopfenberg, V.T. Stannett, 'Transport of Fixed Gases in Radiation-Stabilized Poly(vinyl chloride)', *J. Appl. Polym. Sci.*, **15**, 231 (1971).
30. R.W. Baker, I.C. Roman, K.L. Smith, H.K. Lonsdale, 'Liquid Membranes for the Production of Oxygen-Enriched Air', *Energy Tech.*, 505 (1982).
31. T. Kajiyama, 'Aggregation Structure-Permselective Function Relationships of Hybrid Organic Thin Films', *MRS Int'l. Mtg. on Adv. Mats.*, **1**, 241 (1989).
32. O. Nishimoto, I. Terada, T. Kajiyama, 'Separation Characteristics to Pentane Gas Isomers through Polymer/Liquid Crystal Composite Membrane', *Rept. Progr. Polym. Phys. Jpn.*, **29**, 271 (1985).
33. T. Kajiyama, H. Kikuchi, I. Terada, M. Katayose, A. Takahara, 'Permeation Properties of Polymer/(Liquid Crystal) Composite Membrane', *Current Topics in Polym. Sci.*, **2**, 320 (1987).
34. T. Kajiyama, 'Molecular Filtration of Hydrocarbon Isomers through Polymer/(Liquid Crystal) Composite Membranes', *J. Macromol. Sci., Chem.*, **A25**, 583 (1988).

35. S. Oda, T. Kajiyama, M. Takayanagi, 'Diffusion and Permselectivity of Alkali Metal Ions through Polymer/Liquid Crystal/Crown Ether', *Rept. Progr. Polym. Phys. Jpn.,* **25**, 307 (1982).
36. S. Shinkai, T. Nakaji, T. Ogawa, K. Singematsu, O. Manabe, 'Photoresponsive Crown Ethers, 2. Photocontrol of Ion Extraction and Ion Transport by a Bis(crown ether) with a Butterfly-like motion', *J. Am. Chem. Soc.,* **103**, 111 (1981).
37. S. Shinkai, M. Ishihara, K. Ueda, O. Manabe, 'Photoresponsive Crown Ether Part 14. Photoregulated Crown-Metal Complexation by Competitive Intramolecular Tail(Ammonium)-biting', *J. Chem. Soc. Perkin Trans.,* **11**, 511 (1985).
38. S. Shinkai, O. Manabe, 'Photocontrol of Ion Extraction and Ion Transport by Photofunctional Crown Ether', *Topics Current Chem.,* ***121***, 67 (1984).
39. A. Kumano, O. Niwa, T. Kajiyama, M. Takayanagi, K. Kano, S. Shinkai, 'Photoinduced Ion Permeation through Ternary Composite Membrane Composed of Polymer/Liquid Crystal/Azobenzene-Bridged Crown Ether', *Chem. Lett.,* 1327 (1983).
40. T. Kajiyama, H. Kikuchi, I. Terada, M. Katayose, A. Takahara, S. Sinkai, 'Permeation Properties of Polymer/(Liquid Crystal) Composite Membrane', *Current Topics Polym. Sci.-1984,* **2**, 315 (1987).
41. H. Kikuchi, M. Katayose, S. Shinkai, O. Mamabe, T. Kajiyama, 'Photo-Controlled K^+ Permeation of Polymer/Liquid Crystal Composite Thin Film Containing Photoresponsive Crown Ether', *J. Chem. Soc. Jpn., Chem. Ind. Chem.,* 423 (1987).
42. T. Kajiyama, H. Kikuchi, S. Shinkai, 'Novel Polymer/Liquid Crystal Composite Membrane with Unique Permselective Characteristics', *J. Membrane Sci.,* **36**, 243 (1988).
43. T. Kajiyama, H. Kikuchi, M. Katayose, S. Shinkai, 'Photo-driven Active Transport of Metal Cation through Polymer/(Liquid Crystal)/(Azobenzene Crown Ether) Ternary Composite Thin Films', *New Polym. Mater.,* **1**, 99 (1988).
44. S. Shinkai, H. Kinda, T. Sone, O. Manabe, 'Enforced Ion Release in Proton-driven Membrane Transport', *J. Chem. Soc. Chem. Commun.,* 125 (1982).
45. S. Shinkai, H. Kinda, Y. Araragi, and O. Manabe, 'Photo-driven Ion Transport and Metal-assisted Amino Acid Transport with an Anion-capped Azocrown Ether', *Bull. Chem. Soc. Jpn.,* **56**, 559 (1983).
46. S. Shinkai, S. Nakamura, S. Tachiki, O. Manabe, T. Kajiyama, 'Thermal control of Ion Permeation through Ternary Composite Membranes Composed of Polymer/Liquid Crystal/Amphiphilic Crown Ethers', *J. Am. Chem. Soc.,* **107**, 3363 (1985).
47. S. Shinkai, K. Torigoe, O. Manabe, T. Kajiyama, 'Complete Thermocontrol of Ion Permeation through Ternary Composite Membranes Composed of Polymer-Liquid Crystal-Fluorocarbon-Containing Crown Ethers', *J. Chem. Soc. Chem. Common.,* 933 (1986).
48. S. Shinkai, K. Torigoe, O. Manabe, T. Kajiyama, 'Temperature Regulation of Crown-Mediated Transport through Polymer/Liquid Crystal Composite Membranes. Remarkable Transport Ability of Fluorocarbon-Containing Crown Ether', *J. Am. Chem. Soc.,* **109**, 4458 (1987).

49. S. Shinkai, S. Nakamura, K. Ohara, S. Tachiki, O. Manabe, T. Kajiyama, ' "Complete" Thermocontrol of Ion Permeation through Ternary Composite Membranes Composed of Polymer/Liquid Crystal/ Amphiphilic Crown Ethers', *Macromolecules*, **20**, 21 (1987).
50. E. Shchori and J. Jagur-Grodzinski, 'Permeability to Salts and Water of Macrocyclic Polyether-Polyamide Membranes', *J. Appl. Polym. Sci.*, **20**, 773 (1976).

PHOTOCONTROL OF ION PERMEATION THROUGH MEMBRANE AND MEMBRANE POTENTIAL

Tetsuo Osa and Jun-ichi Anzai
Pharmaceutical Institute
Tohoku University
Aobayama, Sendai 980
JAPAN

ABSTRACT. Principles of photocontrol of membrane characteristics such as ion permeability and membrane potential are explained based on up-to-date experimental examples. The photocontrolled permeation of ions through bulk liquid, solvent polymeric, and bilayer lipid membranes is described with a theoretical consideration. The derivatives of cinnamic acid, azobenzene, and spirobenzopyran are used as photosensitive ion carriers in the membranes. Additionally, polymer membranes which generate membrane potential upon photo-irradiation are reviewed. The photochemically generated or disappeared electric charges on the membrane surface or at the interior of the membrane play a crucially important role in the induction of membrane potential. Finally, this chapter is also concerned with the application of the photosensitive membranes to construction of ion- and bio-sensors.

1. INTRODUCTION

The development of high performance membranes is currently being carried out at an ever increasing rate. Though, until recently, many kinds of artificial membranes have been used to do works that otherwise required more cumbersome and often more energy-intensive processes, the rapid progress in membrane science has shown us the further possible applications of membranes to, for example, separation and purification of chemicals, electric devices such as sensors and transducers, energy conversion and storage, and artificial organs.

Among the high performance membranes recently developed, photo-sensitive membranes, which change the ionic permeability and membrane potential reversibly on photoirradiation, are interesting materials with possible application in many diverse fields in science and technology. In addition to the usefulness of such membranes, the mechanistic study on the photosensitive behavior of the membranes would provide us with a valuable information in connection with the photo-reception and the following signal transduction processes in

T. Osa and J. L. Atwood (eds.), Inclusion Aspects of Membrane Chemistry, 157–189.

some biomembranes.

The purpose of this chapter is to describe 1)photocontrol of ion permeation through liquid, polymer, and bilayer membranes and 2)photocontrol of membrane potential across dye-doped and crown ether membranes, including their application to sensor materials.

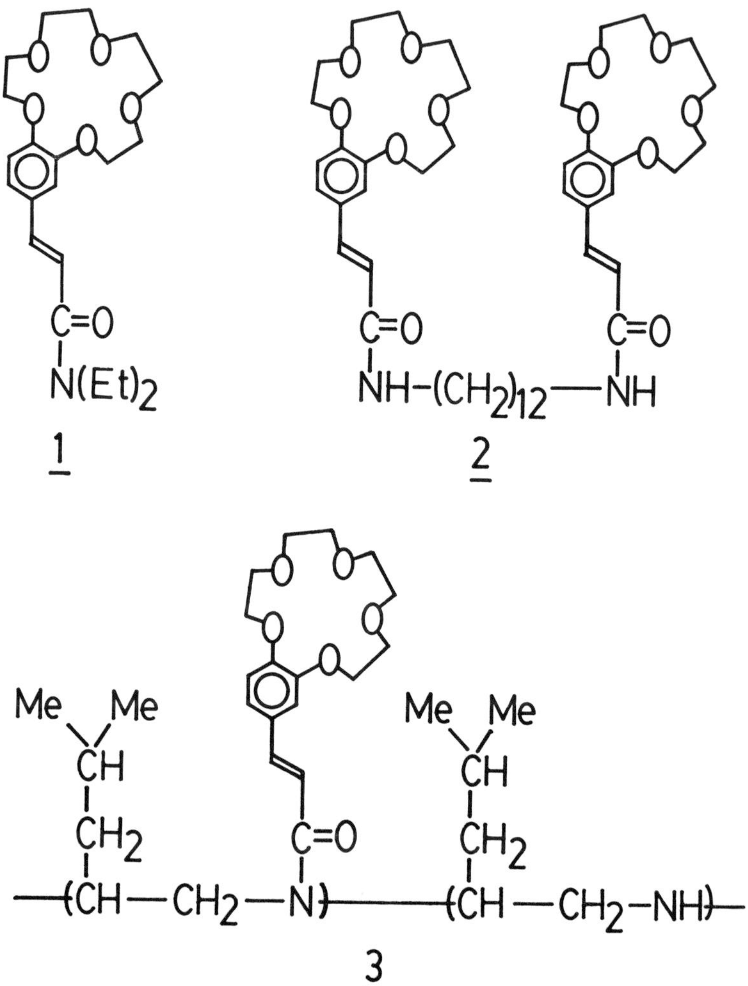

Fig. 1: Photosensitive crown ethers modified with cinnamoyl residue 1, 2, and 3.

2. PHOTOCONTROL OF ION PERMEATION

2.1 Liquid membranes

Ion permeation through liquid membranes has been made subject to photocontrol by several researchers. Crown ethers bearing cinnamoyl moiety have been developed as photosensitive ion carriers to control permeability of alkali-metal picrates across a dichloromethane liquid membrane.[1,2] The monomeric, dimeric, and polymeric crown ethers (1, 2, and 3) were prepared (Fig. 1). It can be monitored by measuring UV spectra that the cinnamoyl residues of 1, 2, and 3 undergo photochemical dimerization to give the mixture of possible stereoisomers, such as neotruxinate, α-truxillate, and β-truxinate types of cycloaddition products. The effects of photoirradiation on the ion transport rates are shown in Table I. Monomeric crown ether 1 transports K^+ ion most efficiently before irradiation and its transport selectivity is fairly compatible with the order of cation concentration in the membrane($K^+ > Na^+ > Rb^+ > Cs^+$). These properties of 1 are very similar to those previously-reported 15-crown-5 derivatives. [3,4]. Photodimerization of 1 in the presence of KCl enhances the transporting efficiency for all cations, especially the transport rate for Rb^+ being accelerated appreciablly. The effect of KCl is also found in bis(crown ether) 2. The Rb^+ selectivity is observed for 2 irradiated in the presence of KCl, while the cation selectivity disappears when 2 is irradiated in the absence of cation. This may arise from the preferential formation of cycloaddition products orienting crown ether moieties to the same direction from a cyclobutane ring (Fig. 2). The enhanced selectivity for the larger cations, Rb^+ and Cs^+, is observed for 3 when irradiated in the presence of KCl. It should be noted that the cation concentrations in the membranes are always higher than those of 1 and 2, and this fact indicates the strong binding ability of 3. The transport efficiency of 3 may be

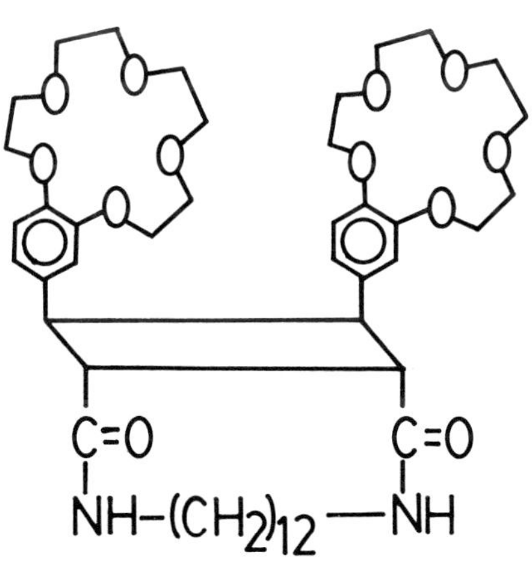

Fig. 2: Cycloaddition Product of 2.

Table I. Cation Transport Rates with 1, 2, and 3 and Cation Concentration in the Membrane

Crown Ether	Template Ion	Transport Rate/10^{-6} Mh^{-1}				Cation Concentration/10^{-5} M			
		Na^+	K^+	Rb^+	Cs^+	Na^+	K^+	Rb^+	Cs^+
Before Irradiation									
1		2.9	4.3	2.3	1.6	0.3	0.4	0.2	0.1
2		3.4	11	15	10	0.4	3.4	2.7	1.3
3		0.5	3.6	4.4	2.9	9.7	12	10	10
After Irradiation[a]									
1	none	14	15	12	13	0.6	0.8	0.5	0.3
1	Na^+	11	10	24	23	<0.1	<0.1	<0.1	<0.1
1	K^+	17	19	30	16	<0.1	<0.1	<0.1	<0.1
2	none	19	16	15	10	3.4	2.7	2.8	2.3
2	Na^+	8.7	6.0	12	8.6	<0.1	1.3	1.0	1.2
2	K^+	8.0	6.6	16	11	1.5	2.2	3.3	3.3
3	none	0.6	1.7	2.3	0.9	15	16	17	16
3	K^+	0.5	1.0	5.2	4.8	17	15	12	9.9

[a]Dimer fraction is always about 75%.

suppressed by the strong binding of cations in the membrane phase.

Azobenzene-modified crown ethers have been used to control the binding equilibrium for alkali-metal cations by photoirradiation.[5-7] Shinkai et al. have studied intensively the photocontrol of binding and transporting properties of azobenzene modified crown ethers such as 4, 5, and 6 (Fig. 3). The azo linkage of these compounds isomerizes to cis form on the irradiation of UV light, and then, the original trans isomers are recoverd by visible light irradiation. They have demonstrated using a liquid membrane (o-dichlorobenzene) in a U-shaped glass cell that the transport rate of K^+ ion with 4 across the membrane is accelerated by photoirradiation when a relatively hydrophilic counteranion (2-nitrodiphenylamine-4-sulfonate) is used. On the contrary, the transport of K^+ ion is suppressed upon photo-irradiation in the case that the counteranion is more hydrophobic (ex., picrate). These results are explained based on the idea that the ion transport is accelerated by light only when the ion transfer at rate-limiting step in the overall transport is facilitated on irradiation.[8] Ion carriers 5 and 6 were also effective for photo-acceleration of alkali- or alkaline earth-metal ions through liquid membranes.[9,10]

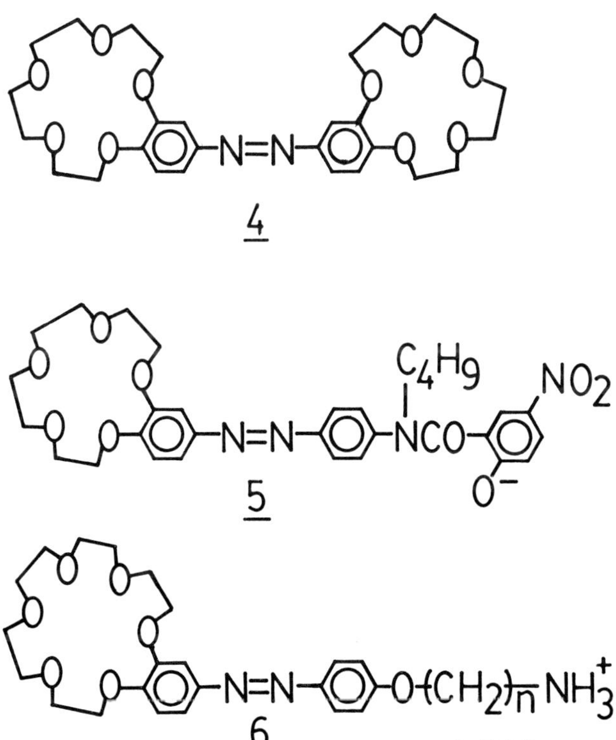

Fig. 3: Azobenzene-modified crown ethers 4, 5, and 6.

A highly lipophilic spirobenzopyran derivative 7 (Fig. 4) has been employed as a photosensitive ion carrier for the transport of NaCl and KCl through 1-octanol liquid membrane.[11] Liquid-liquid extractions reveal that the open 7 binds NaCl and KCl more strongly than the closed form does. Based on this photoresponse in cation binding of 7, it is possible to transport NaCl actively through liquid membrane by irradiating the membrane/source phase and membrane/receiving phase interfaces with UV and visible light, respectively.

As seen in the above-cited systems, the principal problem in developing photosensitive liquid membranes is how one can design and synthesize carrier molecules so functionally as to change the binding equilibrium in an on-off fashion upon irradiation. Once carrier molecules are synthesized, one can prepare membrane easily only by dissolving it into organic solvent immiscible with water. Because of the feasibility of transport experiments using liquid membranes, many synthetic chemists take part in this field.

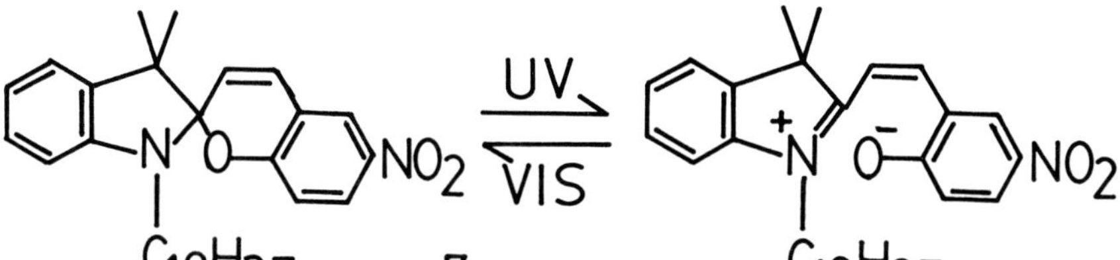

7

Fig. 4: Photochemical reaction of lipophilic benzospiropyran 7.

2.2 Polymer and bilayer membranes

The development of polymer membranes sensitive to light is an important subject from the viewpoint of their practical use to separation process, light energy conversion, etc.

Azobenzene-modified crown ethers can be used for preparing the photosensitive poly(vinyl chloride) (PVC) membranes. Crown ether 8 (Fig. 5) was immobilized into a plasticized PVC membrane.[12,13] PVC/8 membrane is prepared typically by pouring the mixture of 235 mg of PVC, 0.54 ml of dibutyl phthalate (plasticizer), 2.5 mg of 8 in 20 ml of tetrahydrofuran onto a rimmed glass plate (or flat Petri dish) with 9.2 cm diameter and allowing to evaporate the solvent at ambient temperature. This method gives a transparent membrane with 0.1 mm thickness. A U-shaped glass cell (illustrated in Fig. 6) was used for the permeation experiments of alkali-metal picrates.

The permeation rates for alkali-metal picrates before and after UV irradiation are compiled in Tables II and III. From the data, the followings can be concluded;

(a). UV irradiation facilitates the permeation of K^+ picrate, but does not the permeation of the other cations,

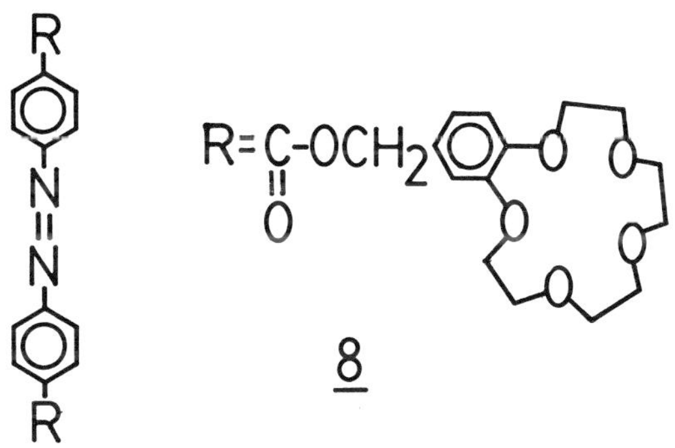

8

Fig. 5: Photosensitive ion carrier 8.

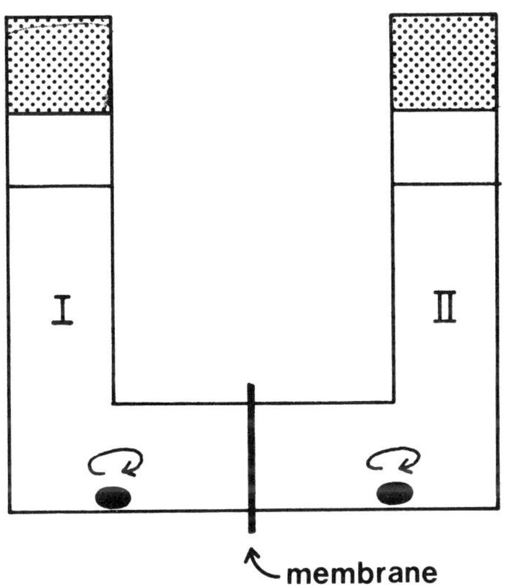

Fig. 6: Schematic representation of U-shaped glass cell for ion transport.

(b). the permeation selectivity of the membrane is in the order of $K^+ > Rb^+ > Na^+ > Cs^+$,
(c). the permeation rate of cations depends almost linearly on the density of 8 in the membrane, and
(d). the permeation rate is inversely proportional to the thickness of the membrane.

It is reasonable to assume that, in this system, a large portion of alkali-metal picrate diffuses across the hydrophobic membrane in an electrically-neutral ternary complex composed of metal cation, picrate anion and crown ether, which forms according to Eq. (1) at the

Table II. Effect of UV Irradiation on the Ion Permeation through PVC/8 Membranes with Various 8 Contents

Membrane	Permeation Rate/10^{-9} mol h^{-1} cm^{-2}			
	Na^+	K^+	Rb^+	Cs^+
Before Irradiation				
PVC/8-A[a]	3.5	9.4	4.6	1.7
PVC/8-B[a]	4.7	13	9.4	2.8
PVC/8-C[a]	8.7	20	21	5.9
After Irradiation[b]				
PVC/8-A	4.3	12	4.7	1.9
PVC/8-B	5.7	18	10	2.8
PVC/8-C	10	29	20	6.9

[a] 8 Content in the membranes (g/cm^2): A;4.0x10^{-5}, B;8.0x10^{-5}, C;16x10^{-5}.

[b] About 50% of 8 was converted to cis form after irradiation.

Table III. Effect of UV Irradiation on the Ion Permeation through PVC/8 Membranes with Various Thickness

Membrane	Permeation rate/10^{-9} mol h^{-1} cm^{-2}			
	Na^+	K^+	Rb^+	Cs^+
Before Irradiation				
PVC/8-B1[a]	13	24	26	7.7
PVC/8-B[a]	4.7	13	9.4	2.8
PVC/8-B2[a]	3.2	7.2	4.6	1.2
After Irradiation[b]				
PVC/8-B1	13	39	24	5.8
PVC/8-B	5.7	18	10	2.8
PVC/8-B2	4.2	12	4.6	1.5

[a] Membrane thickness (mm): B;0.1, B1;0.05, B2;0.2.

[b] About 50% of 8 was converted to cis form after irradiation.

membrane/source phase interface

$$M^{+} + Pi^{-} + CR \rightleftharpoons (CR\cdot M^{+}\cdot Pi^{-}) \quad (1)$$

where M^{+}, Pi^{-}, and CR denote metal cation, picrate anion, and crown ether, respectively, and $(CR\cdot M^{+}\cdot Pi^{-})$ is the ternary complex. The total flux of alkali-metal picrate in the membrane, J_p^T, can be given in the following equation

$$J_p^T = J_p + J_{comp} \quad (2)$$

where J_p and J_{comp} are the fluxes of alkali-metal picrate uncomplexed and those complexed with crown ether, respectively. Equation (2) is the expression taking into account the small portion of the flux of picrate uncomplexed. The J_p value seems small in the present case, because alkali-metal picrates scarcely dissolve into the hydrophobic membrane. In order to estimate the values for picrates, the permeation rates of picrate through the PVC membrane without the crown ether were elucidated. The values for alkali-metal picrates are estimated to be ca. 0.6×10^{-9} mol h^{-1} cm^{-2} (initial concentration of picrate = 5×10^{-3} mol l^{-1}), which is not influenced by photoirradiation. Therefore, one can eliminate the J_p term from the consideration of the photoresponse of the ion permeation.

The flux of ion-piar complex in the membrane is given by

$$J_{comp} = D(C^{0}_{comp} - C^{1}_{comp})/L \quad (3)$$

where D represents the diffusion coefficient of the complex in the membrane, and C^{0}_{comp} and C^{1}_{comp} denote the concentration of complex at the membrane/solution interfaces facing the aqueous phases I and II, respectively. L is the thickness of the membrane. Equation (3) implies that J_{comp} depends linearly on the D value and the concentration gradient of the complex in the membrane.

Under the conditions that the fraction of crown ether participating in complex formation is so small as compared with the total concentration of the crown ether in the membrane, C^{0}_{comp} and C^{1}_{comp} are given, according to Eq. (1), by

$$C^{0}_{comp} = Kk_Mk_PC^{I}_{M}C^{I}_{P}C_{CR} \quad (4)$$

and

$$C^{1}_{comp} = Kk_Mk_PC^{II}_{M}C^{II}_{P}C_{CR} \quad (5)$$

where K denotes the formation constant of ternary complex in Eq. (1), k_M and k_P are the partition coefficients of metal cation and picrate anion, respectively, between the membrane and solution, C^{I}_{M} and C^{I}_{P} are the concentrations of cation and anion in the aqueous phase I, C^{II}_{M} and C^{II}_{P} are the ones in the aqueous phase II, and C_{CR} represent the concentration of crown ether in the membrane. By substituting Eqs. (4) and (5) for C^{0}_{comp} and C^{1}_{comp} in Eq. (3), one has Eq. (6)

$$J_{comp}=DKk_{M}k_{P}C_{CR}(C^{I}_{M}C^{I}_{P}-C^{II}_{M}C^{II}_{P})/L \quad (6)$$

Since $C^{II}_{M}C^{II}_{P}$ is negligibly small under the present experimental conditions, Eq. (6) becomes

$$J_{comp}=DKk_{M}k_{P}C_{CR}C^{I}_{M}C^{I}_{P}/L \quad (7)$$

Equation. (7) means that the flux of the ternary complex between alkali-metal picrate and crown ether is proportional to its diffusion coefficient, formation constant of complex, the concentration of alkali metal picrate in the aqueous phase I, and the partition coefficient of ions between the membrane and water, and is inversely proportional to the thickness of the membrane. Based on these considerations, it is clear that, in the present case, UV irradiation altered the K value of 8 for K^{+} ion.

A polymer/liquid crystal composite membrane has been used to facilitate ion permeation by photoirradiation.[14] Kumano et al. immobilized azobenzene-modified crown ether 4 in the PVC/4-cyano-4'-n-pentyl biphenyl or N-(4-ethoxybenzylidene)-4'-n-butylaniline membrane, and found the photo-accelerated permeation of cations.

The effects of photoinduced pH change on the hydrodynamic permeability through porous membrane with adsorbed poly(L-glutamic acid) (PGA) have been studied by Kinoshita et al.[15] They have measured the filtration coefficient, L_p, of porous Millipore and Nucleopore filters with adsorbed PGA under irradiation. The L_p values change with the pH changes induced by the photo-oxidation reaction of leuco malachite green, which is placed in a carbon tetrachloride solution at the bottom of the cell. The results were explained in relation to the conformational changes of the PGA at the pore wall.

Bilayer membranes have been also made subject to photoirradiation by several researchers.[16-18] Kano et al. have prepared photosensitive bilayer membranes by embedding amphiphatic alkylammonium salts containing azobenzene chromophore 9 (Fig. 7) in dipalmitoyl phosphatidylcholine liposomes. The trans-cis photoisomerization of azobenzene residue in 9 provides perturbation to the ordered structure of the bilayer membrane, and induces the increase of the permeability for water and bromothymol blue. Sunamoto et al. have demonstrated that spiropyran can be used for transporting amino acids across liposomal bilayer of egg phosphatidilcholine. The zwitterionic form of 10 forms an ionic complex with a zwitterionic amino acid, as is not the case for closed-10 (Fig. 8).

Liposomal bilayers are of considerable interest in conjunction with the biomembranes, but they have a serious disadvantage in developing the practically-usable membranes; i.e., easily breakable nature against the osmotic pressure difference. This drawback of bilayer membranes can be overcome by coating the lipid bilayer on the surface of nylon capsules composed of porous membrane. When an azobenzene-

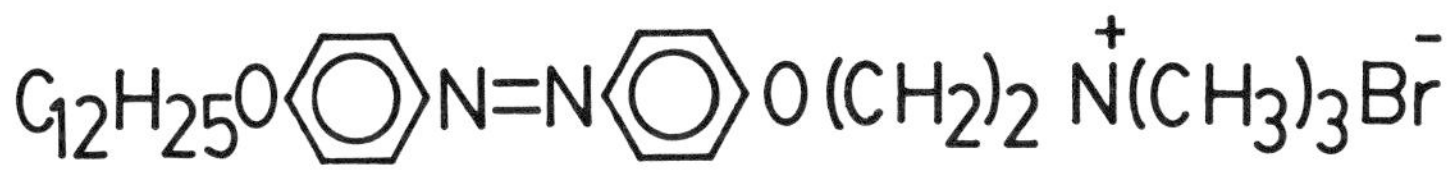

9

Fig. 7: Photosensitive amphiphile 9.

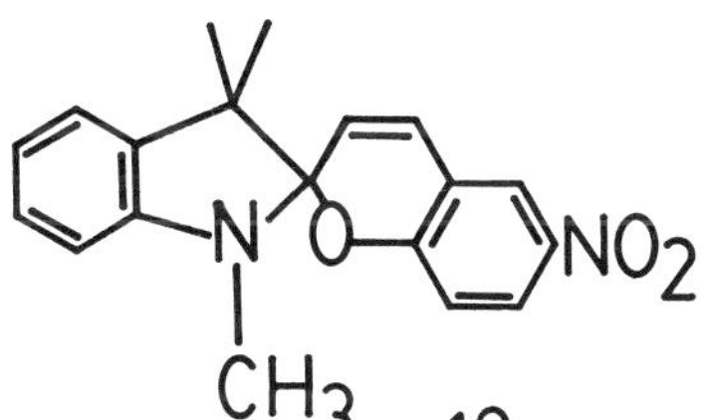

10

Fig. 8: Nitrobenzospiropyran 10.

modified synthetic lipids are coated on the nylon capsule, the permeability of NaCl from the inner aqueous phase to the outer phase is accelerated by UV irradiation.[19,20] The results are explicated based on the distorted configuration of cis-azobenzene moiety in bilayer membrane. Okahata et al. also developed different kinds of stimulireceptive capsule membranes, being sensitive to temperature, solution pH, metal ions, electric field, etc.[21]

Rhodopsin-containing liposomes were constructed as a model of photoreceptor cell membranes.[22-24] It was found that photoirradiation changed the membrane permeability for ions and molecules.

3. PHOTOCONTROL OF MEMBRANE POTENTIAL

Photocontrol of membrane potential across artificial membranes are of considerable interest in connection with the photo-reception and the subsequent signal transduction in photoreceptor cell membranes in plants and animals, as well as the development of signal-transducing artificial membranes that transduce light signal into electric form. Among the naturally-occurring photoresponsive systems, the most carefully studied process is vision. A key point of the mechanism by which eyes receive light and transduce it to electric signal can be summarized as follows;

Light signal

↓

Photoisomerization of 11-cis retinal to all-trans form

↓

Conformational change of rhodopsin

↓

Change in ionic permeability of cell membrane
↓
Membrane potential change
↓
Nerve
↓
Brain

It should be noted here that all processes above occurs on the membrane of the photoreceptor cell. The membrane/solution interfaces, obviously, are essentially required for transducing the light signal to the electric form (i.e., membrane potential).

3.1 Dye-doped membranes

With the intention of mimicking the photo-electric behavior of the photoreceptor cell membrane, many researchers studied the pigmented bilayer lipid membranes.[25-28] A variety of photosensitive molecules including retinal, rhodopsin, and bacteriorhodopsin were utilized to endow the membrane with photoresponse.

Recently, attempts have been made to prepare photosensitive polymer membranes that change the membrane potential in response to photoirradiation. Polypeptide membranes prepared from poly(glutamic acid) having azobenzene side chains 11 (Fig. 9) have been studied.[29, 30] About +4 mV of the potential change was induced by UV irradiation on the membrane (azo group, 3%) interposed between aqueous KCl solutions (1 mM and 0.25 mM, pH 6.2). When the membrane was irradiated in pH 9.0, the membrane potential remains unchanged. The results can be explained in terms of co-operative effects arising from the cis formation of azobenzene residues and the dissociation of carboxylic acid moieties accelerated by the enhanced water uptake in the irradiated membrane. Since the cis-trans photoisomerization of azobenzene residue is inherently not accompanied by the increase or decrease of electric charge, it is necessary for inducing the potential changes by using azobenzene-modified membrane that photo-reaction of azobenzene moiety can be coupled with such ionic phenomena as ion dissociation, adsorption/desorption of ions, and ion transport.

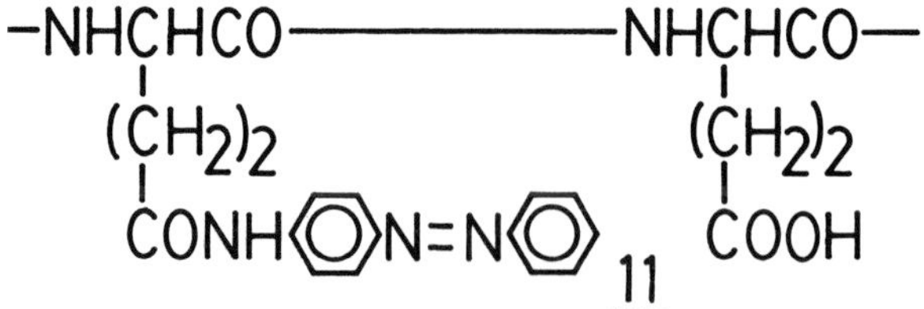

11

Fig. 9: Azobenzene-substituted poly(glutamic acid) 11.

A benzospiropyran derivative was first utilized by Kato et al. for preparing the photosensitive cellulose membranes.[31-33] They entrapped nitrobenzopyran 10 into an acetyl cellulose membrane, and reported that the membrane potential changed upon UV light irradiation. Recently, cellulose 2,4-diacetate membranes incorporating 6-X-1',3',3'-trimethylspiro-(2H-1-benzopyran-2,2'-indoline), with X=H, OH, Br, NO_2, and 1',3',3'-trimethylspiro-(2H-1-naphthopyran-2,2'-indoline) have been investigated with the intention of constructing photoresponsive devices for optical data storage.[34,35] They have asserted based on the consideration of fixed charge concentration in the membrane, that the charge "seen" in the potential measurements corresponds to that present at the oxgen site of spiropyran molecules.

The photoinduced membrane potential across poly(methacrylic acid) membrane having spirobenzopyran pendant groups was also measured.[36] The magnitude of the photoinduced potential was 10 mV or less.

More recently, photoresponse of the spiropyran-containing membrane has been highly improved by the use of a plasticized poly-(vinyl chloride) (PVC) as a membrane matrix.[37-40] Figure 10 shows a typical photoresponse of the membrane potential across the PVC/1-hexadecyl-3,3'-dimethyl-6-nitrospiro(2H-1-benzopyran-2,2'-indoline) 12 membrane (Fig. 11). The composition of the electrochemical cell for the membrane potential measurements is as follows; SCE|NaCl solution (c_1)| PVC/12 membrane|NaCl solution (c_2)|SCE (Fig. 12). After being reached a steady-state potential difference in the dark, a remarkable negative shift of the potential was induced by UV light irradiation.

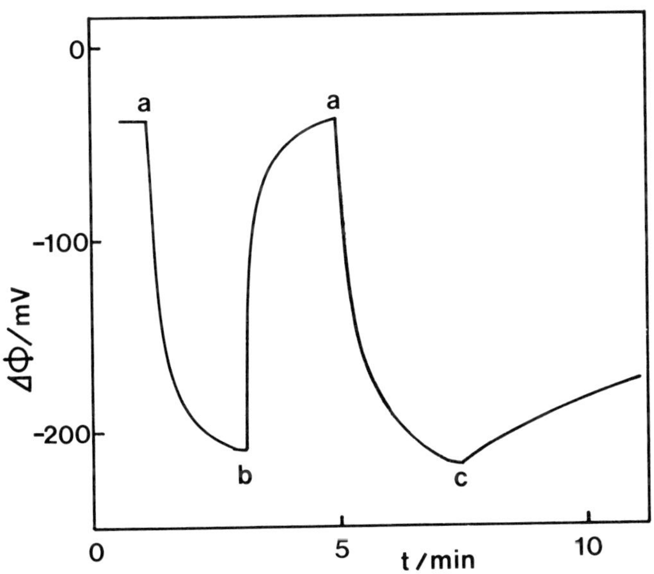

Fig. 10: Photoresponse of the membrane potential across the PVC/12 membrane. UV light (a) and visible light (b) were irradiated from C_2 side, and light was switched off at c. NaCl concentration was 10 mM (C_1) and 100 mM (C_2). pH 6.01.

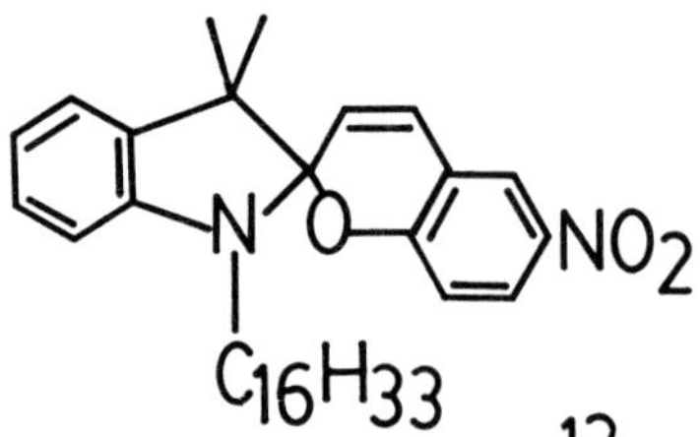

12

Fig. 11: Hexadecylated benzospiropyran 12.

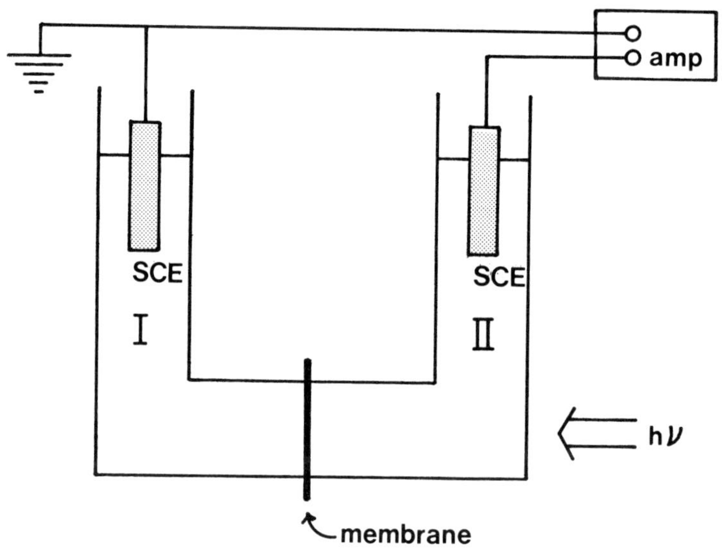

Fig. 12: Schematic representation of the cell for measuring membrane potential.

The photoinduced potential of more than 100 mV were attained with rapid response time (ca. 1 min). By switching the UV light to visible one, the initial potential was recovered rapidly. On the other hand, the photoinduced potential decayed with a half-life of ca. 12 min under dark conditions. The potential changes arise from photochemical reaction of 12 in the membrane.

Figures 13 and 14 show the effects of NaCl concentration in the aqueous solutions on the membrane potential across the PVC/12 membrane. In all cases the photoresponse is high (100 mV or more) and rapid (1-2 min). The absence of any fixed charge except for open-12 in the membrane, which is not the case for cellulose[31-35] and poly-

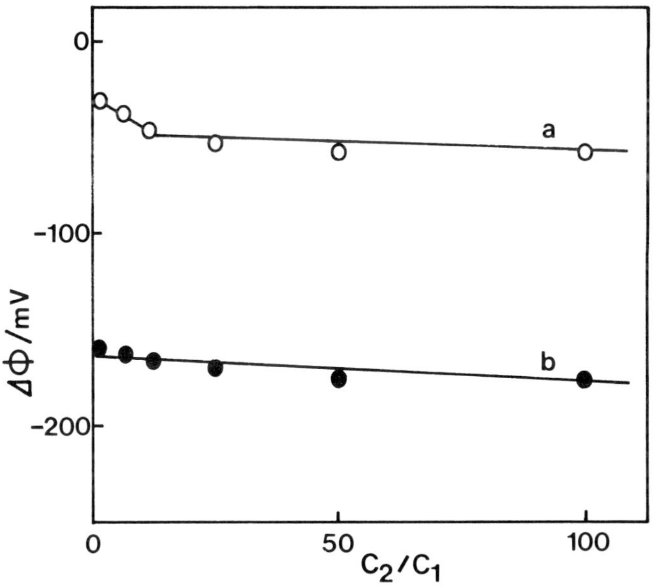

Fig. 13: Effects of NaCl concentration on the membrane potentials at constant C_1 concentration (=10 mM). a;potential under visible light, b;potential under UV light.

(methacrylate) membranes[36], may be favorable for the enhanced photoresponse.

From the viewpoint of mechanistic study of the photoresponse of the PVC/12 membranes, it is of considerable importance to analyse the data obtained under the conditions that NaCl concentrations of solutions in both sides are equal with each other, i.e., $c_1=c_2$. Figure 15 illustrates the photoresponse observed when UV and visible light are irradiated from both sides of the membrane alternately. Even under the conditions of $c_1=c_2$, the membrane potential shifted upon UV irradiation positively or negatively, depending on the direction from which the membrane is irradiated. The membrane potential always has a value near to zero under visible light. These results suggest the following mechanism of the membrane-potential generation by photoirradiation. The membrane potential, $\Delta\Phi$, of a charged membrane is generally given by Eq. 8 when a univalent electrolyte (c_1 M) is contacted with one side of the membrane, and the same electrolyte (c_2 M) is placed in the opposite side of the membrane at equal pressure and temperature;

$$\Delta\Phi = \frac{RT}{F}\left\{-\ln\gamma + \ln\frac{-\theta+\sqrt{\theta^2+4c_2^2}}{-\theta+\sqrt{\theta^2+4c_1^2}} + (2t_- -1)\ln\frac{(1-2t_-)\theta+\sqrt{\theta^2+4c_2^2}}{(1-2t_-)\theta+\sqrt{\theta^2+4c_1^2}}\right\} \quad (8)$$

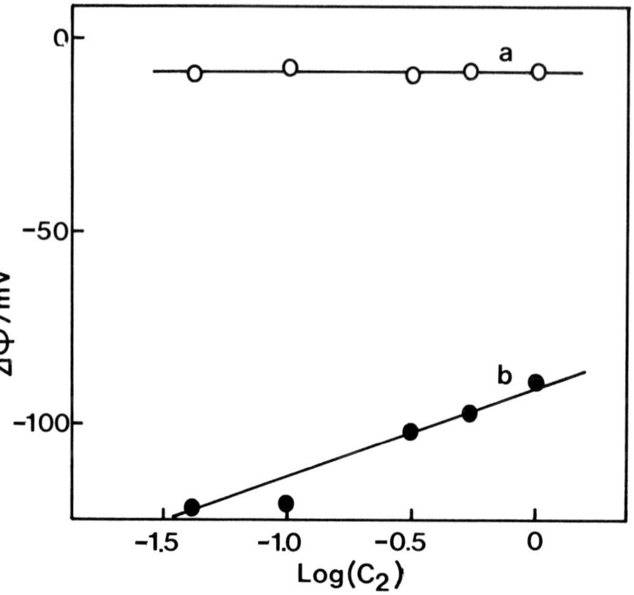

Fig. 14. Effects of NaCl concentration on the membrane potential at constant $C_2/C_1=2$. a;potential under visible light, b;potential under UV light.

where $\gamma=c_2/c_1$, θ and t_- are charge density and transport number of the anion in the membrane phase, respectively, and R, T, and F have their usual meanings. The first and second terms arise from the surface potentials, while the third term is attributed to the diffusion potential. In the case of $c_1=c_2$ in the Eq. (8), the membrane potential should be always zero, $\Delta\Phi=0$, irrespective of θ value. This consideration conflicts with the fact that membrane potentials of ca. 100 mV are induced by UV light irradiation as shown in Fig. 15. The discrepancy can be reasonably solved by introducing θ_1 and θ_2 terms as charge densities under UV light at the membrane surfaces facing c_1 and c_2 side solutions, respectively, into the Eq. (8);

$$\Delta\Phi = \frac{RT}{F}\left\{-\ln\gamma + \ln\frac{-\theta_2+\sqrt{\theta_2^2+4c_2^2}}{-\theta_1+\sqrt{\theta_1^2+4c_1^2}} + (2t_--1)\ln\frac{(1-2t_-)\theta_2+\sqrt{\theta_2^2+4c_2^2}}{(1-2t_-)\theta_1+\sqrt{\theta_1^2+4c_1^2}}\right\} \quad (9)$$

Equation 9 assumes that the charge densities at both surfaces of the membrane are not equal with each other when irradiated with UV light. The uniformity of charge distribution between both surfaces of the membrane is due to that the open form of 12 generates more efficiently at the irradiated surface than at the opposite surface. Namely, the PVC/12 membrane has an asymmetric structure only under UV light with

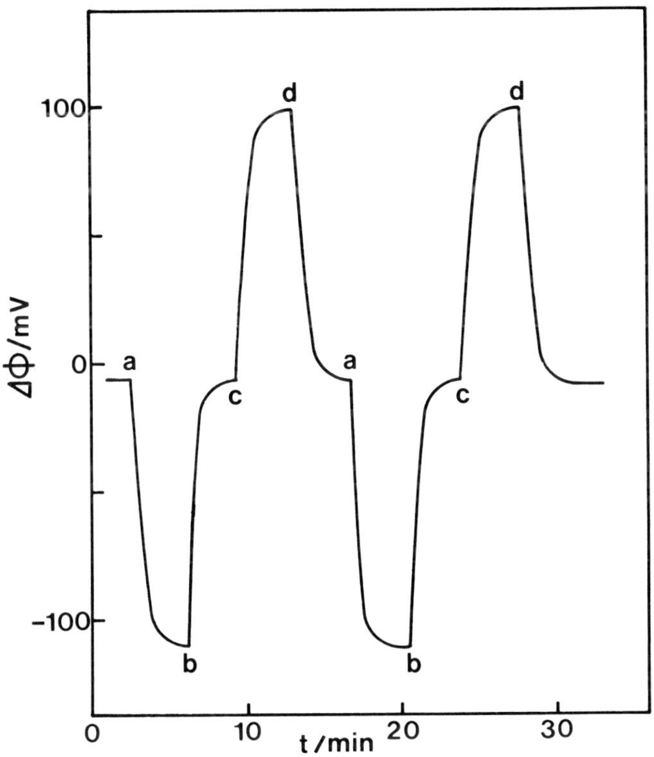

Fig. 15. Photoresponse of the asymmetric membrane potential. UV light was irradiated from C_2 side (a) and C_1 side (c), and visible light was irradiated from C_2 side (b) and C_1 side (d). NaCl concentration in both solutions was 10 mM.

respect to the charge density (i.e., content of open-12) in the membrane. The membrane potentials observed under such asymmetric conditions are called "asymmetric membrane potential".[41,42]

It is evident that the generation of membrane potential can be attributed to the photo-produced open form of spiropyran. However, the exact structure of the open form, which is directly concerned with the potential generation at the membrane/solution interfaces, is not fully obvious. Some authors refer to the significant role of the protonated form of open-spiropyran. In this regard, effects of pH of the solution were examined (Figs. 16 and 17). The magnitude of photoresponse was enhanced in the acidic regions (Fig. 16). Figure 17 shows that photoresponse depends drastically upon the pH value of aqueous solutions in both sides of the membrane, and that the higher response can be obtained when the membrane surface facing acidic solution is exposed to UV light. These data also support the importance of the protonated form of open-12 in inducing the membrane potential (Fig. 18).

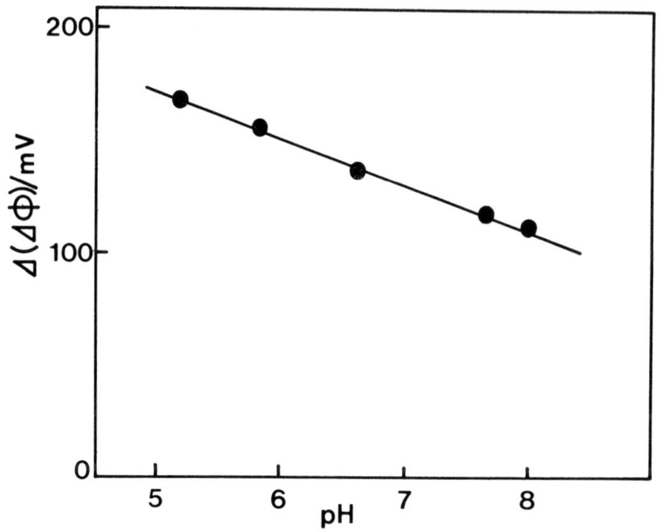

Fig. 16. Effects of pH of the solutions on the magnitude of potential changes. NaCl concentration in both solutions was 500 mM.

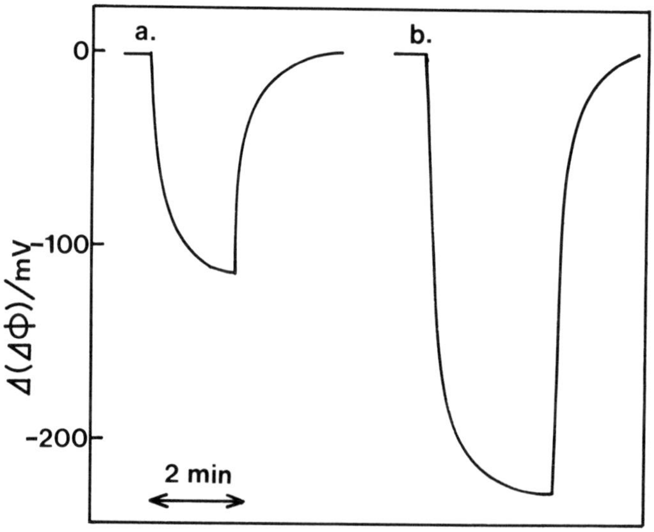

Fig. 17. Effects of pH of the solutions on the photoresponse. a;photoresponse measured at pH 5.3 (C_1) and pH 8.0 (C_2), and b;photoresponse measured at pH 8.0 (C_1) and pH 5.3 (C_2). Membrane surface facing C_2 side solution was exposed to light. NaCl concentration in both solutions was 500 mM.

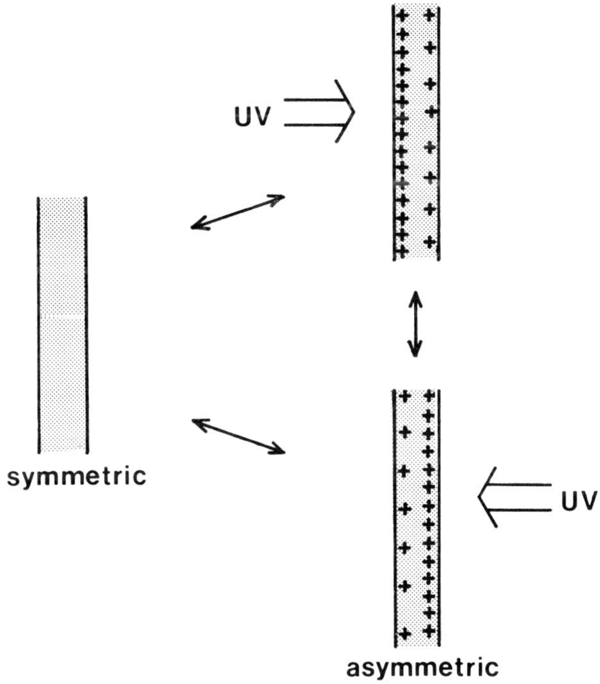

Fig. 18. Schematic representation of the symmetric and asymmetric membranes.

The PVC/12 membrane is applicable to construct photosensitive electrodes and cells which change their electromotive force on irradiation. The devices exhibit photoresponse of the potential changes of 100-200 mV.[43,44] These devices would be useful as photosensors and phototransducers.

It is possible to apply the photosensitive membrane to a sensitive layer of ion-sensors and bio-sensors. By using the PVC/12 membrane doped with nonactin, an ammoniun ion sensor can be fabricated.[45] The configuration of the sensor is as follows; reference electrode internal solution PVC/12 membrane sample solution reference electrode. The photoinduced potentials were measured under UV-irradiation from the sample solution side in the presence of varing amounts of ammonium ion in the samples. Figure 19 shows that an approximately linear relationship exists between the photoinduced potentials, $\Delta(\Delta\Phi)$, and concentration of ammonium ion over the range of 5×10^{-5}-1×10^{-4} M. This sensor seems useful for measuring ammonium ion at μM level.

Another example of PVC/12 membrane-based sensor is an enzyme sensor for determining urea.[46] The urea sensor can be constructed by covering the sensitive layer of the above-mentioned ammonia-sensor with an immobilized urease membrane. The following electrochemical cell was used; reference electrode|internal solution|PVC/12 membrane|

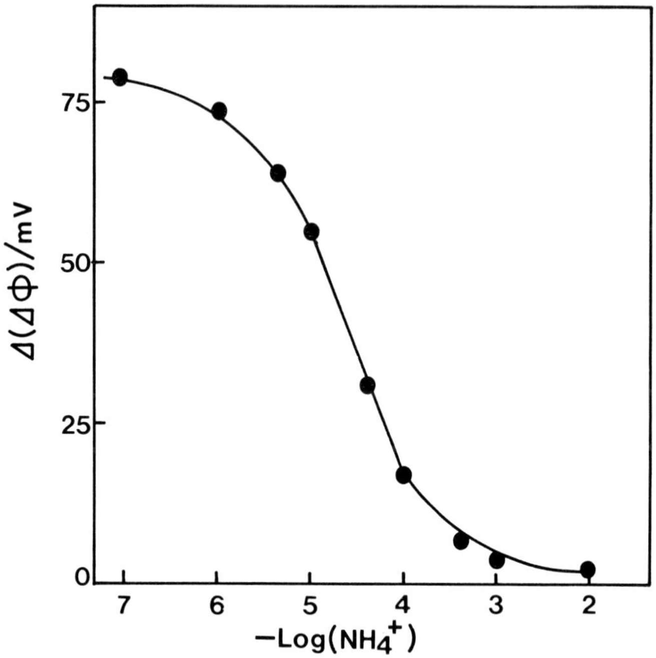

Fig. 19. Effects of ammonium concentration on the magnitude of photo-induced potential across the PVC/12 membrane doped with nonactin.

urease membrane|sample solution|reference electrode. A typical response curve is illustrated in Fig. 20. When urea was added to the sample solution, the membrane potential shifted, with the progress of enzyme reaction, in the negative direction gradually and reached a steady-state value in the dark. This potential shift originates from the selective adsorption of ammonium ion, which is produced by the urease-catalysed decomposition reaction of urea in the enzyme layer, on the membrane surface by forming complex with nonactin. Urea is decomposed to ammonia and carbon dioxide by urease as follows;

$$(NH_2)_2CO + H_2O \xrightarrow{\text{urease}} 2NH_3 + CO_2$$

A further negative shift of potential was induced by UV irradiation. The $\Delta(\Delta\Phi)$ values are plotted as a function of urea concentration in Fig. 21. The calibration graph means that the urea sensor can be used for determining urea at μM level. It is noteworthy that dynamic range of the sensor is significantly improved as compared with those of the conventional types of potentiometric urea sensors developed so far.

These ideas may be extended to the construction of immuno-sensors by immobilizing antigen or antibody molecules on the surface of the

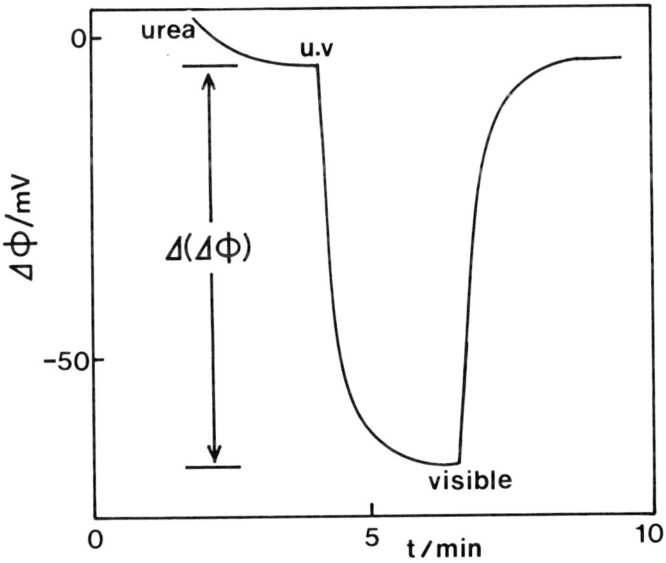

Fig. 20. A typical photoresponse of the PVC/12/nonactin membrane coated with urease layer in the presence of 0.1 mM urea.

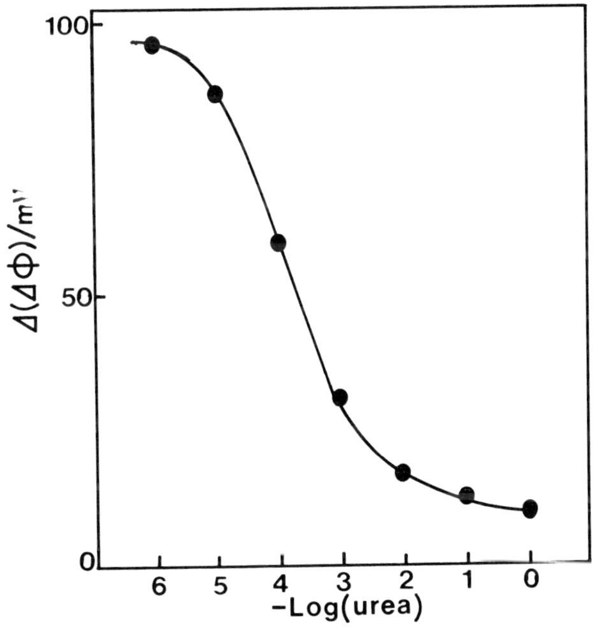

Fig. 21. Effects of the urea concentration on the magnitude of the photoresponse of the urease-coated PVC/12/nonactin membrane.

photosensitive membrane.

3-2 Crown ether membranes

Crown ethers are well-known synthetic ionophore with strong affinity to alkali- and alkaline earth-metal ions. Owing to their selective binding, crown ethers are finding many scientific and practical use including solubilizer of metal salts in organic solvents, catalyst in organic syntheses, chelating agent for solvent extractions, and ion binder of ion selective electrodes. Membranes containing crown ethers are also attracting much attention from the view point of developing selective separation systems. Comparatively few works, however, have appeared on the membrane potential of the crown ether membranes.

We have found that photoinduced membrane potential can be induced by irradiation to a plasticized PVC membrane doped with azobenzene-modified crown ethers 8, 13, and 14 (Fig. 22).[47-54] As discussed already, trans-cis isomerization of azobenzene moiety occurs upon photoirradiation, and therefore their conformations change in an on-off fashion. More important is the fact that their binding equilibria for cations are perturbed by the conformational changes induced by photoirradiation. UV irradiation enhances the binding ability of 8 for larger cations, K^+, Rb^+, and Cs^+, in cis form (bis-crown effects), while the binding abilities of 13 and 14 are reduced. It is reasonable to consider that, if adsorption/desorption equilibria of ions at a membrane/solution interface alter on photoirradiation, the photoinduced surface potential can be realized, due to changes in the structure of electric bouble layer at the membrane/solution interface. Indeed, photoresponse of the membrane potential could be explained successfully in terms of surface potential changes associated with adsorption and/or desorption of ions on irradiation.

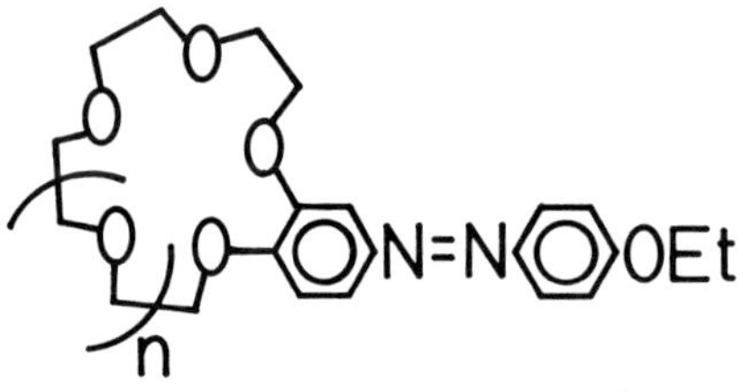

13 (n=1)
14 (n=2)

Fig. 22. Azobenzene-substituted crown ethers 13 and 14.

Figure 23a illustrates the time-response of the potential change of the PVC/8 membrane induced by alternating UV and visible light irradiation. After being obtained a steady-state potential in the dark, a negative shift of the membrane potential was induced by UV irradiation. About 3 min sufficed to attain a steady-state potential

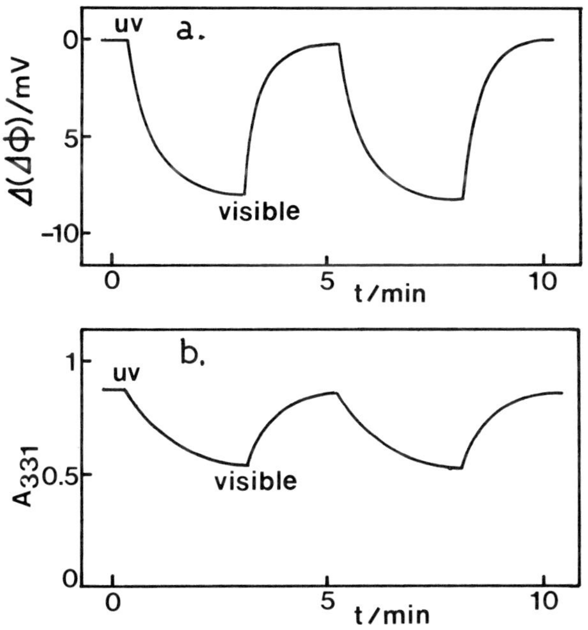

Fig. 23. (a) Photoresponse of the membrane potential across the PVC/8 membrane. C_1=10 mM and C_2=500 mM KCl.
(b) Change in absorbance of the membrane at 331 nm.

Table IV. Photoinduced Potential Changes (mV) across PVC/8 Membrane[a]

Electrolyte	8 Content (g cm^{-2})	C_2 Concentration (mM)				
		1	10	100	500	1000
LiCl	2.0×10^{-5}	-0.5	-0.5	-1.5	-2.0	-2.5
NaCl	2.0×10^{-5}	0	0	0	0	0
KCl	2.0×10^{-5}	-2.0	-6.5	-10	-12	-13
RbCl	2.0×10^{-5}	-2.5	-6.5	-11	-13	-14
CsCl	2.0×10^{-5}	-1.5	-4.5	-6.5	-8.0	-8.5

[a] The concentration of C_1 was constant at 1 mM. The membrane surface facing C_2 compartment was exposed to UV light. The average values of three independent measurements were listed.

under UV light and the initial potential was recovered rapidly by visible light irradiation. The membrane potential change must come from the trans-cis isomerization of 8 in the membrane in view of the fact that the potential change is roughly synchronized with the absorption change at 331 nm of the membrane, i.e., trans-cis isomerization of 8 in the membrane (Fig. 23b).

Table IV describes the effects of the kinds of cation species and the ionic concentration in the solutions on the photoinduced potential change. A marked dependence of the magnitude of photoresponse on the cation species is observed. UV irradiation resulted in large $\Delta(\Delta\Phi)$ values in the presence of KCl and RbCl, while they were only -2.5 mV or less in the presence of LiCl and NaCl. These observations strongly suggest that the potential changes stem from the enhanced binding ability of 8 for the larger cations at the membrane solution interface. A 2:1 sandwich-type complex formation between cis-8 and the cations seems to be the cause of the enhanced ion binding.

In general, the potential difference across a charged membrane (membrane potential, $\Delta\Phi$)is given as an algebraic sum of the surface potential differences at the membrane/solution interfaces and the diffusion potential in the membrane.[41,42] In the case of the PVC membrane, the contribution from the diffusion potential on the $\Delta\Phi$ value is considered to be always small irrespective of the electrolyte concentration on the basis of the following two reasons. First, the PVC/8 membrane has low permeability for alkali-metal chlorides compared with the diffusion of the salts in the aqueous phase, though electrically conductive. Secondly, an ion-pair complex between the crown ether and alkali-metal chloride, which is electrically neutral, is the predominant species in the hydrophobic interior of the membrane (dielectric constant of DBP is 6.42 at 25°C). Thus it is safe to say that one can scarcely find any charged species in the hydrophobic interior of the membrane and, therefore, that the potential gradient in the membrane phase is approximately zero.

In the PVC/8 membrane system, it is reasonable to assume that the distribution equilirium holds with regard to the alkali-metal cation at the membrane/solution interface and that the main factor governing the equilibrium constant is the binding ability of 8 for the cations. Since there is apparently no fixed charge in the PVC membrane, the charge density at the membrane surfaces is inevitably determined by the amount of alkali-metal cation bound to 8 at the membrane surface. As is obvious from the above discussion, trans-cis isomerization of 8, by which the binding ability of 8 for larger cations is enhanced, increases the charge density, that is, surface potential. This mechanism of the potential change reasonably agrees with the results that high $\Delta(\Delta\Phi)$ values are observed in the high salt concentration regions. Another important aspect is that, even for $c_1=c_2$ (1 mM), potential shift are observed. It is well established that the membrane potential should be zero when the charged membrane is placed between two 1:1 electrolyte solutions having the same activity.[41,42] Indeed, in the present case, the membrane potential is almost zero at $c_1=c_2$ under dark and visible light, where 8 assumes wholly the trans form. This discrepancy may be explained in terms of an asymmetric membrane formation under uv light irradiation. In other words, the cis-8:trans-8 ratios of the two surfaces of the membrane are different from each other under UV light. It seems that the trans-cis photoisomerization reaction of 8 proceeds more efficiently on the irradiated surface of the membrane than that on the opposite surface and, therefore, that the cis-8 content of the irradiated surface is higher than that of the non-irradiated surface.

Figure 24 shows the photoinduced potentials observed under the conditions $c_1=c_2=10$, 100, 500, and 1000 mM KCl using the membranes 0.1, 0.2, and 0.4 mm thickness. The magnitude of $\Delta(\Delta\Phi)$ values is independent of the ionic concentration over the range of 10-1000 mM. On the other hand, the $\Delta(\Delta\Phi)$ values depend remarkably on the membrane thickness, the thicker membrane resulting in higher values. This dependence on the membrane thickness is compatible with the interpretation that an asymmetric membrane is formed under UV light. It is reasonable to assume that the cis-8 content of the non-irradiated side of the thicker membrane is suppressed at a lower level than that of the thinner membrane.

The photoresponse of the membrane is closely related to the direction from which the irradiation is performed. Figure 25 shows the potential changes induced by UV light irradiation onto the membrane surface facing the higher or lower concentration compartment.

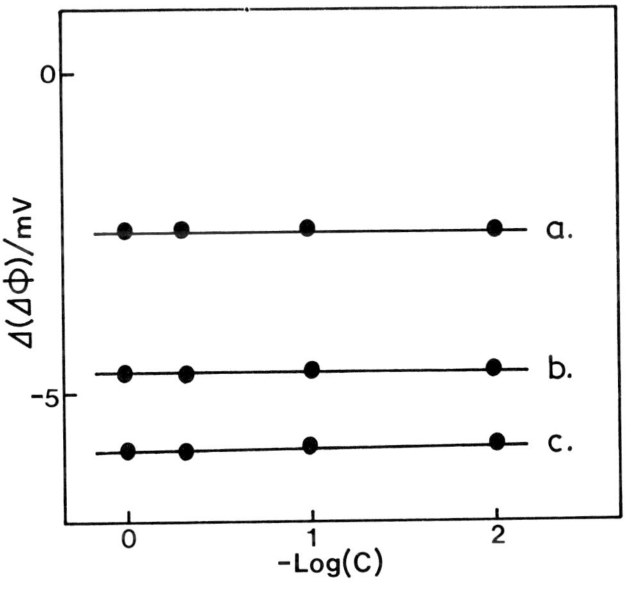

Fig. 24. Effects of membrane thickness and KCl concentration on the asymmetric membrane potential. The KCl concentration in both solutions is equal ($C_1=C_2$). Membrane thickness; 0.1 mm (a), 0.2 mm (b), and 0.4 mm (c).

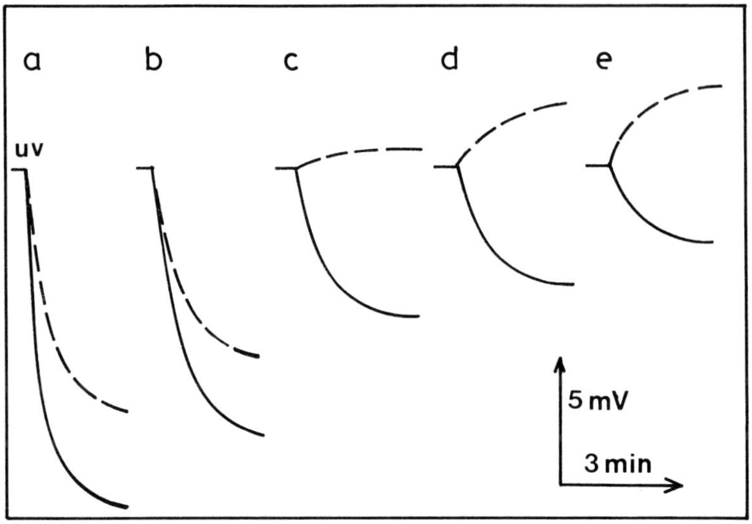

Fig. 25. Photoinduced potential changes across the PVC/8 membrane; irradiated from C_1 side (---) and C_2 side (——). C_1=1 mM (a), 10 mM (b), 100 mM (c), 500 mM (d), 1000 mM (e), and C_2=1000 mM KCl.

When the membrane surface facing the higher concentration compartment was exposed to UV light, negative shifts of the potential were always observed. On the contrary, photoresponse when irradiated from the opposite side of the membrane is somewhat complicated. The negative shift of the potential was observed for 1 and 10 mM whereas the sign of $\Delta(\Delta\Phi)$ values was positive in the higher concentration regions. These photoresponsive behavior of the membrane potential when irradiated from the direction of the lower concentration compartment may be reasonably explicated based on the difference of cis-8:trans-8 ratios between two surfaces of the membrane under UV light irradiation. The cis:trans ratio of the irradiated surface seems to be much higher than that of the other surface. For c_1=1 and 10 mM, the negative shift of the potential resulted from the fact that K^+ uptake by cis-8 at the non-irradiated surface surpasses that at the irradiated surface in spite of the low cis-8 content because of the high c_2 concentration. However, in low c_2/c_1 regions, the difference of the amount of K^+ uptake by cis-8 at two membrane surfaces depends on the cis-8 content rather than the K^+ concentration in two solutions. For c_2/c_1=1 (1 mM), the potential shift read 2.0 mV, the same absolute value as that obtained by irradiating from the side of the higher concentration compartment (-2.0 mV). These results also strongly suggest the asymmetric membrane formation under UV light irradiation.

Figure 26 shows the photoresponse of the membrane potential across the PVC/14 membrane. The membrane potential was near to zero in the dark, which was quite reasonable because of the identical ionic concentration of both solutions. Upon UV light irradiation the membrane potential showed a positive shift and declines, to reach a steady-state value after ca. 2 min. When UV light was switched off, the membrane potential decayed gradually. The membrane potential changes induced by alternating UV and visible light irradiation are also shown in Fig. 26. An on-off cycle of the potential change was observed. The value of the membrane potential under visible light was identical that recorded in the dark. This on-off switching of the membrane potential was fairly reversible and reproducible.

Table V lists the values of photoinduced potential changes across the PVC/13 and PVC/14 membranes. The most remarkable feature is that the potential shifts positively in all cases when irradiated from the c_2 compartment side. We can find the reverse tendency in the PVC/8 membrane (Table IV). The photoinduced uptake of cations from the solution onto the membrane surface is the origin of the negative shifts of the potential across the PVC membrane doped with the bis(15-crown-5) 8. On the other hand, the present positive shifts of the potential can be explained reasonably on the basis of photoinduced cation release from the membrane surface into the solution, in view of the fact that the cation binding abilities of 13 and 14 decrease upon UV light irradiation (Table VI).

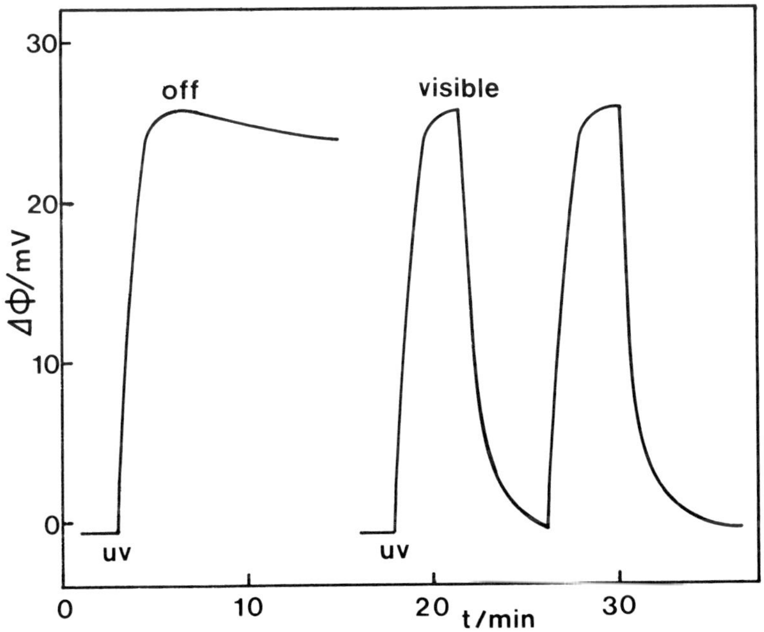

Fig. 26. Photoresponse of the membrane potential across the PVC/14 membrane. C_1=C_2=1 mM KCl.

Table V. Photoinduced Potential Changes (mV) across PVC/13 and PVC/14 Membranes[a]

Membrane	Electrolyte	C_2 Concentration (mM)			
		1	10	100	1000
PVC/13	LiCl	19	14	11	9
PVC/13	NaCl	18	12	10	10
PVC/13	KCl	21	20	21	23
PVC/13	RbCl	15	15	13	15
PVC/14	LiCl	20	18	16	14
PVC/14	NaCl	20	22	23	31
PVC/14	KCl	25	25	30	40
PVC/14	RbCl	20	20	22	21

[a]The concentration of C_1 was constant at 1 mM. The membrane surface facing the C_2 compartment was exposed to UV light. The avarage values of two independent measurements are listed.

Table VI. Liquid-liquid Extraction of Alkali-metal Picrates with 13 and 14[a]

Crown Ether	Extracted Picrate (%)			
	Li^+	Na^+	K^+	Rb^+
trans-13	2.8	25	34	18
irradiated-13[b]	1.4	22	20	10
trans-14	2.0	23	91	78
irradiated-14[c]	1.5	19	88	74

[a]Organic phase (dichloromethane):[crown ether]=0.5 mM; aqueous phase:[picric acid]=0.1 mM and [metal hydroxide]=500 mM.

[b]Cis form, 65%.

[c]Cis form, 72%.

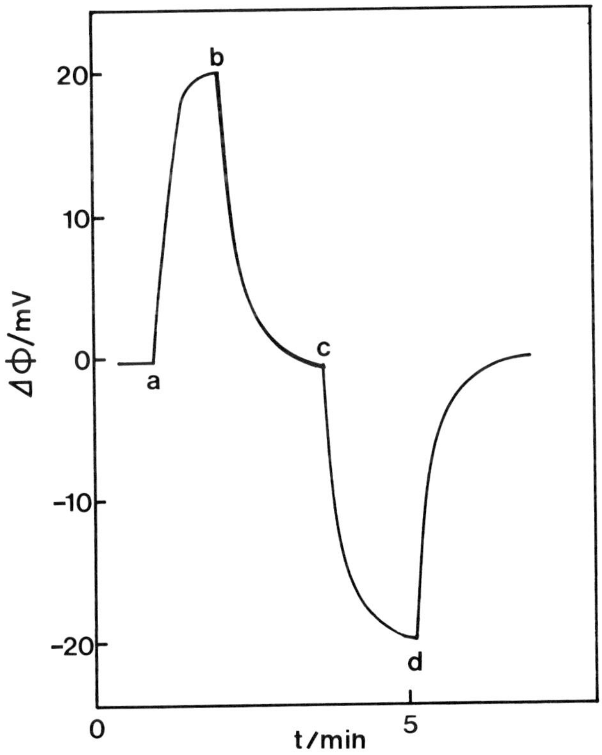

Fig. 27. Photoresponse of the membrane potential across the PVC/13 membrane. UV light was irradiated from C_2 (a) and C_1 (c) sides, and visible light was irradiated from C_2 (b) and C_1 (d) sides. $C_1=C_2$=100 mM RbCl.

Cation selectivity in the photoinduced potential changes is rather small bacause the photoresponse of 13 and 14 does not involve a mechanism with cation-size dependence such as "bis-crown effect", which operates in the case of the PVC/8 membrane. The dependence of the magnitude of $\Delta(\Delta\Phi)$ value on the ionic concentration is somewhat complicated, depending on the cation species.

Figure 27 depicts the photoresponse of the PVC/13 membrane when irradiated with UV and visible light from both sides alternately. About +20 mV of photoinduced membrane potential generated when the membrane surface facing the c_2 solution was exposed to UV light, and the original value of the potential was recovered upon visible light irradiation. UV light irradiation from the c_1 side induced a negative shift of the potentential of the same magnitude as observed upon irradiation from the c_2 side. Figure 28 shows the photoresponse when both sides of the membrane were exposed to UV light alternately. After being reached a steady-state value under UV light irradiation from the

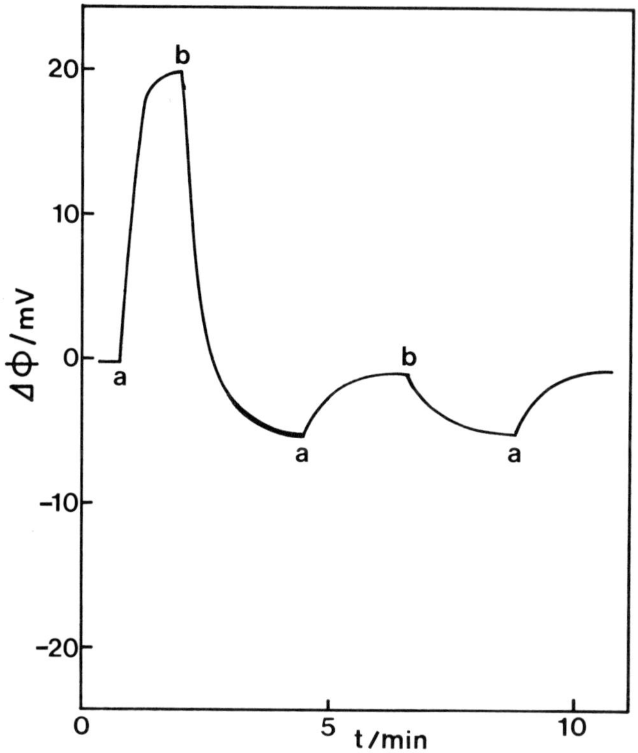

Fig. 28. Photoresponse of the membrane potential across the PVC/13 membrane. UV light was irradiated from C_2 (a) and C_1 (b) sides alternately. C_1=C_2=100 mM RbCl.

c_2 side, the potential shifted negatively, through 0 mV, to reach ca. -6 mV upon UV light irradiation from the c_1 side. This negative shift induced by UV light irradiation has a different mechanism from that observed on irradiation with visible light. The potential shift from +20 mV to 0 mV illustrated in Fig. 27 can be ascribed to the photoisomerization of 13 from cis- to trans-form under visible light irradiation, which means that 13 is wholly in the trans form when the potential exhibits 0 mV. In the case of the negative shift of the potential under UV light irradiation, 13 does not isomerize from cis to trans photochemically. It is considered that UV light irradiation from the c_1 side induces a further isomerization of 13 from trans to cis at the membrane surface facing the c_1 compartment side and, consequently, that the difference in cis-13 content between the two surfaces of the membrane is reduced. When the membrane potential recorded 0 mV, the cis-13 content of the irradiated surface is equal to that of the opposite surface of the membrane, i.e., the membrane is symmetric. Then, with prolonged irradiation, the membrane potential becomes negative in sign, suggesting the formation of an asymmetric membrane with reversed polarity. This is presumably due to the gradual progress of thermal decay of cis-13 to the trans form at the non-irradiated surface as well as to photochemical cis-13 formation at the irradiated surface. The magnitude of the potential shift decreased to 5-6 mV upon repeated irradiation with UV light; this is ascribable to the reduced difference in trans-13:cis-13 ratios between the two surfaces of the membrane.

Photoresponse of the PVC/crown ether membrane can be highly improved by the use of lipophilic crown ether 15 (Fig. 29). Reusability of the PVC/15 membrane is fairly good.[55]

A spiropyran-crown ether compound 16 (Fig. 30) is also employed as a membrane additive for preparing photosensitive PVC membrane. Photoresponse of the PVC/16 membrane depends on the binding-ability changes of the crown ether moiety of 16 as well as on the electric charge produced by photochemical reaction of the spiropyran group.[56]

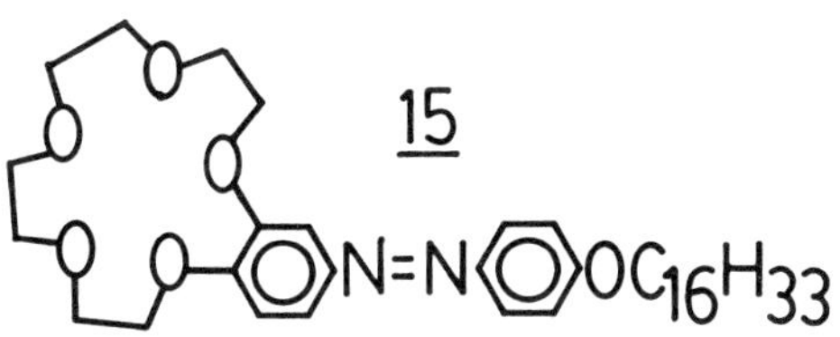

Fig. 29. Azobenzene-substituted crown ether having a lipophilic side chain 15.

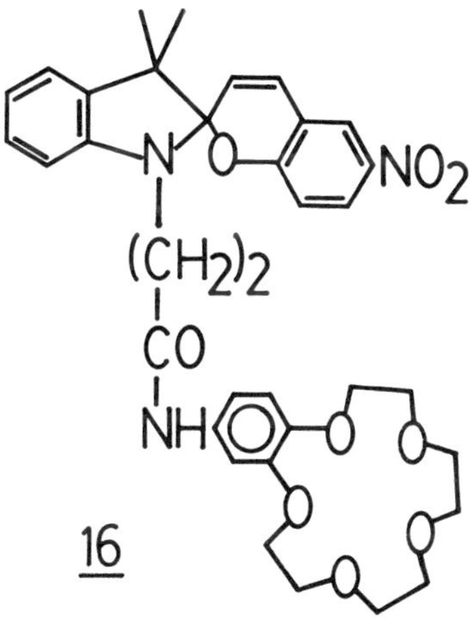

Fig. 30. Crown ether-modified benzospiropyran 16.

4. SUMMARY

Photocontrol of ion permeability of artificial membranes is described, with emphasis on the permeation mediated by synthetic ion carriers sensitive to light. A second section refers to the photoresponse of the membrane potential. Fundamental works of the photoinduced membrane potential are reviewed, together with their applications. It is hoped this article will aid in further developing photosensitive membranes.

ACKNOWLEDGEMENTS:The author's research programs have been supported in part by a Grant-in-Aid from the Ministry of Education, Science and Culture of Japan (Nos. 59390012, 61127003, 62117003 and 63108002). We are grateful to colleagues in this laboratory, especially to associate professor A.Ueno, for their helpful discussion and assistance.

REFERENCES

1)J.Anzai, Y.Suzuki, A.Ueno, T.Osa, Polym.Commun., **25**, 254(1984).
2)J.Anzai, Y.Suzuki, A.Ueno, T.Osa, Israel J.Chem., **26**, 60(1985).
3)J.Anzai, Y.Sakata, Y.Suzuki, A.Ueno, T.Osa, Bull.Chem.Soc.Jpn., **56**, 2541(1983).
4)A.H.Haines, I.Hodgkinsson, C.Smith, J.Chem.Soc.,Perkin Trans. I, **1983**, 311.
5)S.Shinkai, T.Ogawa, T.Nakaji, Y.Kusano, O.Manabe, Tetrahedron Lett., **1979**, 4569.
6)A.Ueno and T.Osa, Yuki Gosei Kagaku, **38**, 207(1980).
7)M.Shiga, M.Takagi, K.Ueno, Chem.Lett., **1980**, 1021.
8)S.Shinkai, T.Nakaji, T.Ogawa, K.Shigematsu, O.Manabe, J.Am.Chem.Soc., **103**, 111(1981).
9)S.Shinkai, T.Minami, Y.Kusano, O.Manabe, J.Am.Chem.Soc., **104**, 1967 (1982).
10)S.Shinkai, M.Ishihara, K.Ueda, O.Manabe, J.Chem.Soc.,Chem.Commun., **1984**, 727.
11)T.Shimizu and M.Yoshikawa, J.Membr.Sci., **13**, 1(1983).
12)J.Anzai, A.Ueno, H.Sasaki, K.Shimokawa, T.Osa, Makromol.Chem.,Rapid Commun., **4**, 731(1983).
13)J.Anzai, H.Sasaki, K.Shimokawa, A.Ueno, T.Osa, Nippon Kagaku Kaishi, **1984**, 338.
14)A.Kumano, O.Niwa, T.Kajiyama, M.Takayanagi, K.Kano, S.Shinkai, Chem.Lett., **1983**, 1327.
15)T.Kinoshita, M.Sato, Y.Yada, A.Takizawa, Y.Tsujita, Maku(Membrane), **11**, 45(1886).
16)K.Kano, Y.Tanaka, T.Ogawa, M.Shimomura, Y.Okahata, T.Kunitake, Chem.Lett., **1980**, 421.
17)K.Kano, Y.Tanaka, T.Ogawa, M.Shimomura, T.Kunitake, Photochem.Photobiol., **34**, 323(1981).
18)J.Sunamoto, K.Iwamoto, Y.Mohri, T.Kominato, J.Am.Chem.Soc., **104**, 5502(1982).
19)Y.Okahata, H.Ihara, M.Shimomura, S.Tamaki, T.Kunitake, Chem.Lett., **1980**, 1169.
20)Y.Okahata, H-j.Lim, S.Hachiya, J.Chem.Soc.,Perkin Trans., II, **1984**, 989.
21)Y.Okahata, Acc.Chem.Res., **19**, 57(1986).
22)A.Darszon, M.Montal, J.Zarco, Biochem.Biophys.Res.Commun., **76**, 820(1977).

23)D.F.O'Brien, N.Zumbulyadis, F.M.Michaels, R.A.Ott, Proc.Natl.Acad.Sci., **74**, 5222(1977).
24)D.F.O'Rien, Photochem.Photobiol., **29**, 679(1979).
25)J.M.Mountz and H.Ti Tien, Photochem.Photobiol., **28**, 395(1978).
26)H-W.Trissl, Photochem.Photobiol., **29**, 579(1979).
27)M.Montal, Biochem.Biophys,Acta, **559**, 231(1979).
28)J.R.Lopez and H.Ti Tien, Biochem.Biophys.Acta, **597**, 433(1980).
29)T.Kinoshita, M.Sato, A.Takizawa, Y.Tsujita, J.Chem.Soc.,Chem.Commun., **1984**, 929.
30)A.Takizawa, M.Sato, T.Kinoshita, Y.Tsujita, Chem.Lett., **1984**, 1963.
31)S.Kato, M.Aizawa, S.Suzuki, J.Membr.Sci., **1**, 289(1976).
32)S.Kato, M.Aizawa, S.Suzuki, J.Membr.Sci., **2**, 39(1977).
33)S.Kato, M.Aizawa, S.Suzuki, Kobunshi Ronbunshu, **34**, 793(1977).
34)I.R.Bellobono, S.Giovanardi, B.Marcandalli, S.Calgari, D.Nosari, Polym.Photochem., **4**, 59(1984).
35)I.R.Bellobono, B.Marcandalli, E.Selli, S.Calgari, Photochem.Sci.Eng., **28**, 162(1984).
36)M.Irie, A.Menju, K.Hayashi, Nippon Kagaku Kaishi, **1984**, 227.
37)J.Anzai, A.Ueno, T.Osa, J.Chem.Soc.,Chem.Commun., **1984**, 688.
38)J.Anzai, H.Sasaki, A.Ueno, T.Osa, Chem.Lett., **1985**, 1443.
39)J.Anzai, A.Ueno, T.Osa, Makromol.Chem.,Rapid Commun., **7**, 133(1986).
40)J.Anzai, Y.Hasebe, A.Ueno, T.Osa, Kobunshi Ronbunshu, **43**, 683(1986).
41)R.Schlogl and F.Helfferich, Z.Electrochem., **65**, 644(1952).
42)S.Ohki, J.Colloid Interface Sci., **37**, 318(1971).
43)J.Anzai, H.Sasaki, A.Ueno, T.Osa, Chem.Pharm.Bull., **33**, 5589.
44)J.Anzai, A.Ueno, T.Osa, Makromol.Chem.,Rapid Commun., **7**, 433(1986).
45)J.Anzai, Y.Hasebe, A.Ueno, T.Osa, Bull.Chem.Soc.Jpn., **60**, 1515 (1987); J.Anzai, Y.Hasebe, A.Ueno, T.Osa, Bull.Chem.Soc.Jpn., **60**, 3169 (1987); J.Anzai, Y.Hasebe, A.Ueno, T.Osa, Chem.Pharm.Bull., **37**, 1307 (1989).
46)J.Anzai, Y.Hasebe, A.Ueno, T.Osa, J.Phys.Org.Chem., 1, 309(1988); C-W.Chen, Y.Sakai, Y.Hasebe, A.Ueno, T.Osa, Chem.Pharm.Bull., **37**, 3316(1989).
47)J.Anzai, H.Sasaki, A.Ueno, T.Osa, J.Chem.Soc.,Chem.Commun., **1983**, 1045.
48)J.Anzai, H.Sasaki, A.Ueno, T.Osa, Chem.Lett., **1984**, 1205.
49)J.Anzai, H.Sasaki, A.Ueno, T.Osa, Maku(Membrane), **9**, 277(1984).
50)J.Anzai, H.Sasaki, A.Ueno, T.Osa, Bull.Chem.Soc.Jpn., **57**, 3331(1984)
51)H.Sasaki, J.Anzai, A.Ueno, T.Osa, Nippon Kagaku Kaishi, **1985**, 1194.
52)J.Anzai, H.Sasaki, A.Ueno, T.Osa, J.Chem.Soc.,Perkin Trans. II, **1985**, 903.
53)J.Anzai, H.Sasaki, A.Ueno, T.Osa, J.Polym.Sci.,Polym.Chem.Ed., **24**, 681(1986).
54)J.Anzai, A.Ueno, T.Osa, J.Chem.Soc.,Perkin Trans. II, **1986**, in press.
55)J.Anzai, Y.Hasebe, A.Ueno, T.Osa, J.Polym.Sci., Part A: Polym. Chem., **26**, 1519(1988).
56)H.Sasaki, A.Ueno, J.Anzai, T.Osa, Bull.Chem.Soc.Jpn., **59**, 1953(1986)

ELECTROCHEMICAL AND PHOTOELECTRICAL PROPERTIES OF BILAYER LIPID MEMBRANES (BLM)

H. Ti Tien
Membrane Biophysics Laboratory
Department of Physiology
Michigan State University
East Lansing, MI 48824 (USA)

ABSTRACT. The use of artificial bilayer lipid membranes (BLM) as an experimental model of biomembranes has provided much insight into the structure and function relationship of natural membranes. Particularly relevant has been the advance in the theoretical and experimental investigations of light energy transduction and that of transport mechanisms via membrane channel reconstitution. This paper will focus on two aspects of BLM studies in relation to biomembranes: namely, membrane electrochemistry and membrane photochemistry. Some attention will be paid to the basic principles as applied to membrane research. Recent experiments with BLMs will be summarized and their potential applications in areas of biosensor development, biomolecular electronic devices, (BED) and solar energy conversion discussed.

Portions of this paper on membrane photochemistry of BLM were presented as an invited lecture at the VIII School on Biophysics of Membrane Transport, Mierki, Olsztynek, Poland, 4-13 May, 1986.

T. Osa and J. L. Atwood (eds.), Inclusion Aspects of Membrane Chemistry, 191–274.

I. INTRODUCTION

The reconstituted planar bilayer lipid membrane (BLM) separating two aqueous solutions has been proven to be an excellent model for biomembranes, which permit charge separation and translocation to be measured directly by modern electrochemical techniques (Davison, 1985; Blank, 1986). Charge separation and translocation are the main concern of basic electrochemistry. Translating these phenomena into membranes, the principal focus of research deals with the mechanisms of reactions at, as well as across, solution/membrane interfaces. The electrical double layer at the interfaces plays a crucial role. In this paper we shall first of all describe the basic principles of membrane electrochemistry and photochemistry, with emphasis on redox reactions and light-induced charge separation and electron transfer. In Section III, a detailed description of relevant methodology for formation and investigation of planar BLM will be presented. Special attention will be paid to experimental materials and procedures common to many studies with BLMs, including the BLM-forming solutions, chamber designs, electrodes, light sources, and electrical measurements. Section IV provides a summary of recent literature on the study of BLM, whereas in Section V their potential practical applications will be given. In these connections, the large number of papers dealing with methods for incorporating membrane proteins and for studying single channel conductance will only be mentioned in passing, owing to space limitation. The reader concerned with membrane channel reconstitution experiments and other topics not covered here, such as thickness determination, interfacial chemistry, permeability to water and non-electrolytes, may consult several comprehensive reviews of the subject matter (Blumenthal and Klausner, 1982; Akabas et al., 1984; Benga, 1985; Blank, 1986; Davison, 1985; Benga, 1986; Harris and Etemadi, 1987).

II. BASIC PRINCIPLES

A. Elements of Membrane Electrochemistry

The principles of membrane electrochemistry may be discussed in terms of two types of measurements. The first is called membrane potentiometry in which the potential is measured under conditions when no current is detected across the membrane. The second is called membrane amperometry in which a current is measured when a potential difference is applied across the membrane. One can also define a third type of measurements, known to electrochemists as voltammetry, in which a scanning voltage is impressed across the membrane and the resulting current is recorded. Of voltammetry, which is a combination of the first two types, the so-called linear sweep voltammetry (LSV) and cyclic voltammetry (CV) are practiced. In the former, a linear voltage scan is used, whereas in the latter the potential is scanned back and forth between two limits. Applications of these methods to membrane research, in particular to the BLM system, will be discussed in Section IV.

1. Potential Differences Between Two Phases and Electrical Double Layer

A separation of charge occurs at interfaces; i.e., almost all interfaces are electrified. These charges may be free charges such as ions, electrons; they may be dipolar molecules such as phospholipids. Separation of charge is usually as a result of: (1) preferential adsorption of ions, (2) adsorption and orientation of dipolar molecules, (3) transfer of charge from one phase to another, and (4) deformation of polarizable atoms or molecules (anisotropy).

a. Anisotropy of an interface. The very existence of an interface between two immiscible liquids necessarily implies a basic anisotropy in the forces operating on the species in the interfacial region, a potential difference (p.d.) will develop and an electrical double layer is formed. There are two aspects of the electrical double layer: (1) the electrical aspect, which is concerned with the magnitude of the excess charge densities across the interfacial region, and (2) the structural aspect, the arrangement of constituents (ions, electrons, dipoles, neutral molecules). These two are obviously intimately related.

b. Electrical double layer (EDL) defined. An EDL is an interfacial region between two homogeneous phases, in which (1) the charge constituents have been separated so that (2) each side of the interface is electrified, (3) the charges on the two sides are equal in magnitude but opposite in sign, (4) the interfacial region as a whole is electrically neutral, and (5) it is impossible to measure the absolute value of the EDL potential difference (p.d.) but changes in p.d. may be determined. The processes ocurring at membranes are closely related to the structure of the region of the solution that is near to the membrane, which is the electrical double layer. Theories about the electrical double layer have been developed over the past 100 years (Blank, 1986; Milazzo, 1983; Koryta, 1982; Tien, 1974). It can be said that an appreciation of the EDL is a 'must' in order to understand the properties of biological membranes both at, as well as across, their solution/membrane interfaces. For general and mathematical descriptions of EDL, a number of excellent papers are available (see Blank, 1986).

c. Electrical double layer theory and BLM. Aspects of the electric double-layer theory have been applied to the BLM system (Blank, 1986; Milazzo, 1983 Tien, 1974). The central idea is based on the Boltzmann distribution of the ions in a varying potential field. One important application of the electric double-layer theory is the estimation of ion concentrations, such as H^+, at the BLM/solution interface. The interfacial and bulk pH can be related by the following equation

$$[H^+]_i = [H^+]_b \, e^{-F\psi/RT} \tag{1}$$

By definition, pH = -log $[H^+]$, we have

$$[pH]_i = [pH]_b + \frac{F\psi^o}{2.3RT} \tag{2}$$

where ψ_o is the interfacial potential at x = 0 (or x = t_m, see Fig. 1). For a negatively charged BLM, it can be readily seen that the pH at the interface is lower than the pH in the bulk solution. In general, ψ_o may be estimated by measuring the zeta (ζ) potential in the plane of shear. For example, Papahadjopoulos and Watkins (1967) have found a reasonable agreement between the values calculated from Eq. (2) and the electrophoretic measurements on spherical BLMs (liposomes). Vaz et al. (1976) using a fluorescent pH indicator (dimyristoyldansylcephalin) estimated the pH adjacent to the BLM interface. They have found that the hydrogen ion concentration is electrostatically enhanced, the enhancement being dependent on the interface charge density and bulk phase salt concentration.

Another approach in estimating ψ_o, is by the use of the Gouy equation which may be simplified

$$\psi_G = 50.4 \sinh^{-1} \frac{134\sigma}{\sqrt{C}} \tag{3}$$

C is the univalent ion concentration and σ the interface charge density (charge/Å^2). It can be readily shown that, at a fixed salt concentration in the bulk phase for a negatively charged BLM, the interfacial pH should be lower by about 0.6 to 0.7 pH units. By changing the charge density on the BLM and by changing the pH or composition of bathing solutions, it should be possible to investigate molecular interactions and ion binding in the BLM. In this connection, mention should be made that photochemical reactions may be studied by monitoring a change in $\Delta\psi$ ($=\psi_i - \psi_o$). This topic will be further considered in Section II B.

2. The Nernst-Planck Equation and Transmembrane Potentials

a. The Nernst-Planck equation. In a system consisting of a BLM separating two aqueous solutions, the law of electroneutrality must be obeyed. That is, the sum of positively charged species is equal to the sum of negatively charged species for the system as a whole. However, as pointed out by Guggenheim (1950), charge separation on a microscopic scale may occur, for example, at interfaces. To develop a potential difference (p.d.) of 50 mV across a BLM of 50 Å thick, the amount of charged species involved is far too small (< 10^{-10} M) to be detected chemically, or by any other means, except electrically. Therefore, the p.d. across the BLM (or biomembrane for that matter) cannot be simply related to the composition of bulk solution concentrations. In order to treat the BLM system thermodynamically, one can, nonetheless, use the concept of electrochemical potential ($\bar{\mu}$), defined in the usual manner

$$\bar{\mu} = \mu_j + z_jF\psi = \mu_j^o + RT\ln a + z_jF\psi \tag{4}$$
$$\bar{\mu} = \mu_j^o + RT\ln [C_j] + z_jF\psi$$

where j is the jth chemical species, μ is the chemical potential, z is the valence, μ^{o} the chemical potential at constant temperature and pressure, a is activity, which is expressed as the product of coefficient (f) times concentration C. In this case f is assumed to be unity. ψ is the electrical potential, F, R, and T have the usual significance.

The driving force for moving the jth charged species across the BLM in the x direction can be expressed as an electrochemical potential gradient

$$\frac{d\bar{\mu}_j}{dx} = \frac{d\mu_j}{dx} + z_j F \frac{d\psi}{dx}$$

$$\simeq RT \frac{d\ln[C_j]}{dx} + z_j F \frac{d\psi}{dx} \qquad (5)$$

The $\bar{\mu}$ is a fundamental measure of irreversibility as can be readily deduced from the first and second laws of thermodynamics;

$$dS \geq 0 \qquad (6)$$

where S = entropy, and dS is the change in S for the reversible and irreversible process respectively, equal to and greater than zero. With an ultrathin membrane interposed between two solutions designated (in or i) and (out or o), it can be shown that, for electrical work (W_{ele}) only,

$$\begin{aligned} TdS &= -\sum\mu\, dn - dW_{ele} \\ &= -(\mu_i - \mu_o)dn - \{ZF(\psi_i - \psi_o)\} \\ &= -(\bar{\mu}_i - \bar{\mu}_o) \end{aligned} \qquad (7)$$

From Eq. (6) and Fig. 1, it may be seen that transport of ions and other species is caused by a force equivalent to $-d\bar{\mu}/dx$.

Since central to membrane transport phenomena is the Nernst-Planck equation, it is instructive to derive it here. The starting point is to consider the molecular flux (J) across a membrane, which is equal to velocity (v) times concentration (C), i.e. J = vC. The flux is also proportional to the mobility (U), concentration (C) and the driving force of the species involved, one obtains

$$J = kUC\left(-\frac{d\bar{\mu}}{dx}\right) \qquad (8)$$

where k is a constant to be evaluated. By substituting Eq. (4) into Eq. (8), we have

$$J = -kUC\left(\frac{RTd\ln C}{dx} + ZF\frac{d\psi}{dx}\right) \qquad (9)$$

The constant k is found by equating

$$kvCF = - \frac{kvC}{-Z(\frac{d\psi}{dx})} \frac{ZFd\psi}{dx} \tag{10}$$

Thus, k = 1/F. Therefore

$$J = - \frac{UC}{F} \left(\frac{RTd\ln C}{dx} + \frac{zFd\psi}{dx}\right) \tag{11}$$

Eq. (11) is the famous Nernst-Planck flux equation, upon which all basic equations relevant to membrane biophysics and electrophysiology depends, as we shall discuss presently.

(1) The Nernst Equation. At equilibrium, there is no net transfer of jth charged species, i.e., $d\bar{\mu}j/dx=0$. We have, from Eq. (5),

$$-z_j F \frac{d\psi}{dx} = RT \frac{d}{dx} \ln [C_j] \tag{12}$$

Integration of Eq. (12), across the BLM from x = 0 to x = t_m, gives

$$\psi_i - \psi_o = (p.d.) = E_m = \frac{RT}{zF} \ln \frac{[C]_o}{[C]_i} \tag{13}$$

where subscripts i and o refer to the inner (right-hand) and outer (left-hand) side of the membrane. The potential difference (p.d.), or E_m, is called the membrane potential. Eq. (13) is known as the Nernst equation. Experimentally, how does one go about measuring E_m? One may place a pair of identical electrodes, such as (i) Ag/AgCl, (ii) K^+-sensitive glass electrodes, or (iii) saturated calomel electrodes (Hg/Hg_2Cl_2) with salt bridges, across the membrane with fixed negative charges. When the concentration ratio of solution (i) and solution (0) is 10:1 (say, 0.1 M and 0.01 M KCl) an E_m of about 116 mV would be measured for case (i), about 0 mV for case (ii), and 58 mV for case (iii). Suffice it to state here that choice of the kind of electrodes used to measure E_m is very important. For BLM studies, calomel electrodes with saturated KCl bridges have been extensively employed (Tien, 1974).

(2) The Goldman Equation. Integration of Eq. (11) has been difficult owing to uncertainty of the events in the membrane phase. A variety of schemes (assumptions) have been proposed (see Lakshminarayanaiah, 1969). The simplest is that due to D. Goldman, proposed in 1943 (Goldman, 1967). He assumed that the membrane potential gradient is linear, i.e.,

$$\frac{d\psi}{dx} = \frac{E_m}{t_m} \tag{14}$$

This so-called constant field theory implies that net charge within the membrane is zero, and that the electric field is governed solely by the charges accumulated at the interfaces between the membrane and its bathing solutions. Upon substituting Eq. (14) into Eq. (13), Goldman obtained

$$J = -\left[\frac{URT}{F}\frac{dC}{dx} + \frac{zFC\,E_m}{RT\,t_m}\right] \tag{15}$$

Under the constant field treatment, J and U are effectively constant. The final expression of the Goldman equation is

$$J = (Py)\left[\frac{C_i - C_o^{-y}}{e^{-y}-1}\right] \tag{16}$$

where $y = \frac{zFE_m}{RT}$ and $P = \frac{D}{t_m}$, the permeability coefficient.

(3) The Hodgkin-Huxley-Katz Equation. Upon introducing $P = D/t_m$, which is the permeability coefficient, Eq. (13) now becomes, for a system with Na^+, K^+ and Cl^- at electrochemical equilibrium,

$$E_m = \frac{RT}{F}\ln\frac{P_k[K]_1 + P_{Na}[Na]_1 + P_{Cl}[Cl]_2}{P_K[K]_2 + P_{Na}[Na]_2 + P_{Cl}[Cl]_1}. \tag{17}$$

Eq. (17) was applied by Hodgkin, Huxley, and Katz with great success to the squid axon membrane (see Scott, 1977; Hughes, 1980). Eq. (17) has also been applied to the BLM system to estimate the permeability ratio (relative permeabilities) between cation and anion (Tien, 1974). Under the condition of zero current, the cation/anion permeability ratio may be evaluated from measurements of E_m. If the BLM, bearing negative fixed charge, is not permeable to anion, Eq.(17) then becomes

$$E_m = \frac{RT}{F}\ln\frac{p[K]_1 + [Na]_1}{p[K]_2 + [Na]_2} \tag{18}$$

where $p = P_K/P_{Na}$. Eq. (18) has been applied to describe a valinomycin-treated BLM separating KCl on one side and NaCl on the other side at the same concentration (e.g., 0.01 M). The membrane potential difference, E_m, thus generated is known as the bi-ionic potential, which can be used to determine ion specificity (Tien, 1974). Since the salt concentrations are equal, and only one kind of ions is present on one side, we can rewrite Eq. (18) as follows:

$$E_m = \frac{RT}{F} \ln p = \frac{RT}{F} \ln \frac{U_{K^+}}{U_{Na^+}} \tag{19}$$

where, as before, U_K and U_{Na} are the mobilities of K^+ and Na^+.

(4) The Ussing Equation. The Goldman equation (Eq. 16) can be separated into two parts as has been done by Ussing in his frog skin membrane experiment (see Scott, 1977). Let

$$J = J_{in} - J_{out} \tag{20}$$

where subscript in and out denote inward and outward fluxes, respectively. Thus,

$$J_{in} = -PC_o \left[\frac{ye^{-y}}{e^{-y}-1} \right] = PC_o \left[\frac{-y}{1-e^{y}} \right] \tag{21}$$

and

$$J_{out} = PC_i \left[\frac{y}{1-e^{-y}} \right] \tag{22}$$

As is typical, y is negative, for E_m is negative. Therefore, J_{in} and $-J_{out}$ are both positive. Upon dividing Eq. (21) by Eq. (22), we have

$$\frac{J_{in}}{J_{out}} = \frac{C_o e^{-y}}{C_i} \tag{23}$$

Eq. (23) is known as the Ussing flux ratio equation which can be shown to be equal to $(\bar{\mu}_o - \bar{\mu}_{in})$ by taking natural log of Eq. (23) (see also Eq. 7).

(5) Other Equations. As a concluding comment on the Nernst-Planck equation, one can also obtain Fick's law of diffusion, $[J=-D(dC/dx)]$ by letting z=0 in Eq. (11), and the Einstein equation, $D = URT/F$.

3. Transmembrane Potentials and BLM

a. Transmembrane potentials. By means of thermodynamics, transmembrane potentials of BLM can be related to the distribution coefficients and electrochemical potentials of species present in aqueous solution and oily membrane phases. Fig. 1 shows an aqueous solution/BLM/aqueous-solution system, in which a transmembrane potential E_m is measured. The Galvani potential, ϕ, is plotted on the vertical axis. The BLM thickness is denoted by t_m, starting from x = 0 at the left (outer) or solution (1)/BLM interface to x = t at the right (or inner) BLM/solution (2) interface on the horizontal axis. By convention, the aqueous solution on one side of the BLM is taken to be zero potential. Let D stand for ionic distribution coefficient, C and A for cation and anion, respectively. Superscripts 1 and 2 denotes aqueous phase 1 (or 2) and m the membrane phase. At equilibrium, for the cation

$$ {}^{1}\bar{\mu}_{c} = {}^{m}\bar{\mu}_{c} = {}^{2}\bar{\mu}_{c} \tag{24} $$

and

$$ {}^{1}\bar{\mu}_{A} = {}^{m}\bar{\mu}_{A} = {}^{2}\bar{\mu}_{A} \tag{25} $$

for the anion, where $\bar{\mu}$ is the electrochemical potential. Since each phase must be electroneutral, we have

$$ [C]_1 = [A]_1; \quad [C]_m = [A]_m; \quad [C]_2 = [A]_2 \tag{26} $$

where species concentrations are enclosed in brackets, subscripts denote various phases as before. For the time being, let us focus our attention on the left solution/BLM interface (Fig. 1). The difference in the aqueous solution and membrane phases can be expressed as follows:

$$ D_c = \exp \frac{{}^{1}\mu^{o}_{c} - {}^{m}\mu^{o}_{c}}{RT} $$

and

$$ D_A = \exp \frac{{}^{1}\mu^{o}_{A} - {}^{m}\mu^{o}_{A}}{RT} \tag{27} $$

where μ^{o} stands for standard chemical potential. In terms of electrochemical potentials,

$$ {}^{1}\bar{\mu}_{c} = {}^{1}\mu^{o}_{c} + RT \ln [C]_1 \tag{28} $$

$$ {}^{1}\bar{\mu} = {}^{1}\mu^{o}_{A} + RT \ln [A]_1 \tag{29} $$

$$ {}^{m}\bar{\mu} = {}^{m}\mu^{o}_{c} + RT \ln [C]_m + F\,\phi_1 \tag{30} $$

$$ {}^{m}\bar{\mu}_{A} = {}^{m}\mu^{o}_{A} + RT \ln [A]_m - F\,\phi_1 \tag{31} $$

It should be noted that, in writing these equations (Eqs. 24 - 31), we have chosen monovalent salts (z = 1) and used their concentrations, rather activities, for simplicity (i.e., $a \simeq [\,]$, where activity coefficient is assumed to be unity). Subtracting Eq. (31) from Eq. (30), and remembering Eqs. (24 - 26), the following is obtained

$$ 2F\phi_1 = ({}^{1}\mu^{o}_{c} - {}^{m}\mu^{o}_{c}) - ({}^{1}\mu^{o}_{A} - {}^{m}\mu^{o}_{A}) \tag{32} $$

By substitution of Eq. (27) into Eq. (32), we have, upon rearrangement

$$ \phi_1 = \frac{RT}{2F} \ln \frac{D_c}{D_A} \tag{33} $$

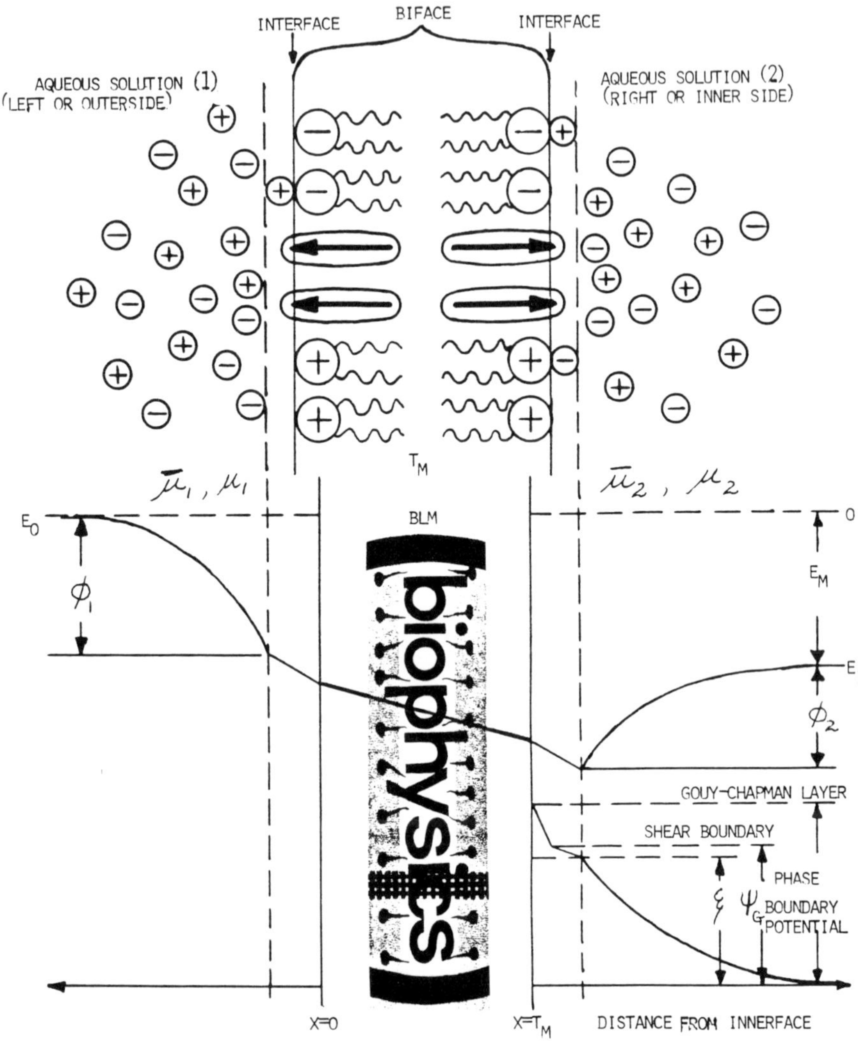

Fig. 1. Transmembrane potential profiles across a bilayer lipid membrane (BLM) with charged and dipolar lipids. Also shown are various interfacial potentials and electrical double layers, for detail see text (adapted from Tien, 1974. p. 139).

where $\emptyset$ is the inner potential at x = 0 and is not measurable. However, as we shall see shortly, the differences in $\emptyset$ may be detected.

Since single ion distribution coefficients can be related to the classical partition coefficient, K, defined by

$$K = \frac{[\mathrm{Salt}]_m}{[\mathrm{Salt}]_1} \tag{34}$$

That is, the ratio of salt concentration in the membrane phase $[\mathrm{salt}]_m$ to that in the aqueous solution, $[\mathrm{salt}]_1$, Eq. (33) may be rewritten as

$$\emptyset_1 = \frac{RT}{F} \ln \frac{[K]}{[D_A]_1} \tag{35}$$

since $K = [D_c]_1^{1/2} \times [D_A]_1^{1/2}$

Referring to Fig. 1, $\emptyset_1$ and $\emptyset_2$ at each interface are not measurable, owing to a lack of suitable contact with the membrane, we nonetheless can measure $\emptyset = (\emptyset_2 - \emptyset_1)$ by placing a reference electrode (e.g., calomel electrode) in each of the bulk solutions bathing the membrane. For example, KCl is introduced in the outer aqueous solution (1) and NaCl in the inner aqueous solution (2). From Eq. (35), we obtain

$$\emptyset = \emptyset_2 - \emptyset_1 = \frac{RT}{F} \ln \frac{[K_{KCl}]}{[Cl^-]} - \frac{RT}{F} \ln \frac{[K_{NaCl}]}{[Cl^-]}$$

or

$$= \frac{RT}{F} \ln \frac{[K_{KCl}]}{[K_{NaCl}]} = \frac{RT}{2F} \ln \frac{D_{Na}}{D_K} \tag{36}$$

where K_{KCl} and K_{NaCl} are partition coefficients defined by Eq. (34). Of interest to note is that, Eq. (36), the observed E_m ($\Delta\emptyset$) is independent of the common anion, Cl^-. Experimentally, Eq. (36) may be readily tested by forming BLM containing valinomycin separating NaCl and KCl solutions.

The Nernst equation (Eq. 13), albeit exceedingly useful in accounting for the transmembrane potential, E_m, does not reveal very much about the underlying mechanism responsible for the observed potential. In the following paragraphs, we will discuss the origin of membrane potentials that have been detected in the BLM systems.

b. BLM Potentials. The unique aspect of BLM, similar to biological membranes, is the ultrathin lipid bilayer separating two aqueous solutions (see Fig. 1). Membranes of these types have two interdependent interfaces. The creation of these two interfaces (or a biface) is the result of two immiscible phases with very different dielectric constants (ϵ_{H_2O} = 80; ϵ_m = 2-5). At each interface, as well as across the ultrathin bilayer (~50 Å), powerful electric fields may exist, which facilitate charge separation and transport, as will be discussed later. In this segment, we will consider the origins of BLM potentials.

An observed transmembrane potential in a BLM system may be due to one or a combination of the following potentials:

(1) Diffusion (concentration) potential, ϕ_d
(2) Adsorption (surface or interfacial) potential, V
(3) Distribution (outer) potential, ψ
(4) Galvanic (inner) potential, Φ (= $V + \psi$)
(5) Gouy (Donnan) potential, ψ_G or ψ_{DON}.
(6) Electrokinetic potential, ψ_k
(7) Thermoelectric potentials, E_t
(8) Redox (electrostenolytic) potential, E_{RX}
(9) Photoelectric potential, $E_{h\nu}$

In the classification given above, redox potential, thermoelectric, and photoelectric potential in BLM are discussed in Section IV. Here, we shall describe the first five potentials that are relevant to the BLM systems.

The nature and origin of adsorption, distribution, and Galvani potentials are best described in the language of interfacial chemistry, and have been treated in detail (Davies and Rideal, 1961). In order to apply these concepts to the BLM systems, it is informative, first of all, to define the terms and to summarize some of the general conclusions (Tien, 1974).

Adsorption potentials, also known as surface or interfacial potentials (designated by symbol V), are opposite to that observed with the diffusion potential. For this interfacial potential to appear, a difference in salt concentrations across the BLM is not required; it depends upon the sorbed species and/or the dipoles. With cationic interface-active compounds, such as hexadecyltrimethylammonium bromide (HDTAB) and phosphatidylethanolamine (PE) at low pH, a negative potential results and vice versa with anionic interface-active compounds, such as dodecyl acid phosphate (DAP) and phosphatidylserine (PS). The magnitude of V decreases with increasing salt concentration. Two other caveats should be noted about this potential: (i) it may have only transient existence, however, the time for its decay can take a long time, and (ii) the law of electroneutrality at the interfaces is not obeyed.

Distribution potentials, also called outer potentials, are symbolized by ψ. These potentials arise as a result of differential distribution of the oppositely charge species across the interface, the magnitude of which is influenced by the diffusion of ions and their concentrations, and increases with increasing salt concentration, but is independent of the nature of the common ion.

The sum of interfacial potential and distribution potential is termed the inner or Galvani potential, designated by Φ (= $V + \psi$). It will be discussed below that the observed transmembrane potential difference, E_m, is due either to ΔV, $\Delta\psi$, or to ΔV plus $\Delta\psi$, as defined above.

It should be remembered that there are two components of ϕ (= V + ψ). Eqs. (33,35,36) are concerned with only the distribution potentials. If compounds, such as phospholipids and interface-active agents, are preferentially adsorbed at the interface, the so-called adsorption potential (V) may also develop. As shown in Fig. 1, both adsorbed fixed charge species and dipoles may contribute to the observed potential. The nature and origin of adsorption (or interfacial) potential can be discussed in terms of the classical electric double-layer theory of Gouy-Chapman-Stern-Graham (Davies and Rideal, 1961).

c. Potential difference across BLM. An unmodified oxidized cholesterol or phosphatidylcholine BLM presents a very high energy barrier for the transport of common salts, such as NaCl and KCl. This high energy barrier arises mainly from the difference of the electrostatic energies of the ions in the two phases: aqueous solution and oily membrane (Finkelstein and Cass, 1968). The energy necessary to transport an ion from aqueous solution to the bilayer lipid phase can be estimated using the Born equation

$$E\ (a) = - \frac{z^2 q^2}{8\pi\epsilon_o a} \left(\frac{1}{\epsilon_m} - \frac{1}{\epsilon_w} \right) \qquad (37)$$

where a is the ion radius, z the valence, q the charge, ϵ_o = air permitivity, ϵ_m and ϵ_w are the dielectric constants of the membrane and the aqueous solution, respectively. For z = 1, q = 1 electronic charge, ϵ_o = 8.85 x 10^{-6} μF/m, a = 2 Å, ϵ_m = 2.0 to 3.5, and ϵ_w = 80, E(a) is calculated in the order of 1 - 1.5 eV (about 40 kT), which is much larger compared with kT at room temperatures. One can conclude, with confidence, as supported by experimental conductivity data, that an unmodified BLM contains few dissolved common ions (eg. Na^+ or K^+). This being the case, then how does one explain the observed E_m, when such BLM are perturbed by interface-active compounds, such as $HDTA^+$-Br^- and SDS (Na^+DS^-)? Here again, classical interfacial chemistry provides a ready answer.

According to Gibbs' adsorption equation, any substance that has the ability to lower the interfacial free energy tends to be spontaneously adsorbed at the interface. Thus, for example, when $HDTA^+Br^-$ is introduced to the outer side of the BLM, the cation (C) and anion (A) concentrations can be described by the Boltzmann equation:

$$[C]_m = [C]_1 \exp \left[- \frac{F\ (V - \psi)}{RT} \right] \qquad (38)$$

$$[A]_m = [A]_1 \exp \left[\frac{F\ (V - \psi)}{RT} \right] \qquad (39)$$

where subscripts m and 1 as before, refer to the membrane and aqueous solution (1), respectively. V is the interfacial potential at the solution/BLM interface at $x = 0$ and ψ is the distribution potential at a distance in the interior of the lipid bilayer, where in principle at least, $[C]_m = [A]_m$. However, at the interface ($x = 0$ or $x = t$) the electroneutrality conditions are not observed, Eq. (33) becomes

$$\phi = RT \ln \frac{D_C}{D_A} = RT \ln \frac{[C]_m}{[A]_m} \tag{40}$$

Substituting Eqs. (38) and (39) into Eq. (40), one obtains at the condition where ϕ at $x = 0$

$$\phi_1 = \psi - (\psi - V) = V \tag{41}$$

Eq. (41) shows that at the solution/BLM interface, the inner (or Galvani) potential is determined only by the interfacially adsorbed species and their dipoles (Davies and Rideal, 1961; Tien, 1974).

The dipolar components of interfacial potentials, V, can be analyzed using BLM forming molecules, such as phospholipids (e.g., phosphatidylcholine) and interface-active agents (e.g., HDTABr). These molecules are oriented at the solution/BLM interface in the manner shown in Fig. 1. Together with water dipoles (μ_w) associated with the fixed charge and polar-head group (μ_p), the terminal hydrocarbon chain (μ_t) constitutes another dipolar component. The overall contribution to the interfacial potential is given by

$$V = \psi_G + 4\pi n \mu_d \tag{42}$$

where ψ_G is the Gouy double-layer potential due to the fixed charges at the interface, n is the number of dipoles per cm^2, and $\mu_d = \mu_w + \mu_p + \mu_t$, where μ_d is the overall dipole moment of the constituent molecules in the BLM. Techniques for measuring interfacial potentials as a result of these dipoles have been discussed (Tien, 1974; Blank, 1986).

The Guoy potential, ψ_G in Eq. (42), is due to the presence of fixed charges associated with BLM-forming constituents and is given by (for uni-univalent salts)

$$\psi_G = \frac{2RT}{F} \sinh^{-1}\left(\frac{2\pi\sigma}{\epsilon_w \kappa RT}\right) = \frac{2RT}{F} \sinh^{-1}\left(\frac{\sigma}{A\sqrt{C}}\right) \tag{43}$$

where σ = interface charge density, ϵ_w = dielectric constant of water, $\kappa = [8\pi Ce^2/\epsilon_w kT]$, the reciprocal of which is defined as the Debye length, A = the area occupied per ionogenic group, and C = bulk salt concentration. The considerations given are for the left-hand or outer side interface. For the entire BLM system,

$$E_m = \Delta\psi_G + E_p \tag{44}$$

where E_p is the polarization potential as defined in Eq. (42). For a symmetrical BLM (i.e., the constituent molecules are the same at the two interfaces), E_p will be nullified. However, if the composition is not identical, as is known to be the case for many biomembranes, the contribution due to E_p should not be neglected. The potential due to $\Delta\psi_G$ can be tested. According to Eq. (43).

$$\left(\frac{\partial \Delta\psi_G}{\partial \log C}\right)_{A = \text{constant}} = \pm 58 \text{ mV} \tag{45}$$

for a symmetrical BLM at 20°C. Experimentally, either by holding one side of the BLM constant or changing the concentration of both sides, a positive slope should be obtained for negatively charged BLM. For positively charged BLM, minus 58 mV should be observed for a 10-fold change in concentration, assuming of course, all the conditions are ideal (see Fig. 1).

More often than not, ideal conditions are rare. When a concentration gradient is imposed across the BLM, a diffusion potential, ϕ_d, may also be present. This potential, ϕ_d, also known as <u>concentration potential</u>, is frequently used to estimate transmembrane ionic permeabilities, and is given by

$$\begin{aligned} E_m &= \frac{RT}{F} \frac{U_C - U_A}{U_C + U_A} \ln \frac{[CA]_1}{[CA]_2} \\ &= \frac{RT}{F} (2t_C - 1) \ln \frac{[CA]_1}{[CA]_2} \end{aligned} \tag{46}$$

where U_C and U_A are the mobilities of cation and anion, respectively, in the BLM. t's are <u>ionic transference numbers</u> ($t_C + t_A = 1$), and are defined by

$$t_C = \frac{U_C}{U_C + U_A} \; ; \quad t_A = \frac{U_A}{U_C + U_A} \tag{47}$$

Further, the ionic transference number is related to the membrane conductance (G),

$$t_C = \frac{G_C}{G_m} \; ; \quad t_A = \frac{G_A}{G_m} \tag{48}$$

here G is the total membrane conductance ($1/R_m$). If there is a difference in the interfacial potentials on the two sides of a BLM, G_m, hence G_C and G_A, can be related by the following

$$\frac{{}^2G_C}{{}^1G_C} = \exp -\left(\frac{F\Delta\psi_G}{RT}\right) \tag{49}$$

$$\frac{{}^{2}G_{A}}{{}^{1}G_{A}} = \exp \frac{F\Delta\psi_{G}}{RT} \tag{50}$$

Thus, from measurements of E_m and/or membrane conductivity under experimental conditions, various membrane parameters, such as transference numbers, ionic mobilities and their related quantities, permeabilities ratios and ion selectivity, may be evaluated. Also, a BLM is said to be <u>ion-selective</u>, if it can select ions of one type of charge (e.g., cation over anion), which means, in this case, $t_C = 1$. Upon modification, a BLM can also be made <u>ion-specific</u> in that ions of similar charge are discriminated (e.g., K^+ over Na^+). In both cases, interaction with, as well as transport across, the BLM is involved.

d. <u>Relationship among various potentials</u>. Of immediate concern here we have been discussing the observed membrane potential, E_m, which is the sum of all interfacial potentials and diffusion (concentration) potentials at the interfaces and within the membrane. Thus, we have the appropriate name, the so-called <u>transmembrane potential</u>, E_m, which is given by

$$E_m = \Delta\phi + \phi_d \tag{51}$$

or

$$E_m = \Delta V + \Delta\psi + \phi_d \tag{52}$$

or

$$E_m = \Delta\psi_G + E_p + \Delta\psi + \phi_d \tag{53}$$

For most of these individual potentials, the final expression takes the form of the Nernst equation (Eq. 13). Experimental conditions or circumstances can be such E_m measures one or a combination of these potentials. Specifically, (i) for a symmetrical BLM, the polarization potential, E_p, is unimportant, (ii) for an unmodified BLM, whose interior is liquid-hydrocarbon like, is not measured, (iii) for BLM with low interfacial charge density (e.g., $\sigma < 1$, net charge per 100 $Å^2$), $\Delta\psi_G$ contributes very little, (iv) for charged species with substantial solubility in BLM, $\Delta\psi$ may be neglected, and (v) for a BLM with high interfacial charge density (> 1e/100 $Å^2$), the diffusion potential, ϕ_d, is inconsequential.

It should be emphasized, however, that the above statements are offered as a general guidance only. Even for a relatively simple BLM, the measured transmembrane potential (E_m) could be a result of various potentials. The extent of contributions of these potentials to the overall E_m is dependent upon the nature and properties of BLM-forming constituents, whose charge densities, dipole orientations and relative permeabilities to ions are some of the factors that should be taken into consideration for a careful analysis of the observed membrane potential.

In concluding this section, it should be mentioned that there are other types of membrane potentials under specific names, such as Donnan membrane potential and ion-exchange membrane potential, both of which can be applied to the BLM system (Lakshminarayanaiah, 1969). The equations describing these potentials can be derived either by a thermodynamic or a kinetic treatment using Eq. (11). Mention should also be made concerning the dipoles of membrane constituents in the BLM. The dipole flip-flop could be important for the voltage-dependent conductance in BLM (e.g., alamethicin-modified BLM. Derzhanski et al. (1981) have derived an equation for the current as a function of frequencies. Pastushenko and Chizmadshev (1981) have considered the energetic profile of dipole molecules in biomembranes in general and in BLM in particular. This topic has been reviewed in detail (Shchipunov and Drachev, 1982; de Levie, 1978).

4. Electronic Processes

Up to now, we have used mainly the classical concept of electrochemistry in describing the various phenomena associated with membranes, involving only hydrated ions and ions surrounded by other ions (ionic cloud). There is no rest for ions in the solution (ceaseless motion or random walk). Ions get nowhere by random walk alone (their time average displacement is zero). A net transport of ions is possible by (a) the presence of an electrical field, and (b) chemical potential gradient. Thus, a net drift is superimposed on the random walk. When ions reach the interface, properties change abruptly; the anisotropy of the forces compel ions to adopt new configurations unknown in bulk solution. Here, a variety of phenomena occurs such as charge separation, potential gradient, adsorption, and orientation of H_2O dipole, etc. At present, modern electrochemistry is focused on the transfer of charges across the interface, in particular, the electrified interface. The involvement of the electrical double layer, therefore, is self-evident.

The charges that we are concerned with here are the electronic charges (electrons and holes). For charges of this type to be transported across the interface, electrochemical reactions must take place. In the presence of a membrane that is impermeable to ions, what will happen then? Here, the membrane must serve at least two functions: (1) pathway for electronic charges, and (2) electrode surface for chemical transformation (reduction and oxidation or redox reactions).

a. Evidence for electronic processes in biomembranes. More than 40 years ago, Albert Szent-Gyorgyi suggested the idea of the role of solid-state electronic processes in biology (see Tien, 1974; Weller, 1973; Lewis, 1982; Taylor and Mahboubian-Jones, 1982; Pethig, 1979; Zon and Tien, 1986). Many attempts have been made during the intervening years to demonstrate that such electronic processes can occur in biomembranes and their constituents such as proteins and lipids (see Milazzo, 1983). The origin of the concept of electronic processes in membranes and related systems was first reviewed in 1971 (see Tien, 1976), in which the phenomenon known as "electrostenolysis" was stressed. In this connection, it should be mentioned that the term "electrodics" has been proposed (Bockris, 1980). In the language of membrane electrochemistry, electrostenolysis simply means that a reduction reaction takes place on one side of the membrane (or barrier) where the positive electrode is situated and the oxidation occurs on the other side of the membrane. Although electronic processes in BLM in the dark had been discussed since 1970 (see Milazzo, 1983), no conclusive evidence was shown. This is because of the fact that an unmodified BLM is an excellent insulator (resistivity $>10^{15}$ ohms) incapable of either ionic or electronic conduction. In order to demonstrate electronic conduction in BLM it seems that the membrane must, first of all, be modified to function as a redox or semiconductor electrode. Secondly, an appropriate method must be found for studies of electronic processes that can be applied to the BLM system. Our recent attempt at demonstrating the electronic processes in membranes in the dark is presented in the following paragraphs.

b. Basics of cyclic voltammetry. Cyclic voltammetry (CV) is an elegant and simple electrochemical technique for studying redox reactions at the electrode/solution interfaces and has become increasingly employed in all fields of chemistry (Tien, 1984; 1985). Until very recently, however, the technique has apparently not been applied to any membrane systems including the BLM separating two aqueous solutions. It may be informative, therefore, to give a brief account of the fundamentals of CV and to show how the technique can aid us in membrane research.

In conventional CV, one uses immobile working electrode in unstirred solutions and applies a triangular potential waveform between it and a reference or an auxiliary electrode. The working electrode (WE) may be made of Pt, Au, carbon paste, glassy carbon, or semiconductor, SnO_2. A silver/silver chloride or a saturated calomel electrode (SEC) is used as the reference electrode (RE). In the present day practice of CV, a Pt wire or a second reference electrode is employed as auxiliary electrode (AE), the purpose of which is to avoid large currents being passed through the reference electrode during the potential scan, which could change its potential. The potential scan (or sweep) is carried out between two potential values of interest (eg., from about 1.2 to -0.8 V vs. SCE). The scan rates can be anywhere from 0.1 mV to 100 V or more per second but values between 10 to 400 mV/s are frequently used. During the potential scan, the current is measured at the working electrode. The graph with the current plotted on the vertical axis versus the potential on the horizontal axis is called the voltammogram, which is characterized by several parameters. These are the peak potentials (E_{pc} and E_{pa}, for cathode and anode, respectively), the peak current (I_{pc} and I_{pa}), and the half-peak potential ($E_{p/2}$) and the half-wave potential ($E_{1/2}$). The last parameter is related to the standard redox potential ($E^{o\prime}$) by the following equation

$$E_{1/2} = E^{o\prime} + \frac{RT}{nF} \ln\left(\frac{D_{RED}}{D_{OX}}\right)^{1/2} \tag{54}$$

where D_{RED} and D_{OX} are the diffusion coefficients, n is the number of electrons involved and the other terms have the usual significance. Since $D_{RED} \simeq D_{OX}$, therefore $E_{1/2} \simeq E^{o\prime}$, which is located exactly halfway between E_{pa} and E_{pc}. Under a given set of conditions, the criteria of reversibility of electron-transfer reactions at the electrode surface is given by

$$E_{pa} - E_{pc} = E_p = \frac{59mV}{n} \tag{55}$$

For a reversible reaction, the ratio of I_{pa}/I_{pc} should be unity and the I_p (peak current in amperes) is expressed as follows:

$$I_p = kn^{3/2}AD^{1/2}C\,\nu^{1/2} \tag{56}$$

where A = the electrode area (cm^2), D = the diffusion coefficient (cm^2/s), C = the molar concentration, ν = the scan rate (volt/s), and k (= 269) is the so-called Sevick-Randles constant (see Tien, 1984). In Eq. (56), it can be seen that I_p increases with C and is directly proportional to the square root of ν. For a general redox reaction

$$OX + ne \underset{k_b}{\overset{k_f}{\rightleftharpoons}} RED \tag{57}$$

where k_f and k_b denote the forward and backward rate constant, respectively, the reaction is said to be reversible if $k_f \simeq k_b$. The k's are related to k_s, the standard heterogeneous electron-transfer rate constant, which may be estimated from the value of E_p (Eq. 55), according to the method worked out by Nicholson (see Tien, 1985). The final results of Nicholson's analysis is that the voltammograms (or current/voltage curves) of CV depend on the three variables, from which a practical relationship between k_s, E_p, and ν is obtained, i.e.,

$$\psi \simeq 28.8\, k_s\, \nu^{1/2} \tag{58}$$

where ψ as a function of E_p is available. Thus, known ψ and from experimentally determined E_p values, k_s can then be estimated.

Regarding the cyclic voltammograms of electron-conducting BLM, the following remarks should be made. First, mass transfer by migration of electroactive species under the influence of applied voltage is rendered negligible by using a high concentration of supporting electrolyte (e.g. 1 M KCl). Second, it is assumed that the principal mode of mass transport of electroactive species at the surface of BLM is by diffusion, which occurs as a result of the electrochemical potential difference developed between the BLM/solution interface and the bulk solution. Third, electron-transfer reactions can be directly related to the voltammograms. This is owning to the fact that the redox products generated near or in the BLM are available for further redox reaction as the direction of scan is reversed. Perhaps, the best way to explain these factors is by referring to Fig. 2. During the cathodic scan (-E or increasingly negative), more and more electrons are available for reduction to the species adjacent to the BLM. This is the BLM becomes a stronger reducing electrode. Conversely, as the potential is increased toward the positive direction, the BLM becomes a better oxidizing

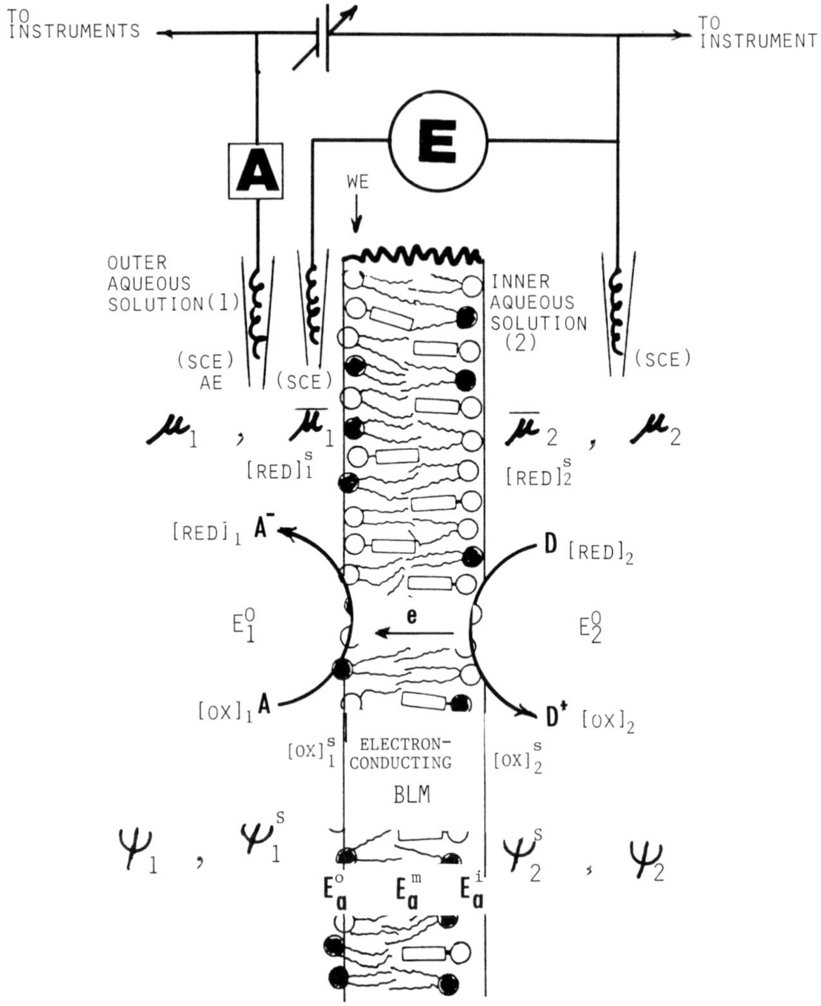

Fig. 2. Transmembrane electron movement and redox reactions. Also shown schematically are electrodes and circuit diagram for cyclic voltammetry. WE = working electrode, SCE = saturated calomel electrode, AE = auxiliary electrode. μ and $\bar{\mu}$ are chemical and electrochemical potentials, respectively. Bulk and interface concentrations of reduced [RED] and oxidized [OX] species are indicated by subscripts (1 and 2) and superscript s, respectively (see text for details).

electrode, with positive holes, so to speak, increasingly ready for oxidation. As shown in Fig. 2, a saturated calomel electrode (SCE) is placed in the inner and the outer solution, respectively, which are used to monitor the applied potential (E). The third SCE, also placed in the outer solution, is used as an auxiliary electrode for passing the current. The outer BLM surface functions as the working electrode and is labelled WE. The two redox couples are denoted by A and D, whose reduced and oxidized forms are represented by A^- and D^+, respectively.

c. Electron-conducting BLM. Fig. 2 shows an electron-conducting bilayer lipid membrane (BLM) separating two aqueous solutions containing different redox couples. We will refer to, by subscripts, the left side as the outside "i" (or Aqueous Solution 1) and the right side as the inside "o" (or Aqueous Solution 2). To facilitate the discussion, let's hold the inside constant and consider the interfacial electron transfer reaction on the outside, i.e.,

$$D^+ + ne \underset{k_b}{\overset{k_f}{\rightleftharpoons}} D \tag{59}$$

where D and D^+ denote an electron donor in its reduced and oxidized form, respectively, whose concentrations are $[Red]_2$ and $[Ox]_2$ in the bulk phase. The corresponding interfacial concentrations are $[Red]_2^s$ and $[OX]_2^s$. The forward and backward heterogeneous rate constants are k_f and k_b, respectively.

The observed current is then given by

$$I_{net} = nFA\,(k_f[OX]_2 - k_b[Red]_2) \tag{60}$$

where I_{net} is the net current passing over the interface.

In order for an electron to move from the inside originating from an electron donor (D) to the outside accepted by an electron acceptor (A), it has to transverse at least three energy barriers, two of which are situated at the two membrane/solution interfaces and the third one in the membrane phase, which are labelled E_a^o, E_a^i, and E_2^m, respectively, in Fig. 2. Clearly, one of the tasks is to evaluate these energies of activation. According to the Arrhenius equation, one can relate the heterogeneous rate constant, k_s to E_a, which can be related in turn to the free energy of activation (ΔG^*), we have

$$k_s = Ae^{-E_a/RT} \tag{61}$$

and

$$k_s = Ae^{-\Delta G^*/RT} \tag{62}$$

Since the observed current is a measure of the rate of an electrochemical reaction, which may consist of non-faradaic processes (capacitive or changing currents), several equations are available for the evaluation of various parameters including the rate constant (see Tien, 1985). One of the general expressions is the Eyring equation

$$I_{net} = Fk[OX]_2 e^{-F(E-E^o)RT} - [Red]_2 e^{(1-\alpha)nF(E-E^o)/RT} \quad (63)$$

where α is the so-called transfer coefficient, E^o is the standard redox potential of the couple and E is applied voltage. In this simplified Eq. (63), we have assumed activity coefficients, n (= 1) and A (1 cm^2) to be unity. It has been shown (see Tien, 1985) that the Eyring equation (Eq. 63) can be expressed in another form known as the Butler-Volmer equation

$$I_{net} = I_o \left[\exp \frac{\alpha F \eta}{RT} - \exp \left(\frac{(1-\alpha)F\eta}{RT} \right) \right] \quad (64)$$

where I_o is the so-called exchange current at $\eta = 0$, α is the transfer coefficient as before (under ideal situation $\alpha = 0.5$), and $\eta = E - E^o$, the so-called overpotential. Eq. (64) has two limiting cases: for small η (<10 mV)

$$I_{net} = \frac{I_o F(E-E^o)}{RT} \quad (65)$$

since for small value, exponential terms may be linearized. For large (> 100 mV), it can be shown that the second exponential term in Eq. (64) is negligible, thus

$$I_{net} = I_o \exp - \frac{\alpha F(E-E^o)}{RT} \quad (66)$$

Expressing Eq. (64) in logarithmic form, we have

$$\ln I_{net} = \ln I_o - \frac{\alpha F(E-E^o)}{RT} \quad (67)$$

which is known as the Tafel equation. By plotting log I vs. E, various parameters such as I_o and α may be obtained from the intercept and the slope, respectively. As we will see presently, all afore-given equations, which are well known to physical electrochemists, are highly useful to membrane bioelectrochemistry studies (Blank, 1980; 1986; Milazzo, 1983).

To explain the observed membrane potential (E_m), it is best to refer again to Fig. 2 in which the membrane is assumed to behave like an ideal electron conductor (ie., Pt). From thermodynamics, the overall free energy change associated with an electrical cell reaction is given by

$$\Delta G = -nF\Delta E \quad (68)$$

where E is the electromotive force (emf) of the cell. Consider the BLM system as an electrical cell shown in Fig. 2, each solution/membrane interface is assumed to behave as a redox electrode. From Eq. (68) the individual free energy terms are denoted by μ. Thus, as indicated on the left side of the BLM $\mu_A + \mu_e = \mu_{A^-}$. Similarly, on the right side

$$D - e \longrightarrow D^+ \tag{69}$$

and

$$\mu_D - \mu_e = \mu_{D^+} \tag{70}$$

In writing these equations the electrons involved are considered as one of the reactants. Since $\mu_i = \mu^o + RT \ln a_i \simeq \mu^o + RT \ln [i]$, where μ^o is a constant, R and T have the usual significance, [i] = the concentration of the redox species when the activity coefficient is taken to the unity. The free energy of the electron μ_e, is given by

$$\mu_e = -FE_{out} = \mu_{A-} - \mu_A \simeq \mu^o_{A-} - \mu^o_A + RT \ \ln \frac{[A^-]}{[A]} \tag{71}$$

for the outer (left) solution membrane interface, where E_{out} = the electrode potential. For the inner (right) solution/membrane interface, we have

$$\mu_e = -FE_{in} = \mu_D - \mu_{D^+} \simeq \mu^o_D - \mu^o_{D^+} + RT \ln \frac{[D]}{[D^+]} \tag{72}$$

At equilibrium, the free energy difference of the electron across the BLM is equal to

$$-\frac{\Delta\mu_e}{F} = E_m = (E^o_{A/A^-} - E^o_{D+/D}) + \frac{RT}{F} \ln \frac{[A][D]}{[A^-][D^+]} \tag{73}$$

where $[A^-]$ - [A] amd $[D^+]$ = [D], we have (74)

$$E_m = E^o_{A/A-} - E^o_{D+/D} = E^o_{redox}$$

Eq. (74) can be obtained in any way by letting I_{net} = 0, therefore Eq. (63) is reduced to the Nernst equation

$$E = E^o + \frac{RT}{F} \ln \frac{[OX]}{[Red]} \tag{75}$$

The redox reaction of each compartment may be treated as an electrical cell, thus, one can describe, the electrical potential of each side of the membrane relative to a reference electrode such as to calomel electrode used in our measurement. Therefore

$$E_1 = E_1^o + \frac{RT}{nF} \ln \frac{[OX]_1}{[Red]_1} \qquad (76)$$

and

$$E_2 = E_2^o + \frac{RT}{nF} \ln \frac{[OX]_2}{[Red]_2} \qquad (77)$$

where E_1 and E_2 are the potentials of Cell 1 and Cell 2, while E_1^o and E_2^o are standard redox potentials for the two redox couples. If the membrane is electron-conducting, as in the case depicted in Fig. 2, we have in essence two electrochemical cells connected in series. The overall potential generated by transfer of electrons from D and A should then be

$$E = E_2 - E_1 = E_m = E_2^o - E_1^o + \frac{RT}{nF} \ln \frac{[OX]_2[Red]_1}{[Red]_2[OX]_1} \qquad (78)$$

In principle, E_2^o and E_1^o, in each compartment should be measurable between the BLM and calomel electrodes, which should be equal to the algebraic sum of E_2 and E_1, if $[OX] = [Red]$ according to Eq. (78). Unfortunately, this cannot be done with the BLM separating two aqueous redox solutions. However, the difference can be easily measured. According to Eq. (74), we have

$$\Delta E = E_m = E_2^o - E_2^o = E_{op} = E_{rx}^o \qquad (79)$$

where E_m or E_{op} is the membrane potential measured under open-circuit condition ($I = 0$). This, incidentally, is the proper situation where Eq. (78) applies. In carrying out our cyclic voltammetry of BLMs, which involves non-equilibrium conditions, we have to rely on other than the Nernst equation.

In cyclic voltammetry of electron-conducting BLMs, a voltage is applied across the membrane and the resulting current recorded. If redox reactions occur, there are changes of concentration of [OX] and [Red] in the immediate vicinity of the membrane (i.e., $[Red]_2 \neq [Red]_2^s$, etc.). For most experiments described here, we have plotted the logarithm of current as a function of applied voltage, essentially straight lines have been obtained. The results are in accord with Eq. (67). Of interest to note is that Eq. (67), when written in exponential form, is used in solid-state physics as the diode equation. In this connection, it should be stressed that as a result of ultrathinness (< 100 Å), an electric field gradient of 10^5 volts cm^{-1} is easily developed across the BLM (i.e., in the presence of $E_m \geq 10$ mV). Under high field conditions, two mechanisms may be envisioned for electron translocation across the BLM system; (i) by tunneling, and (ii) by hopping (Tien, 1984). A straightforward but highly significant experiment would be to measure the electrical conductivity of the doped-BLM system as a function of temperature (also as a function of applied voltage to assess high field effects). With the aid of standard semiconduction equations (e.g., Eqs. 61 and 62), the temperature-dependent pre-exponential factor may be determined, from which the nature of electronic processes in the doped-BLM and its related systems may be elucidated (Krysinski and Tien, 1986, to be published).

B. Elements of Membrane Photochemistry

1. Background and Perspective.

Solar energy in the form of visible light is what makes life possible on Earth. Light is used by living systems for two main purposes: (i) as a source of free energy, and (ii) as a source of information. To achieve either purpose, light must be first converted into an appropriate form of energy that can be used by the organism. To effect light transduction, certain "machineries" have been evolved in nature; they are membrane-bound. The details of these machineries (or transducers) in biological membranes for light transduction are largely uncertain but they are definitely made of lipids, proteins, and pigments. These constituents are believed to be organized in special arrays with lipids in the form of a bilayer, with proteins "floating" about or spanning across the bilayer, and with the pigments in some ordered lattices. Clearly, until our knowledge of biomembrane transducers is vastly improved, alternative and conflicting explanations can be offered for almost any experimental findings in work with intact biological membranes. If, on the other hand, one can obtain similar observations in a sufficiently well-defined artificial membrane system, the interpretation of experimental results would be much less ambiguous. It is along this line of reasoning that we wish to discuss membrane photochemistry with emphasis on pigmented bilayer lipid membranes (BLM) in relation to natural photoactive membrane systems. One way to begin this exciting story of light and life, it seems that an overview is in order. This is presented in Table 1, in which the major events are listed in chronological sequence.

For photochemical studies, pigments and dyes may be incorporated directly into the bilayer forming solution, or may be solubilized in the aqueous bathing media and permitted to adsorb onto the lipid membrane after formation. Spectroscopic methods which are useful to photochemical studies are available (Tien, 1976; Antolini et al., 1982); however, for electrical measurements of detection, one only needs a light source and a monochrometer. It is also recommended that a heat filter be placed between the light source and the membrane chamber. Electrical changes can be readily made using a high impedance electrometer and ordinary calomel electrodes. If the measurement of transient responses is desired, one can use an oscilloscope to record the parameter of interest (Tien, 1974; Davison, 1985).

Table 1. Major Events in Membrane Photoscience Research

Date	Event
Genesis:	"Without light there can be no life" (visible light range 390-760 nm)
1660s:	Robert Boyle: phosphorescence, chemiluminescence, bioluminescence (needs O_2)
19th Century:	Photography
1900:	The quantum theory (Planck); Grothus-Draper's Law - only light absorbed causes chemical reactions
1908:	Einstein - Law of Photochemical Equivalence
1920:	Warburg - quantum yield concept
20th Century:	Ciamician - Organic photochemistry (many interesting photochemistry reactions are discovered by him)
1920s-1940s:	Photosynthesis - Molecular spectroscopy; Direct fluorescence lifetime measurement; Flash photolysis; Visual processes
1950s-1960s:	Photosynthesis: primary events, photophosphorylation and oxygen evolution; structure of photoactive membranes
1968:	Photoelectric effects in pigmented bilayer lipid membranes
1970s:	The purple membrane of Halobacterium halobium; energy transduction
1980s and beyond:	Molecular details of light transduction in photoactive membranes; membrane-based electronic devices activated by light; water photolysis via ultrathin pigmented membrane for solar energy utilization

The interpretation of light induced phenomena are based on the theory of photochemical reactions and the quantum theory of radiation and matter. There follows a discussion of the general principles of photophysics and photochemistry and their relationship to reactions at membrane/solution interfaces.

2. The Nature of Light and Matter as Particle-Waves.

The spectrum of electromagnetic radiation of which ultraviolet (UV) rays, visible light, infrared (IR), and microwaves are examples, extends through 10^{16} different wavelengths. Visible light from about 390 nm to about 760 nm, constitutes a tiny segment, on which life as we know it depends. The basic equation of electromagnetic radiation due to Planck (1900) is given by

$$E = h\nu \tag{80}$$

where E = energy (Kcal/mol or eV), h = Planck's constant (6.6×10^{27} erg-sec), ν = the frequency (cycles per sec). Eq. (80) expresses the quantitization of energy, i.e., any atom or molecule has available to it only a discrete set of energy levels. The particle-like nature of radiation is the basis of the photoelectric effect, in which shining light on a metal causes electrons to be ejected from it. The kinetic energy of the emitted electrons depends linearly on the frequency of light, but not on the light intensity. Since $\nu = c/\lambda$ where c is the speed of light (3×10^{10} cm/sec) and λ is the wavelength of light (nanometer, nm; 1 nm = 10 Å), by substituting into Eq. (80), we have

$$E = h\frac{c}{\lambda} \tag{81}$$

The duality of particle-wave nature of light was dramatically demonstrated by de Broglie (1924), who combined the Einstein equation ($E = mc^2$) with the Planck equation (Eq. 80) resulting

$$\lambda = \frac{h}{p} \tag{82}$$

in which the momentum, p, of the particle of mass (m) and velocity (v) is p = mv (note: for photons v = c). Thus, there is a characteristic wavelength associated with each particle (for a ping pong ball traveling at 45 miles/hr, the de Broglie wavelength is about 1.6×10^{-26} nm, which is exceedingly small, exhibiting essentially no wave character. As we will discuss later, the wave-nature of particle is essential in the understanding of electron transfer via tunneling in certain biomembranes.

The particles of light are called photons (or quanta). From Eqs. (80) and (81) we can see that the frequency of radiation is inversely proportional to wavelength; the shorter the wavelength, the greater the energy. To bring about a photochemical change, Einstein postulated that all the energy of a single photon is transferred to a single electron of matter. This is known as Einstein's law of photochemistry (Barltrop and Coyle, 1978; Atkin, 1982).

3. States and Transitions.

The concept of "state" is used to describe the condition of a given particle (or collection of particles) at some given instant of time. In general, we can describe a state with a mathematical function which provides information on all important variables which may affect the condition of the particle. In principle, the function can supply us with information needed to predict the dynamics of the system in question. Atoms and molecules are described by wave function (usually designated) which are functions of nuclear and electronic coordinates. The wave function is obtained by the solution of a differential equation known as the Schrodinger equation. It is frequently found that the solution is not unique; that is, there exists a set of solutions that solve the equation. The individual members of a set of wave functions each possess an energy which may be unique or common to several members (degeneracy). The members of the set are ordered in increasing energy to give a series of energy levels. These energy levels are populated according to a rule known as the Pauli Exclusion Principle (see below) and then sums the energy.

Suppose we are considering a single particle (e.g., an electron in a potential well), then each level occupied by the electron represents a different total energy of the system and therefore represents a state of the system of one electron. If we have a many particle system (several electrons in a potential well) then we populate the levels and tally the energies. Each unique permutation of the energy level population gives a unique state of the many particle system. It is important to realize that a state is defined by the total configuration of the entire system under consideration. Transitions between states may occur if a given system is provided with sufficient energy to make the transition from initial to final state. For instance, an atom or molecule may absorb a photon which induces an electron to enter a higher energy level and thereby enter a new state. In the case of atoms and molecules, the state of lowest total energy is called the ground state, whereas all others of higher energy are called excited states.

The frequencies of light that have been absorbed by the material constitute a characteristic absorption spectrum of that material and represents the energies of transition between the states. The photon of energy h has been absorbed by the atom and its energy used to undergo a state-to-state transition. This radiation induces transitions between vibrational energy levels within molecules. These levels are separated by only a small quantity of energy and can be regarded as different modes of oscillation about a chemical bond. There are some aspects of IR photochemistry that involve reactions, however for our applications these vibrational states will only be important in the discussion of energy deactivation (or de-excitation). Visible and uv light are of just the right energy to induce a molecule to undergo a transition between states which may be more energetically favorable for a chemical reaction, and at the same time not alter the chemical nature of the system. These transitions occur between electronic energy levels which correspond to what are often called orbitals (Barltrop and Coyle, 1978).

4. Energy Levels and Multiplicity.

The concept of state applies to the configuration of the occupation of energy levels describing a system. We often speak of the energy levels as orbitals. For an atom, the energy levels are separated by a quantity of energy which is very well defined. A molecule may have its energy levels separated by a less precise quantity of energy. The reason for this is the presence in molecular systems of additional energy levels, called vibrational levels, which describe different vibrational modes of the chemical bonds between atoms (analogous to the classical vibrational modes of objects joined by a spring). Vibrational states are not occupied in the manner described above. They are regarded as a subclass of the electronic states. In other words, we first designate an electronic state for a system, then given that information, ask if the system is vibrationally excited.

There is one final concept which is important to the understanding of photochemical processes, and it is a consequence of an empirical rule called the Pauli Exclusion Principle. This law of quantum mechanics states that no two electrons (or particles, in general) may be described by the identical set of quantum numbers. The upshot is that one cannot have more than two electrons in a given energy level. An electron is said to have parity or spin which is assigned a value of $\pm 1/2$. If an electron occupying an orbital has a spin of $\pm 1/2$, then if another electron is to enter the orbital, it must be of opposite parity, or $\mp 1/2$. The multiplicity of a system is defined as the sum of the total parities of the system. If the system consists of paired electrons of opposite parity, then the multiplicity is zero and the state is called a singlet (S). If the system has a pair of electrons of like parity (in different orbitals), the state is called a triplet (T), and has multiplicity of unity. A transition from a state of one multiplicity to another (e.g., $S \rightarrow T$, $T \rightarrow S$) is quantum mechanically "forbidden", meaning that it occurs with a very low probability. The parity of an electron is usually represented by an arrow with direction either up or down. The common transitions encountered in biomolecules are $n \rightarrow \sigma^*$, $n \rightarrow \pi^*$, $\pi \rightarrow \pi^*$, and charge-transfer (CT), where n, σ, and π denote, respectively, non-bonding, sigma, and (pi) orbitals. Their respective anti-bonding orbitals are indicated by $*$. The charge-transfer or CT transition is due to the presence of an acceptor-donor complex whose absorption band is generally broad and structureless (Barltrop and Coyle, 1978).

The absorption spectrum (optical density vs. wavelength or frequency) of most biomolecules show two or more bands, exhibiting that there are two or more distinct electronic excited states. Instead of displaying sharp and well-defined lines as in the case of an atom, the lines of biocompounds tend to fuse and form a broad band because of interactions between the compound and its surroundings. This is shown in Fig. 3 for chlorophyll. At room temperature in thermal equilibrium, most electrons are in the lowest or "ground states" of these respective orbitals. Upon absorption of light, they may be excited or in a higher energy state.

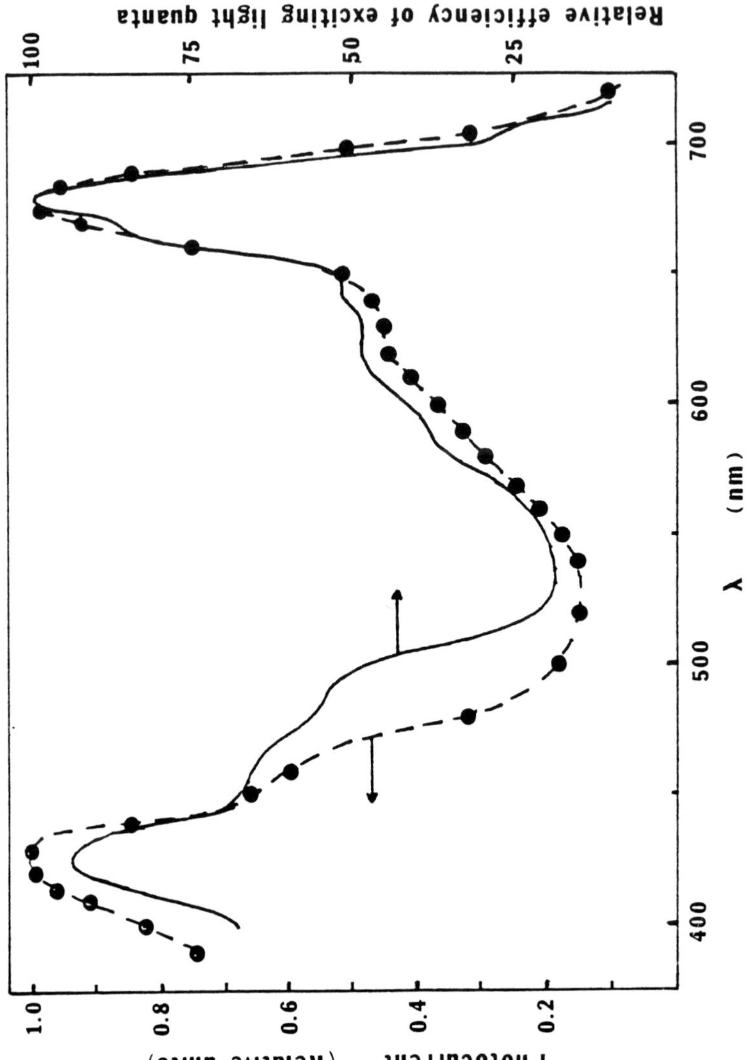

Fig. 3. The absorption spectrum (right ordinate) is from PS-I reaction center shown in solid line. Also shown is the photoelectric action spectrum of a BLM containing PS-I reaction center (taken from Lopez et al., 1984).

5. The Fate of the Excited State

For a collection of atoms or molecules, the overwhelming majority are in the lowest energy or ground state. If a molecule is induced to undergo a transition to an excited state, it will eventually return to its ground state via some energy deactivation mechanism. Photochemical reactions which proceed from these energetically more favorable excited states must compete with these de-excitation processes. The various photophysical chemical processes of energy deactivation are described in terms of transition (state) energy diagram due to Jablonski and extended by Lewis, Kasha, Terenin, and others and are shown in Fig. 4. The singlet (S_0 , S_1, S_2,...) and triplet (T_1,T_2,T_3,...) states are arranged in order of increasing energy. All the physical processes involved in the interconversion of the states can be represented on such a diagram as indicated. The overall efficiency of any particular process is determined by the relative magnitude of a number of individual rate constants and can be expressed in terms of the quantum yield (ϕ) in Eq. (83)

$$\phi = \frac{\text{rate of conversion}}{\text{rate of absorption}} = \frac{dc/dt}{I_0 A \Delta \epsilon} \tag{83}$$

where C = the pigment concentration, I_0= the incident light intensity, A = the fraction of light absorbed, and $\Delta\epsilon$ = a difference in extinction coefficient (units: moles cm^{-3} sec^{-1}/einstein cm^{-3} sec^{-1}).

A molecule, following excitation, might find itself with excess vibrational energy, that is, in one of the higher vibrational energy levels of the excited electronic state (point a of Fig. 4). Unless the molecule is in the gas phase at very low pressure, it is subjected to as many as 10 collisions per second. These collisions release the excess vibrational energy as kinetic energy. The process is known as vibrational relaxation and it is often referred to as an exchange of energy with a thermal or heat sink. Vibrational relaxation almost invariably (depending on the temperature) returns the system to the lowest vibrational energy state. In Fig. 4, this corresponds to the transition from a to b, b to c, and c to d. For example, the molecule has absorbed a photon (f to a) and vibrationally relaxed to the zero vibrational level of S (point b); from this point, one of several processes may occur. The molecule may simply emit a photon of energy $h\nu$, $S_1 - S_0$, and return to the ground state. This transition is allowed by the rules of multiplicity and is called fluorescence to distinguish it from the case where the states differ in multiplicity, phosphorescence. Fluorescence is a rapid process which occurs at a rate of 10^6-10^8 s^{-1}. With fluorescence, as with absorption, there is no assurance that the transition will be to the zero vibrational level. Often the transition is to an excited vibrational level (b to e'), followed by further vibrational relaxation (e to f). Other processes which may occur are the so-called radiationless transition. These transitions are not associated with a loss or gain of radiational energy, as the name implies. They occur as the result of the overlapping of energy surfaces. For a molecule of a given energy (e.g., point b of the Jablonski diagram), there may be vibrational levels of other electronic states which are isoenergetic. Because there are no energy barriers between these levels (and states), a transition may occur. The transition is called internal conversion if

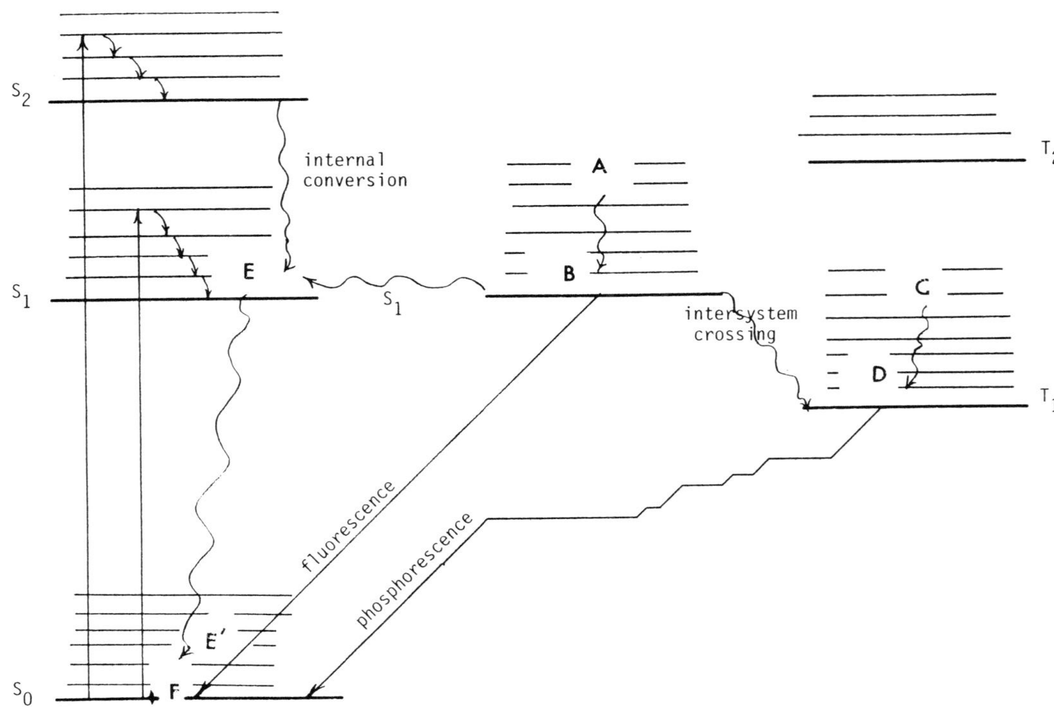

Fig. 4. An Jablonski energy level diagram. S_0 = the ground singlet electronic state. S_1 and S_2 are the first and second excited singlet state. T_1 = the first triplet state (see text for details).

the initial and final states are of the same multiplicity (e.g., S to S, T to T), and intersystem crossing if the states differ in multiplicity (S $\rightleftarrows$ T). Like fluorescence, these processes may be accompanied by vibrational relaxation. If a molecule has attained a triplet excited state (f to a to b to c), it is in essentially the same situation as it would be if it were in an excited singlet state (except for the difference in multiplicity). Here again, the molecule may emit radiation (phosphorescence), or undergo some radiationless transition. The emission and nonradiative decay from a triplet to singlet state is forbidden on the basis of multiplicity and therefore occur with low probability. The low probability of transition translates into a slow kinetic process. This means that if the molecule can be induced into a triplet state, it will remain there for a relatively long period of time. This is desirable from the standpoint of photochemistry since it allows the excited state molecule more time to react chemically (Barltrop and Coyle, 1978; Atkin, 1982).

6. Energy Transfer Processes.

It often happens that, although one molecule is excited, another molecular species reacts photochemically. Such a process is said to be sensitized and is indicated by the reaction, $D^* + A \rightarrow D + A^*$, and $A \rightarrow$ products. The mechanism of such reactions varies and proceeds via radiative and non-radiative (analogous to the intramolecular de-excitation described above) processes. The radiative mechanism is quite simply the fluorescence de-excitation of the donor D^*, followed by the immediate absorption of the emitted photon by the acceptor, A, as follows: $D^* \rightarrow D + h\nu$ and $A + h\nu \rightarrow A^*$.

Since a photon is exchanged between the donor and acceptor, no fluorescence is observed from D . Another consequence of the involvement of a photon is the relatively long range of separation between D^* and A over which the energy may be transferred.

Non-radiative energy transfer has the perturbation theory of intermolecular forces as its basis. When two different molecules interact, the energy of the system can be expressed as a sum of the energies of the individual molecules plus an interaction term. Intermolecular forces are distinguished on the basis of the range of their influence: namely, long range vs. short range forces. Each of these categories is mathematically derived by a different theoretical formalism. Without delving into the details of the theory, energy transfer occurs because a finite probability exists of a state $D\text{-}A^*$ in the presence of state $D^*\text{-}A$. In quantum mechanics, one often describes a system as a sum of all the possible states of its components and then computes the relative contribution of each component to the total. At the risk of oversimplification, one can draw an analogy to the problem of energy transfer. The D and A are in some way associated due to intermolecular forces and can be written as resonance structures as $D\text{-}A \rightleftharpoons D\text{—}A$. Because of this analogy, one often refers to the process as resonance energy transfer; a special case of which is known as Forster energy transfer (Barltrop and Coyle, 1978; Atkin, 1982).

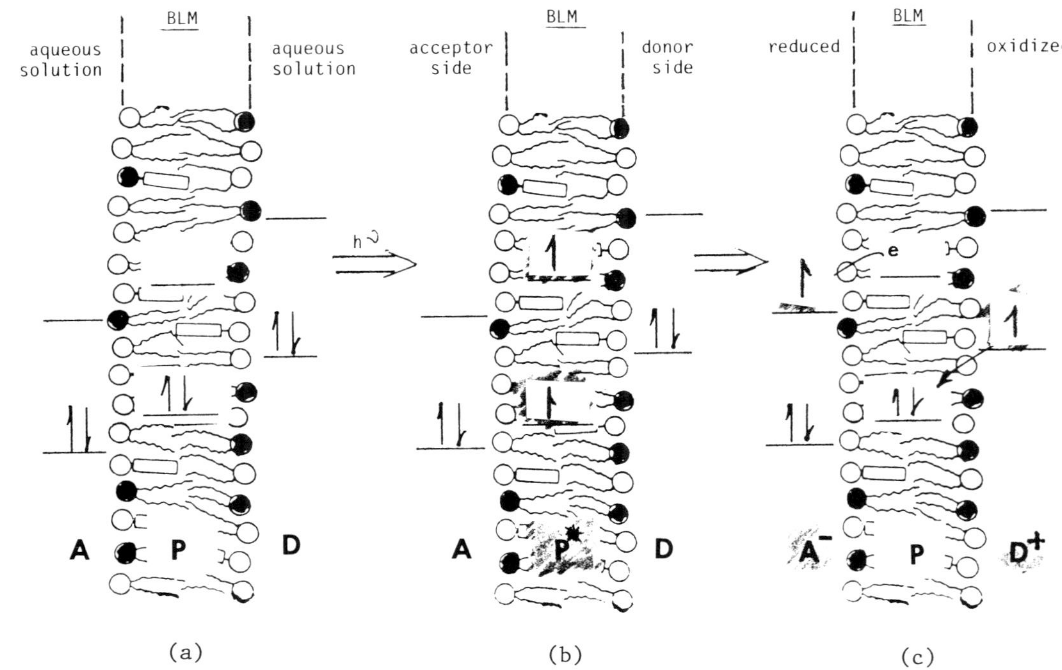

Fig. 5. The redox membrane electrode model of photosynthesis illustrating the mechanism of light-induced charge separation and electron transfer processes in a pigmented BLM separating two aqueous solutions. (a) the system at the ground state, (b) after absorption of photons by the pigment resulting in charge separation, (c) Electron transfer processes (see Nature, 219, 272, 1968; 227, 1232, 1970).

7. Photochemical Redox Reactions.

From the above discussions, there appears to be a great number of very rapid processes occurring from the excited state of a given molecule. The only way for a reaction to occur from an excited state is for it to take place on a time scale very nearly the same as the processes with which it must compete (see Fig. 6). The overriding factor of quantum mechanics and its influence on macroscopic behavior is the law of probabilities. If a process is very slow compared to others, it will certainly not be a dominating process. It is not enough for a process to yield stable products, it must also be very fast to happen photochemically. Electron transfer is among the fastest type of chemical processes, therefore photochemistry consists almost exclusively of charge or electron transfer reactions. Examples include:

a. Dissociation

$$A\text{-}B \xrightarrow{h\nu} A + B \tag{84}$$

The energy of visible light is in the range 40-80 Kcal/Einstein (1 Einstein = 6 x 10^{23} photons or quantum mole).

b. Ionization

$$AB \xrightarrow{h\nu} AB^{+} + e \tag{85}$$

or $AB \longrightarrow A^{+} + B^{-}$ (radical ions)

c. Association

$$A \xrightarrow{h\nu} A^{*} + A \longrightarrow A_2 \tag{86}$$

d. Photosensitization

$$A \xrightarrow{h\nu} A^{*} + B \longrightarrow A + B^{*} \tag{87}$$

e. Photoredox reactions

$$A + D \xrightarrow{h\nu} A^{-} + D^{+} \tag{88}$$

For our purposes of energy conversion, we shall confine our discussion to photoredox reactions, as indicated in e (Eq. 88).

In a photoredox process, an electron is excited to a higher energy level, altering both the electron donor and electron acceptor properties of the molecule (P). The excitation of the electron leaves a positive "hole" in its previous location in the orbital and may permit the excited molecule (P^{*}) to accept an electron (reduction). Alternatively, the excited molecule may now have an electron energetically favorable for transfer to another molecule (oxidation). Reverse reactions, which would result in charge recombination and subsequent de-excitation, must be prevented for a photochemical charge transfer reaction to proceed with any efficiency. One means of accomplishing this is by separating

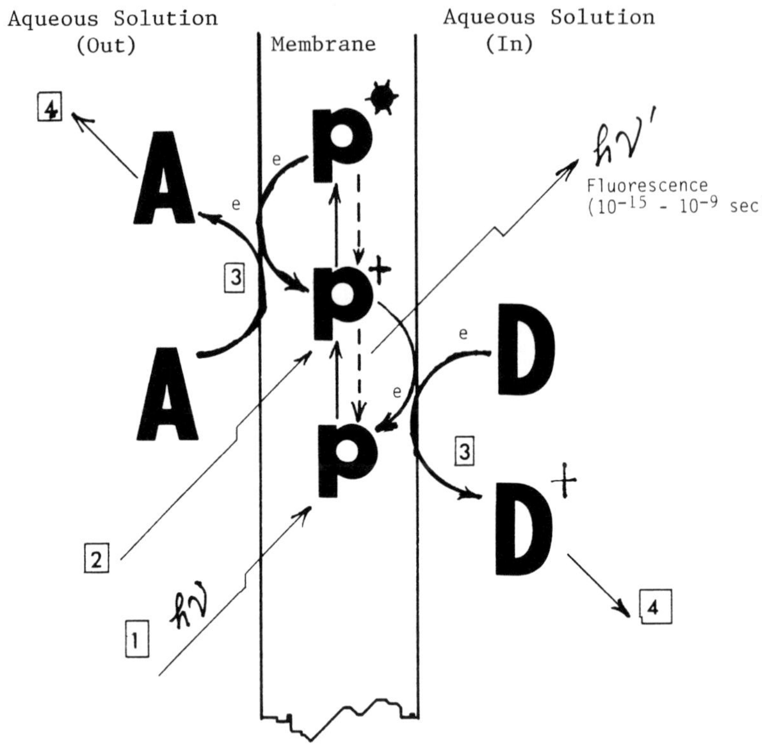

Fig. 6. The time scale of membrane photophysical photochemical processes. (P pigment, ground state; P* excitated state; P+ oxidized state)

(1) Photoabsorption. A photon is absorbed by the pigment. This takes place within 10^{-15} to 10^{-9} sec.

(2) Charge separation and dissociation. This happens in about 10^{-10} sec. The presence of an electric field plays a crucial role.

(3) Electron transfer and ejection. Interfacial electron transfer and ejection is a relatively slow process and may take 0.1 usec to 1 msec.

(4) Diffusion of products. The time necessary for the products of the primary photophysico-chemical process to diffuse away from the membrane/solution interface is about 1 msec.

the products in some way. A simple example of product separation would be the charge separation process which occurs at semiconductor junctions. If one sandwiches p- and n-type semiconductors, one forms an interface between the two, called a junction. When light falls on the interface, an electron-hole pair is formed across the junction resulting in charge separation and a voltage difference. Charge separation is maintained by migration of the electron and hole away from each other on separate sides of the junction (Tien and Karvaly, 1977). The semiconductor model is analogous to processes desired to create with redox reactants, namely allow a redox reaction to occur and to get the products apart as quickly as possible. The processes of diffusional mass transport or convection are the only means of separating the products in solution, and these are slow compared to the mechanism involved in semiconductors. To enhance mass transport, it is possible to introduce intermediates so that oxidant and reductant are separated by fast electron transfer reactions. In this case, the recombination is prevented (to some extent) by the physical separation of oxidant and reductant using intermediary donors and acceptors.

More germane to our discussion is the light-driven photoredox process illustrated in Fig. 5, in which a membrane provides a barrier to the oxidant (A) and reductant (D), preventing their recombination. Photochemical redox reactions are light-driven in that photoactive pigments are involved in electron transfer from donor molecules (D) to acceptor molecules (A). The net result is that D becomes oxidized (D^+) and A becomes reduced (A^-). The membrane consists of an ultrathin lipid bilayer containing pigments (P). On one side of the membrane are electron acceptor molecules, and on the other side are electron donor molecules. Owing to the presence of the lipid bilayer, there is no overlap on the orbitals of the acceptor and the donor molecules, whereas some overlap of the orbitals between the pigment and acceptor (or donor) are assumed. Upon absorption of a photon ($E_{h\nu} > E_g$) by the pigment, an electron in the highest filled orbital (ground state) is promoted to the lowest empty orbital. The excited electron may return to the ground state or move to the empty orbital of an acceptor. Concurrently, an electron can be transferred from a donor to the empty orbital left in the pigment. Thus, a portion of the free energy of the photon has been conserved in the form of an oxidized donor (D^+) and a reduced acceptor (A^-) on opposite sides of the membrane (Fig. 5).

8. Time Scales for Photophysical and Photochemical Processes.

From the absorption of light by a pigment molecule in the membrane to the formation of a stable product in bulk solution, there are at least four distinct processes occurring with vastly different time scales. Fig. 6 shows the time scales in membrane photobiophysical chemistry, beginning with the primary event to charge separation and dissociation, ending with the final products, A^- and D^+.

III. EXPERIMENTS WITH BILAYER LIPID MEMBRANES (BLM)

This section is devoted to a summary of the experimental methods for forming and studying bilayer lipid membranes of planar configuration, hereafter it will be referred to either as planar BLM or simply BLM(s). Detailed procedures, as well as instrumentation and apparatus common to many studies with BLM including the BLM-forming solutions, formation methods, thickness determination, permeability measurements, electrical measurements, bifacial tension measurements, spectroscopy and light-induced characteristics may be found in the monograph on BLM (Tien, 1974).

A. Materials and Procedure

1. BLM-forming Solutions.

Critical to the success of a stable BLM is the proper formulation of BLM-forming solutions. During the past two decades, various lipid solutions have been established for the formation of BLM: the choice depends on the interest and purpose of individual investigators. The widely used materials for BLM formation include synthetic lipids, oxidized cholesterol, interface-active agents, carotenoid pigments, and extracts of brains, eggs, chloroplasts, mitochondria, red blood cells, and of bacteria. In addition to these materials, liquid hydrocarbons (hexane to hexadecane) and/or other organic solvents (chloroform, methanol and butanol) are generally required. Now, synthetic compounds, tailor-made for bilayer formation are available. Kunitake and Okahata (1980) have synthesized ammonium amphiphiles of single-chain (hexadecyl), double-chain (didodecyl and dioctadecyl) and triple-chain (trioctyl), as well as those containing the diphenylazomethine segment. They have demonstrated that some of these compounds possess esterolytic reactivities. Alkyltrimethyl ammonium bromides in conjunction with cholesterol have been reported by Bagaveev et al. (1979). Ringsdorf and associates (Hub et al., 1980; Benz et al., 1982) synthesized a number of polymerizable phospholipid analogs useful in model membrane studies. Benz et al. have been successful in forming BLM by synthetic lipids and lysolipids containing polymerizable acrylic-, butadiene or diacelylene units, either in the hydrophobic part or in the hydrophilic part of the molecule. Pure lipids, such as monoglycerides, diolein and triglycerides, form BLMs in the absence of alkane solvents. These BLMs are characterized by a large capacitance ($0.59-0.96\mu F/cm^2$) and fluidity (Laprade and Lapointe, 1982). For new investigators, oxidized cholesterol is highly recommended. This compound, although commercially available (P-L Biochemicals, Inc., Milwaukee, Wis.), is easily prepared by bubbling oxygen or air through a solution of 4% cholesterol in n-octane under reflux temperature (125 C) for about 5-6 hours. BLMs formed from oxidized cholesterol solutions are known to be stable up to 3 hours, even after one month's storage.

2. BLM Chamber and Formation.

The materials which are suitable for construction of BLM chambers and supports are limited by the requisites of chemical inertness to both aqueous and organic solvents, electrical insulation and optical transparency. Glass components, because they possess these requisites, are useful when they can be incorporated into the system. Cast Lucite (plexiglass) is electrically nonconductive, transparent, and resistant to many organic solvents. Because Lucite may be easily machined and cemented, it has frequently been used in lieu of glass. However, Lucite should not be used when halogenated solvents, such as chloroform and methylene chloride, are used as lipid solvents. Teflon, polyethylene, and polycarbonate have the requisite chemical and electrical properties and may be machined or cast, but lack transparency and rigidity. Teflon has proven particularly useful in O rings and in aperture-containing sheets or cups for BLM support.

BLM separating two aqueous solutions may be formed using a number of methods including the brush technique, the dipping technique, the bubble technique, the syringe technique, and the monolayer technique (for a detailed description see Tien, 1974). Referring to Fig. 7, a 10 ml Teflon cup with a 1 mm hole in the side is held in a Lucite (plexiglass) block, which has two chambers of about the same volume. Both the cup and the chamber are filled with aqueous solution. Next, a small amount of lipid solution is introduced in the hole with the aid of a small syringe or pipette. Since the area of a lipid film is usually small, it is necessary to observe it with a low power (20-60X) microscope in reflected white light.

When a BLM is formed in a single aqueous compartment, the various transmembrane gradients are nulled. The membrane itself, and the Plateau-Gibbs (P-G) border which surrounds and supports it, are variables which cannot be so well controlled. In a complete BLM system, the total quantity of lipid present would ideally be distributed between the P-G border and the BLM. At equilibrium, the quantity of BLM lipid can be estimated theoretically from the membrane volume by taking the product of membrane area and membrane thickness in a manner analogous to that used with the monolayer film balance. Optical and electrical methods are available which permit measurement of BLM area, and estimates of BLM thickness. A combination of the monolayer technique and the dipping method were first described in 1965 by Takagi, et al. (see Tien, 1974). A partition, made from a sheet of Teflon (25-500 μM thick) with a hole (0.25-1 mm) at the center, is placed between the two halves of a trough with the hole above the aqueous phase. The trough is first filled with an aqueous solution. After the surface has been cleared, a monolayer of a suitable lipid solution is spread in the surface in the usual manner. After evaporation of the solvent, the Teflon partition is then lowered slowly into the aqueous solution. A BLM is thus formed over the hole (see Fig. 11-3, p. 477, Tien, 1974). Conversely, with the Teflon partition bearing a hole in place, the aqueous solution on either side is covered by a monolayer. Then, the levels of aqueous solutions are raised either sequentially or simultaneously above the hole thereby forming a BLM from the apposition of the hydrocarbon chains of the two lipid monolayers. Another variation of the Takagi technique is to have a closed chamber as used in water permeability studies (Fettiplace et al., 1980; Tien, 1974). The closed

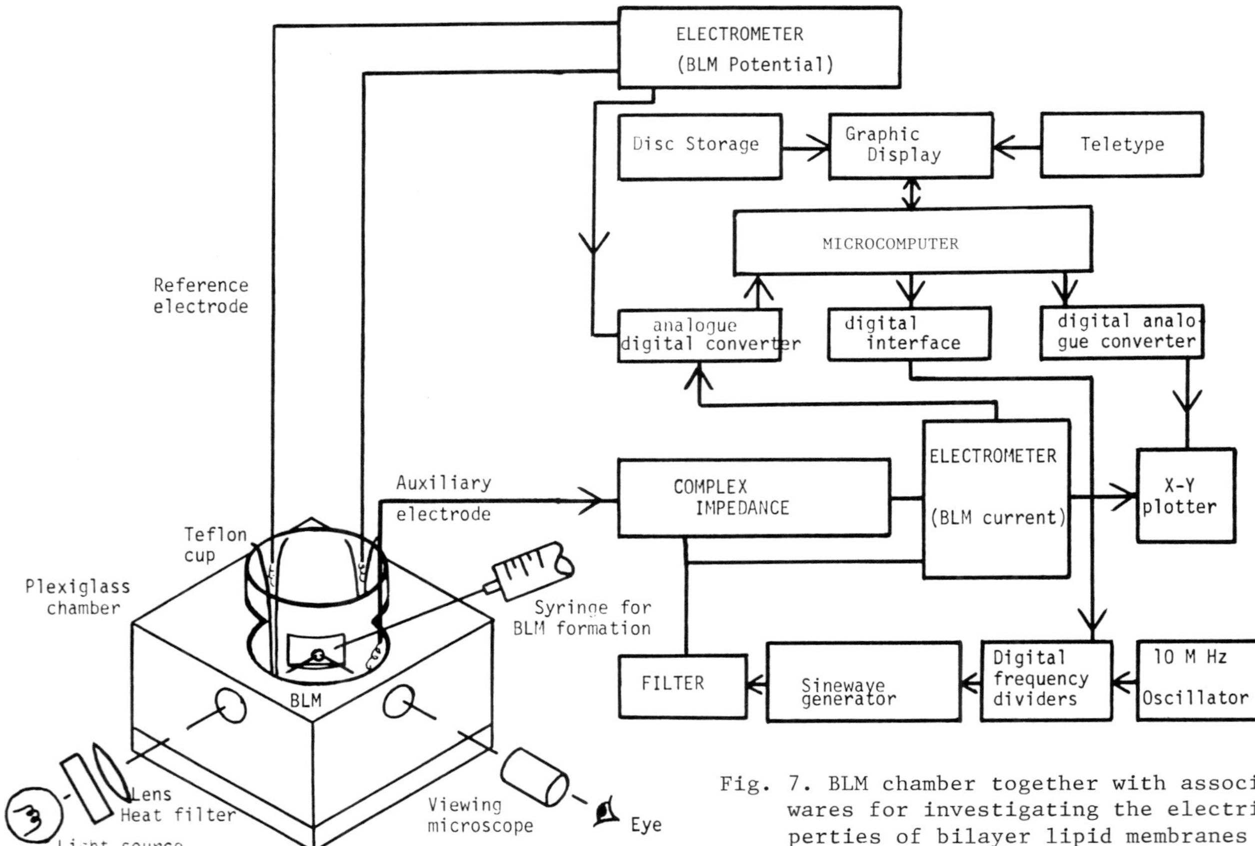

Fig. 7. BLM chamber together with associated hardwares for investigating the electrical properties of bilayer lipid membranes (BLM) being developed in our laboratory.

chamber, with an aperture in its side, is held in place. The chamber, while immersed in aqueous solution below its aperture, is filled completely with aqueous solution. A few drops of BLM-forming solution (n-hexane or benzene as lipid solvent) is introduced onto the surface of the outer aqueous solution. After the solvent has evaporated, the level of outer aqueous solution coated with a lipid monolayer, is slowly lowered below the aperture and then raised carefully above it, thereby forming a BLM. The above procedure may be repeated as many times as necessary until a BLM is obtained (Vodyanoy et al., 1982). The existence of a membrane is easily monitored by observing a change in electrical properties of the BLM. It should be noted that, thus far, even lipid bilayers formed from monolayers according to the Takagi technique are not absolutely solvent-free, since petroleum jelly (vaseline) silicone grease, or squalene in hexane must be used to coat the membrane support (Tancrede et al., 1983).

BLM of spherical configuration can also be produced by blowing a lipid bubble under water at the end of a tubing attached to a syringe filled with aqueous solution. Spherical BLM, up to 1 cm in diameter, have been obtained in this manner (Tien, 1974).

For studying planar BLM, Procopio et al. (1982) have reported a very interesting setup using a quartz cell. In this connection, Yoshida and Okabayashi (1982) in a series of papers reported physiocochemical properties of BLM of several cm^2 in area. Contrasting to these giant BLM, long-lived BLM of μM diameter have been reported (Mountz and Tien, 1978; Sanderman, 1979; Kossi and LeBlanc, 1981; Thompson et al., 1982). These micro BLM are made by filling the smooth circular pores of known diameter and density in a commercially available polycarbonate membrane (Nuclepore filters). Under an ideal situation this new BLM system can be visualized as tens of thousands of micro BLM simultaneously generated in the polycarbonate support. The membranes formed in this manner exhibit far superior stability and manipulability as well as longer lifetimes and larger areas than can be achieved with conventional BLM. Since the new BLM system has the desirable quality of long-term stability, it would be interesting not only to repeat some of the difficult and uncertain experiments reviewed in the following sections, but also to carry out new experiments, such as antigen-antibody-complement interactions, transport and energy transduction studies hitherto plagued by BLM instability problems.

So much for the techniques. We shall now give a brief description of the formation characteristics of a BLM. When first formed, the lipid membrane at the biface is usually thick, and interference colors may not be visible, as viewed through a 10-40 power microscope in reflected white light. Under suitable conditions, the thick membrane begins to thin spontaneously. Rainbow colors soon appear as horizontal bands, at either the top or the bottom of the aperture, depending on the density of the lipid solution used. Initially, only faint pinkish and greenish bands appear. These bands usually increase in breadth and brightness. At other times, depending on the lipid solution, as the lipid membrane thins, bright swirling patterns of interference colors are displayed. Finally the "black holes" are seen, either singly or several at once, and growing at a fairly rapid rate. The appearance and rate of growth

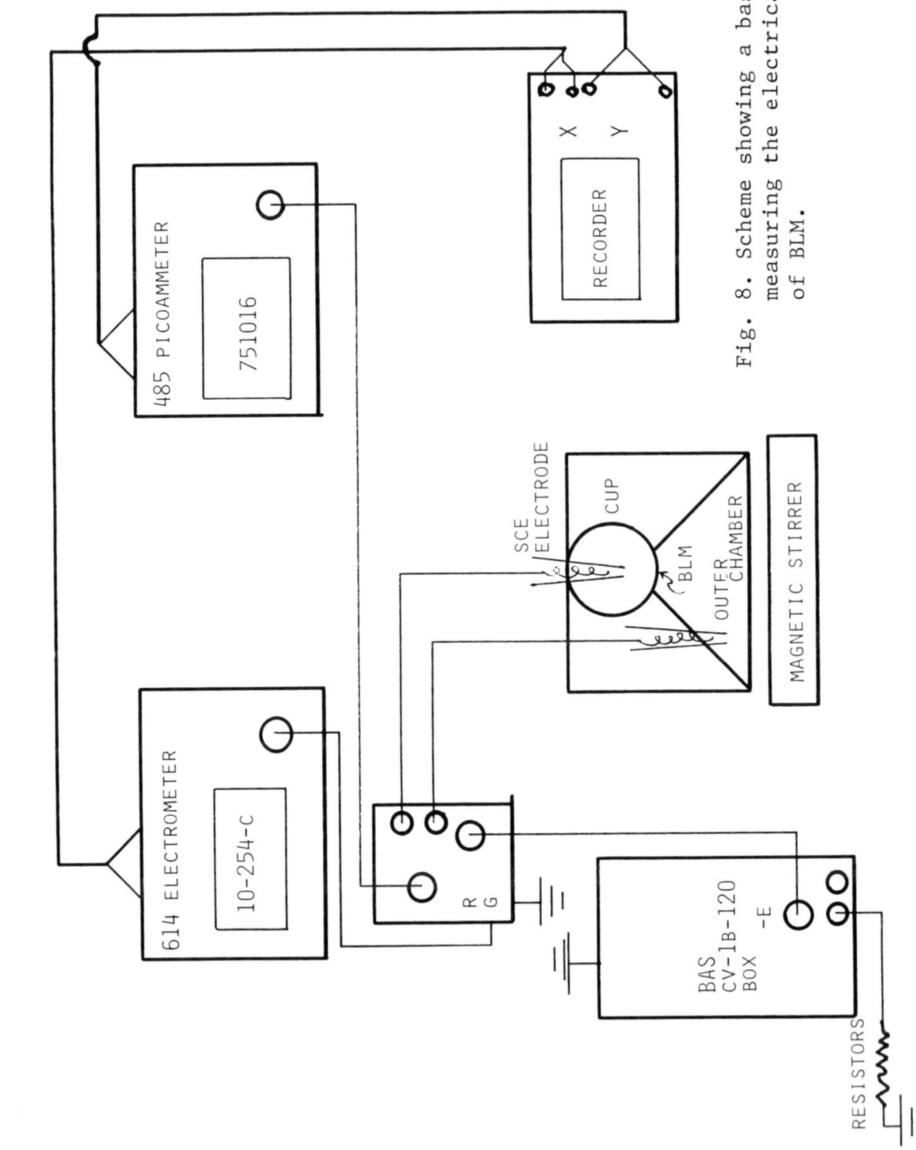

Fig. 8. Scheme showing a basic setup for measuring the electrical properties of BLM.

of black holes depend greatly on experimental conditions. These black holes grow continuously until they cover the entire aperture, resulting in a black or bilayer lipid membrane (hence the acronym BLM). When viewed by reflected light at a large angle of incidence, the BLM appears as a faint gray sheen. At random intervals colored spots swirl in the black membrane then disappear, demonstrating fluidity of the structure.

3. Experimental Setup

a. Basic setup. Fig. 8 illustrates the apparats and circuitry of a basic setup for measuring the electrical properties of BLMs. The setup uses commercial off-the-shelf electronics and circuitry and is quite conventional. The setup does not require any specialized instrumentation and is suitable for several purposes (e.g., undergraduate and graduate lab experiments). To minimize interference by external noise and electrical transients, a Faraday cage made of copper wire mesh is recommended. All cable used are shielded and the shields grounded.

b. Advanced setup. In recent years, many new electrochemical methods have been developed and applied to membrane research. Among them, cyclic voltammetry (CV) turned out to be a very powerful one. Its application to BLM study has been employed in our laboratory since 1982 for investigating membrane mediated redox reactions, properties of dye-doped membranes, and photoeffects in the membrane systems.

A standard circuitry of the setup for BLM cyclic voltametry is available (Tien and Kutnik, 1984). A two or three electrode system may be employed. A voltammetric cell for BLM research consists of a Teflon cup with a small aperture (usually 1.5 mm) diameter) in which the BLM is formed, an outer vessel, and two or three calomel electrodes, which are sufficient for current measurements in the nanoampere range. A voltametric analyzer (e.g. IBM, Model EC/225) in conjunction with x-y recorder can also be used. Another setup consisting of separate measuring devices such as: a fast picoammeter (Keithley, Type 417), a function generator (EG & G PARC, Type 175) and an electrometer (Keithley, Type 610 BR,) is also suitable for two electrode measurements.

c. Setup for studying two BLMs. Recently, our laboratory has initiated a new series of experiments, the aim of which is to investigate the communicative function between membranes (Vassilev et al., 1984; 1986). This entails the formation of two BLMs at close proximity. In one of the arrangements (Fig. 9a), the two holes for BLMs are located against each other while in the other they are in one plane (Fig. 9b). After the formation of the BLMs, square electrical pulses are passed through electrode a between the first and the third compartment and the resulting displacements of the potentials across the two BLMs are measured simultaneously by using the electrodes b and c. The appearance of potential changes across the second membrane are interpreted in terms of electrical coupling between the membranes. In this connection, mention should be made about the experiments of Virysov and Perelygin (1984), who have proposed a mechanism of electric connection between two BLMs.

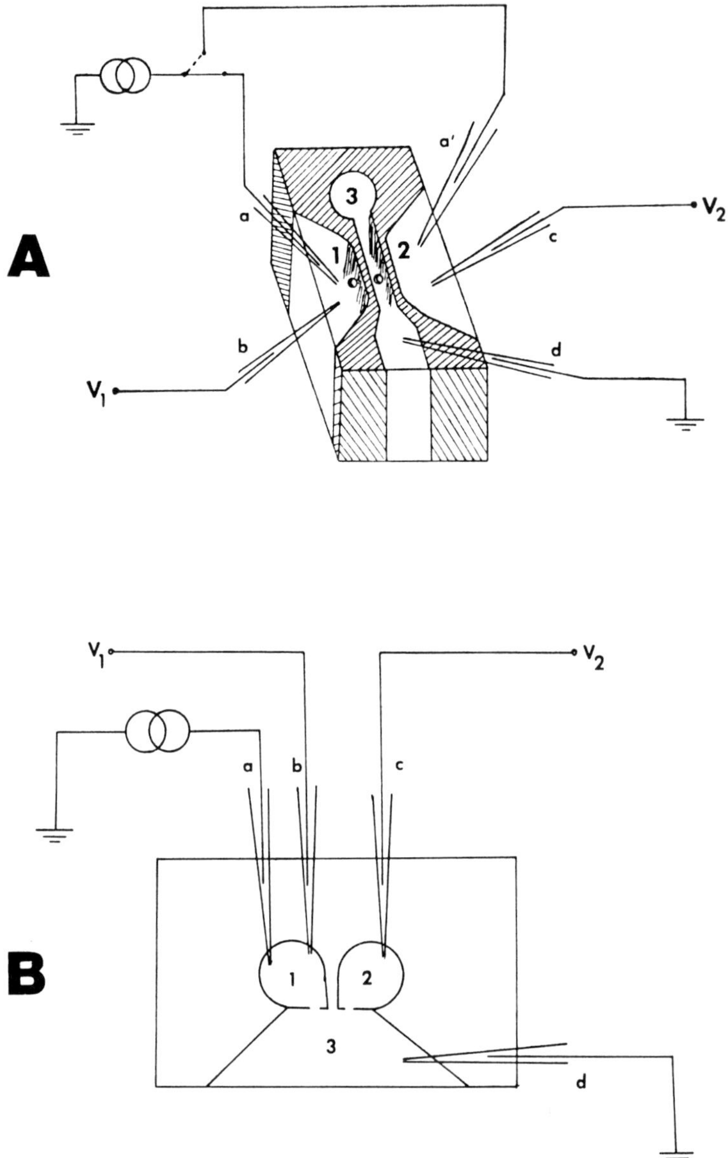

Figs. 9A & 9B. Setups for two BLM experiments. A. Facing each other. B. in the same plane. BLMs are formed in compartments 1,3 and 2,3. a, a', b, c, and d are electrodes. Measurements taken at V_1 and V_2 to ground (Vassilev et al., 1986).

d. Setup for studying light-induced effects in pigmented BLM. In view of the sensitivity and sophistication of spectroscopy techniques, and their utility in elucidating certain aspects of the structure and dynamics of membranes, attempts have been made to apply absorption, and fluorescence spectroscopy to the BLM system (Barber, 1979). Additionally, by combining electrical methods with those of spectroscopy, a photoelectric action spectrum of a pigmented BLM may be obtained. This technique of measuring action spectra of pigmented membranes, termed photoelectrospectrometry, is several orders of magnitude more sensitive then conventional absorption spectroscopy (Lopez and Tien, 1980; 1984; Weller and Tien, 1973; Lopez and Tien, 1984).

Photoelectric effects from BLM in the presence of pigments and dyes may be investigated by two basic types of setups. One type uses the electrometers to detect the so-called open-circuit voltage, while the other monitors transients in feedback circuit that clamp the membrane voltage. The advantage of the latter is large responses at the oscilloscope for photocurrents of short duration. The "open-circuit" voltage method offers advantages of measuring smaller photocurrents of long duration.

Rapid electrical transients of pigment-containing BLM in response to flash excitation can also be obtained. Essentially the same instruments and techniques as those developed for research in photobiology and photocheistry are employed. Huebner (1979) has described an apparatus in detail for recording flash induced voltage transient with 10 ns resolution. The open-circuit method is limited by the resistances of aqueous solution and electrodes used, the electrometer and oscilloscope speed, the flash duration, and the noise superimposed on the open-circuit voltage. With 10^{-3} M NaCl, for example, as the bathing solution, electrometer resolution is limited to about 1 sec, whereas with 1 M NaCl, electrometers follow BLM voltage variations with 100 nsec resolution, and BLM current is proportional to the derivative of the electrometer output.

IV. RESULTS

A. Electrical Properties of BLM in the Absence of Light

We shall review mainly the electrical properties of BLM obtained recently in our laboratory. For literature references prior to 1980, see Martonosi (1982). Here it is sufficient to state that, due to its ultrathinness (< 100 Å), an electric field of 100,000 volts per cm is easily developed across the BLM. This very high field strength greatly modifies, besides charge separation and transport properties, hydrocarbon solubility in BLMs. White (1980) has been studying the physical chemistry of the BLM, and reported how electric fields alter n-alkane solubility in the bilayer, causing it to shift from the bilayer to the Plateau-Gibbs border and microlenses. The solubility of n-alkanes in BLMs has been investigated by Gruen and Haydon (1980). They have reported that the n-alkanes dissolve primarily into the central core of a BLM and alter significantly the BLM's thickness. N-alkanes with increasing chain length have less solubility contrary to the expectation. This phenomenon has been considered at length also by McIntosh et al. (1980) based on the results obtained by the combined use of differential scanning calorimetry, x-ray diffraction, and monolayer techniques. To complete the picture, we should also mention the paper by Sugar and Neumann (1984) who have proposed a stochastic model for electric field-induced pores in membranes (termed electroporation). At subcritical electric fields, the average pore size is stationary and very small. At super critical field strengths, the pore radius increases and, reaching a critical pore size, the dielectric breakdown occurs. If, however, the electric field is turned off before membrane rupturing, the pore apparently completely reseals itself. Hence, the BLM lives. In connection with this discussion, mention should be made about experiments of Dely, Prager, and Puppi (1981) and of Zimmermann and his colleagues. The former group examined the effect of redox compounds and found that the reductants made the BLM stable while the oxidants have the opposite effect. Zimmermann (1982), in his review, states that the maximum voltage across BLMs is about 300 mV. This is contrary to the findings reported by Tien and Diana in 1967, who have observed that a breakdown voltage as high as 550 mV could be set up across BLMs formed from a mixture of cholesterol and dodecyl acid phosphate (see Tien, 1974). The preceding comment notwithstanding, Zimmermann (1982) has presented an excellent discussion on the mechanisms of electrical breakdown of both BLMs and biomembranes and their relevance to cell-cell fusion and channel formation in membranes.

Before discussing our recent results obtained on a BLM system using the powerful CV technique, a brief description of the setup used in our work is in order (see Fig. 7; Tien and Kutnik, 1984). The Teflon cup will be referred to as the inside, and the other chamber will be referred to as the outside. We have used a three-electrode system for obtaining voltammograms in the following configuration: one calomel electrode (SCE) is placed in the Teflon cup and two other calomel electrodes are on the outside. An EC/225 Voltammetric Analyzer (IBM) in the cyclic voltammetry mode in conjunction with an x-y recorder is used throughout in our second setup. The hole of 0.15 cm diameter in the Teflon cup is usually pretreated with the BLM-forming solution to be used later. First of all, in order to demonstrate electronic conduction in BLM in the absence of light, the membrane must be modified to function as a metallic of semiconductor electrode, so that the powerful CV technique can be applied. To impart electronic proerties, we have discovered that, by incorporating an organic semiconductor such as TCNQ (7,7,8,8-tetracyanoquinodimethane) or TTF (tetrathiafulvalene) the BLM becomes electron-conducting (Worthy, 1984).

In order to apply the powerful CV technique to the BLM system, a conceptual effort has to be made, that is to consider one side of the BLM as the working electrode, while the other side is providing the connection to the external circuit. As already mentioned, an unmodified BLM behaves essentially as an excellent insulator (specific resistivity $> 10^{14}$) and does not function as a working electrode. However, upon incorporation of TCNQ into the BLM and in the presence of ascorbic acid in the outer solution and equal molar (0.001 and 0.01)

$K_3Fe(CN)_6/K_4Fe(CN)_6$ in the inside, the observed electrical properties are such (R_m decreasing from 10^8 to 10^6 ohms cm^2, C_m increasing from 0.4 to 0.5μFcm^{-2}, E_m developing to about 180 mV with the ascorbic acid side positive. In the absence of TCNQ in BLM with other experimental conditions the same, the I/V is linear; it practically coincides with the x-axis on the scale used. In the absence of redox couples in the bathing solution, but with TCNQ in the BLM, the I/V curve remains linear. To show more clearly that the redox reactions are occurring across the TCNQ-containing BLM the following experiment was carried out.

To one side of a TCNQ-containing BLM, we added a high concentration of equal molar ferri-ferro cyanide solution, whereas to the other side, aliquots of ascorbic acid of known concentration were introduced (Tien and Lojewska, 1984). Concurrently, the membrane potentials (E_m) after each addition of ascorbic acid were measured. It is most interesting to note that the extrapolated value of E_m at the equal molar concentrations of the two redox couples agrees very well with the difference of the two standard redox potnetials ($E^{o\prime} \simeq 300$mV). The highly asymmetrical current/voltage curves are reminiscent of those of a p-n junction diode which permits an electron current flow in a forward-bias direction only. In our TCNQ-BLM system, oxidation occurs at the membrane-solution interface contacting the ascorbic acid solution and reduction of ferricyanide to ferrocyanide takes place on the other side of the BLM which has a negative polarity. Implicit in this interpretation is the transmembrane movement of electrons via the TCNQ molecules imbedded in the lipid bilayer; that is, the whole system has the properties of a p-n junction with the TCNQ-BLM acting as a rectifier.

To demonstrate that the TCNQ-BLM behaves like a metallic electrode (e.g., Pt which is frequently used in CV), a comparative experiment was carried out recently, in that cyclic voltammograms of quinhydrone were obtained using either Pt or TCNQ-containing BLM under very similar conditions. We have found that in particular the cathodic portions of the voltammograms are quite alike, thus substantiating that the TCNQ-BLM functioned as a working electrode. In this connection, a cyclic voltammogram of horse-heart ferricytochrome c was also obtained similar to that reported by Eddowes and Hill in 1977 (see Tien, 1984), who used a gold working electrode in their experiment. Thus, in order to be consistent with the theory of CV, the TCNQ-BLM must function as a redox electrode in the usual practice of electrochemistry.

The significant advantages of the new technique are, besides its simplicity and the good precision of measurement, the involvement of the lipid bilayer and the capability for future development. Since the electron-transfer chain components are known to be closely associated with the lipid bilayer, the values thus determined hitherto by the usual Pt electrode may be quite different from their actual values in the membrane. Conceivably, therefore, the technique described here offers a new approach to the determination of $E^{o\prime}$ of membrane-bound biomolecules such as the cytochromes and other redox enzymes, using modified BLM as the working electrode. Further, this new type of electronically conducting BLMs coupled with the CV technique may be useful in the study of membrane bioenergetics (see Blank, 1986) and in the designing of molecular electronic devices based on ultrathin films (see Section V A).

B. BLM as a Model for the Photosynthetic Thylakoid Membrane

1. Background.

In green plant photosynthesis, the primary event, together with its subsequent redox reactions, is believed to take place in and across the thylakoid membrane. On the basis of evidence from electron microscopy, the structure of a thylakoid membrane is that of a lipid bilayer with embedded pigment-protein complexes (Olson and Hind, 1976; Metzner, 1978; Akoyunoglu, 1981). To explain the light-driven redox reactions from water to NADP, the so-called Z-scheme, with two photosystems (PS-I and PS-II) was proposed in the 1960s (Barber, 1979). With the establishment of the Mitchell's Chemiosmotic hypothesis in the 1970s, a hypothesis for accounting electron transfer and photophosphorylation via an ion-impermeable lipid membrane, PI-I and PS-II of the Z-scheme have been incorporated into the thylakoid membrane in the late 1970s (Bolton and Hall, 1979; Barber, 1979). Concurrently with afore-mentioned developments, the BLM system was developed in 1962 with the observation of photoelectric effects in chlorophyll-containing BLM in 1968 (Tien, 1976; Antolini et al., 1982). These past accomplishments have enhanced our understanding of light transduction processes in the thylakoid membrane and set the stage for future experiments. In the following paragraphs, I wish first of all to review briefly the recent literature of photoeffects in BLM containing chlorophyll and its related compounds. For references prior to 1980, a number of reviews are available (Tien, 1976; Barber, 1979; Gliozzi and Rolandi, 1984). Secondly, I would like to discuss a redox membrane electrode model, first introduced in 1968, based on our findings and others with pigmented BLM, in relation to the thylakoid membrane of photosynthesis (Robert et al., 1985; Tien and Joshi, 1986).

2. Summary of Recent Experiments.

BLM containing chlorophylls and related compounds have been studied by several groups of investigators (Barber, 1979; Antolini et al., 1982). Elegant techniques and setups have been described by Huebner (1979), Tancrede et al. (1983) and Kutnik and Tien (1984). To elicit appreciable photoelectric effects, asymmetrical conditions across the pigmented BLM are necessary. These conditions can manifest themselves in terms of pH difference, electrical potential difference, different composition or concentration of electron acceptors and/or donors present in the bathing solutions. For electron transfer and charge separation studies, Joshi et al. (1982) have formed BLM from lipid solutions containing Chl or porphyrins with and without added modifiers, such as quinones and carotenes. The additions of quinones and/or carotenes is due to the well-documented evidence that quinones serve as the primary electron acceptor and carotenes as effective light-gathering accessory pigments. Therefore,it seems highly probable that a close proximity between a donor species (e.g., chlorophyll or porphyrin) and an electron acceptor molecule (e.g., quinone) is a prerequisite. To test this hypothesis, several groups of workers have synthesized covalently linked porphyrin-quinone and porphyrin-carotene complexes Tabushi et al., 1979; Moore et al., 1980). Incorporation of these newly available compounds, as a model for the initial photophysicochemical event in reaction centers of photosynthesis, into reconstituted BLM is of interest and has been carried out in our laboratory. The photopotentials obtained on some of these new BLM systems are indeed interesting in that they are larger by at least a factor of three (the highest values hitherto reported) than those of chlorophyll-containing BLMs (see Fig. 10). In this connection, the experiment reported recently by Rich and Brody (1982) is of interest. They found that Chl-BLM in the presence of dihydroxy-carotenoids give rise to much greater photocurrents than either the simple carotenes or the diketocarotenoids. BLM formed using pheophytin (a Mg-free chlorophyll) generate very small photocurrent regardless of the carotenoids used. This is explained in terms of lack of interaction between pheophytin and carotenoids. The negligible photocurrent in pheophytin BLM might also be due to the mismatching of redox potential between pheophytin and ferricyanide, thus preventing the latter to be reduced by the former. The interaction between chlorophyll and β-carotene in BLM has also been studied by Vacek et al. (1982), they observed a decrease of photopotential upon repeated flash excitation, which might be owing to lipid oxidation. The kinetics of photopotential was studied in more detail by Liu and Mauzerall (1985), who explained the decay by a rate constant that exponentially decreases with distance between pigment cation and reduced acceptor (e.g., ferricyanide). The other conclusion from their study is that the pigment cation does not transverse the BLM in less that 10 msec. Instead of using ferricyanide, $FeCl_3$ was used as an electron acceptor in the Kutnik and Lojewska experiment (1983). They measured the photoconductivity of Chl-BLM with added β-carotene as a function of wavelengths, with highest efficiency obtained at 541 nm, in good agreement with a value reported earlier using photoelectrospectrometry (Lopez and Tien, 1980). Photoelectrospectrometry is a technique by combining the classic photoelectric effect with optical spectroscopy. Thus far, only in range of UV and visible wavelengths has been reported (Tien, 1976; Putvinskii, 1977). Putvinskii, for example, has reported the influence of UV light on the stability of BLM formed from mitochondrial lipids and explained his results in terms of photoperoxidation of unsaturated fatty acid chains of phospholipids. Sielewiesiuk (1983) reported the photooxidation of Chl in BLM on the basis of a decrease in photocurrent. Recently, Kadoshnikov and Stolovitsky (1982) reported the results of spectral photoelectrochemical investigations of chlorophyll (Chl) BLM and liposomes. Chl in these membranes has been shown to have monomeric (λ=640 nm) and crystalline (λ=740 nm) forms. The line width of the EPR signal of monomeric Chl cation-radical is within 8-9 G, and that of crystalline is 1.7 G with g-factor = 2.0024. It was also shown that about 25 Chl molecules were incorporated in the aggregates. The BLM, with incorporated Chl and pheophytin, generated photopotentials of opposite signs.

In a novel experiment, Koyama et al. (1982) obtained a spectrum of carotenoid-BLM by resonance Raman spectroscopy, which should be considered as a major advance in BLM spectroscopy. For efficient charge

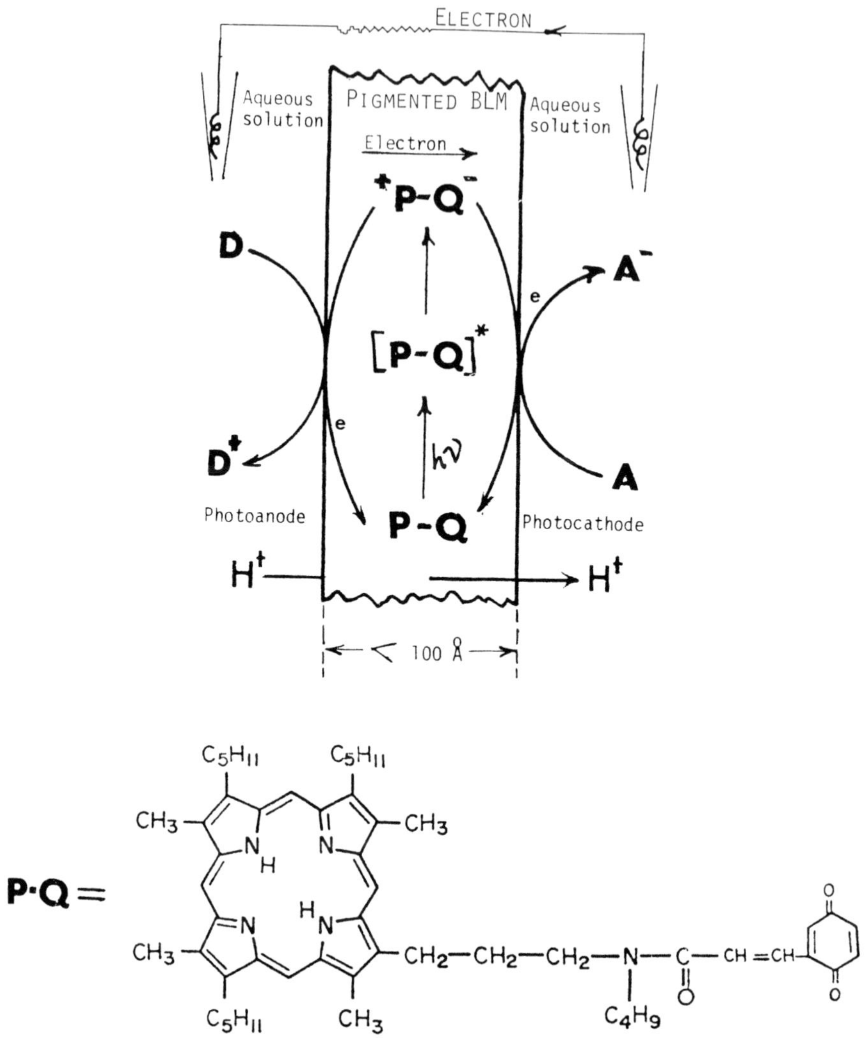

Fig. 10. Light-induced transmembrane charge separation and electron transfer in a BLM containing covalently linked porphyrin-quinone complex (see Joshi et al., 1982 for detail).

transfer, the orientation of chlorophyll molecules at the membrane/solution interface is important. Brasseur et al. (1984) have developed a new procedure for conformation analysis to define the position of chlorophyll in BLM. They found that the porphyrin ring is orientated at an angle of 45 $\pm$ 5° to the plane of the BLM, in excellent agreement with the value reported previously (see Barber, 1979).

In an interesting study, Hattenbach et al. (1982) raised the question, "does phytochrome interact with lipid bilayers?" Phytochrome, a plant chromoprotein, claimed to respond reversibly to red light of different wavelengths (667 and 725 nm), has been incorporated into a BLM. Mueller, Hattenbach and their colleagues (1982) have found no detectable change in membrane conductance during illumination with red light (660 nm). Their finding in BLM, however, does not rule out the regulatory role of phytochrome in plant membranes.

A unique feature of the BLM system is that a coupled photosensitized redox reaction may be independently activated at the two interfaces of the BLM, mimicking the Z-scheme of photosynthesis (1-3,11,12). One way to accomplish this is to use ZnTPP (or MgTPP), as the photoabsorber in the BLM, in conjunction with tris(2,2'-bipyridine) ruthenium ion, $Ru(bpy)_3^{2+}$, methylviologen (MV^{2+}, 1,1'-dihexadecyl-4,4'bipyridium) as electron acceptor. EDTA (ethylenediamine-N,N,N',N'-tetraacetate) as electron donor, and vitamin K (VK_1) as a hydrogen carrier (Tien and Joshi, 1986). In one system, the BLM phase contained VK_1 and MV^{2+} with $Ru(bpy)_3^{2+}$ added to one side of the bathing solution and EDTA to the other side. The second system differed from the above by having only ZnTPP added to the BLM-forming solution. All experiments were performed in the absence of O_2 by bubbling Ar (or N_2) through the bathing solution for at least 30 minutes before forming BLM. Photovoltages with viologen-free side were positive 50 mV and 90 mV, respectively, for the first and second systems. The time course for the photocurrent is as follows: photocurrent rises rapidly, which is followed by a slow increase, then it decays to a steady value. When the light is switched off, the current goes back to the dark value. Since in both systems the photoresponses were observed, $Ru(bpy)_3^{2+}$ must be bound to the BLM. The presence of ZnTPP in the BLM greatly enhanced the response, which implies facilitated electron transport. Photoexcitation of $Ru(bpy)_3^{2+}$ and/or ZnTPP leads to viologen reduction, as evidenced by the polarity of photopotentials. Since acceptor (MV^{2+} or Fe^{3+}) and donor (EDTA) are physically immobilized, some transmembrane redox reactions must take place (see Fig. 11). The most direct interpretation of the data is that reduction of $Ru(bpy)_3^{2+}$ occurs on the side containing EDTA with transmembrane electron tunneling facilitated by ZnTPP and VK_1. The drop in photocurrent may be caused by the diffusion of reduced MV^+, which could become more soluble in the membrane phase. In analogy to the Z-scheme of photosynthesis, this two-step electron-transfer system should be further studied.

Reconstitution with entire reaction centers (i.e. Photosystem I of green plants) has also yielded a photoelectric effect (Lopez and Tien, 1984). Upon illumination, the PS-I reaction center (RC) containing BLM generated transient and steady-state light-induced voltages and currents. The wavelength dependence of the photoresponse matched the

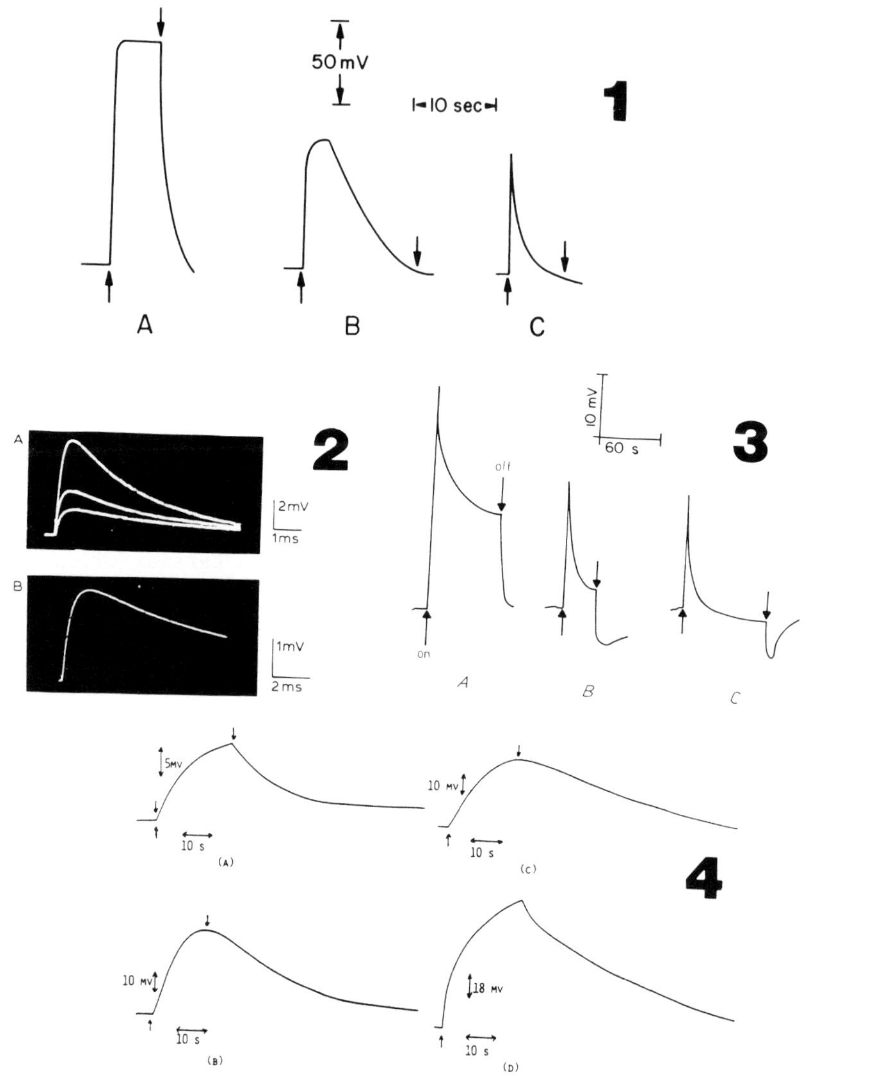

Fig. 11. Photoelectric effects observed in a variety of pigmented BLM systems.
(1) Chlorophyll-containing BLM (see Olson and Hind, 1976. p.114)
(2) Rhodopsin BLM (Liu and Tien, 1982. p. 76)
(3) Bacteriorhodopsin BLM (Higgins et al., 1979. p.513)
(4) ZnTPP-Ru(bpy)$_3^{2+}$ BLM (Tien and Joshi, 1985. p.245)

absorption spectrum of reaction centers (RC). The main features of the photoresponse are explained by the redox membrane electrode model (see below). The observed photosignals are due to the transfer of electrons from a donor to an acceptor located on opposite sides of the BLM (see Fig. 11). This RC-containing BLM provides direct evidence that the RC is an intrinsic membrane protein that span the lipid bilayer. In connection with reconstitution experiments, mention should also be made of pigmented liposomes, reported recently by Chen and Li (1986).

3. The Redox Membrane Electrode Model.

In the first report describing the BLM photoelectric effect, three possible mechanisms were proposed, one of which is the redox electrode mechanism; electronic charges are generated as a result of light excitaation which cause one side of the BLM-solution interface to be reducing and the other side oxidizing (see Tien, 1976; Barber, 1979). Before discussing the redox membrane electrode model, one may recall the usual picture of a BLM separating two electrolyte solutions consisting of a liquid hydrocarbon core (~50 A) sandwiched between two polar regions which contain the sites for ion binding. These sites, if charged, determine the interfacial potentials, and constitute one of the two major factors governing the charge transport, the other factor being the hydrophobic layer. The number of pigment molecules in a BLM is estimated to be about 3×10^{13} molecules per cm^2 with average distance of about 10 ± 5 Å between photoactive molecules (e.g., chlorophyll or Chl). The porphyrin is oriented 45 ± 5° to the plane of the BLM at the interface and most likely located between the polar groups of the lipids (Barber, 1979; Brasseur et al, 1984). Therefore, energy transfer is possible from a donor (D) to an acceptor (A), if the bands of respective D and A are overlapped. It should be noted that a Chl can serve as either a donor or acceptor in the process. When electrons and holes are generated in the BLM, they will polarize the surrounding and attract charges of opposite sign across the lipid bilayer. These charges will, in turn, attract the electrons (or holes) toward the oppositely charged interface, resulting in the reduction of the interfacial potential. Similarly, it can be shown that the interfacial barrier for charge transport can also be reduced by an externally applied voltage across the BLM (Tien, 1974).

If a Chl-BLM separating two different redox couples is illuminated, the basic experimental results are as follows: a change in membrane potential ($E_{op} = E_{h\nu} - E_{dark}$) occurs under open-circuit conditions, and a change in membrane current ($I_{sc} = I_{h\nu} - I_{dark}$) under short-circuit conditions. E_{op} or I_{sc} vs. time curves can take several forms, depending on the experimental conditions. Fig. 11 shows some typical curves obtained in a variety of BLM systems.

Evidence favors a redox electrode mechanism in which the excited pigment molecule is directly involved. It seems reasonable that the excited pigment molecule (P) ejects an electron to an acceptor (A). This is followed by accepting an electron from a donor (D) by the oxidized pigment molecule (P^+). The separated charges are prevented from recombining by the presence of the lipid bilayer core of the BLM.

Thus, an electric field should appear shortly after the absorption of photons by the pigments in the BLM.

Mechanistically, two modes of operation are possible, i.e., via a one-photon or two-photon process (Tien and Joshi, 1986). In the former case, the ion-radicals (P^+ and P^-) generated are reduced or oxidized, respectively, by D and A. In the two-photon process, the reaction of P with A on the acceptor side (left) is to produce A^- and P^+. Being in the membrane phase, this P^+ is highly likely to be reduced by an adjacent P^* on the donor side (right) of the membrane. The resulting P^+ on the donor side is, in turn, reduced by D. The overall process is the sum of the equations given in Fig. 12, namely,

$$A + D \longrightarrow A^- + D^+ \qquad (8)$$

In the two-photon scheme, it is envisioned that transmembrane diffusion of P^+ (or P^-) is unlikely. Rather, the electron translocation across the membrane is explained in terms of a tunneling mechanism between P^+ and P^* . This suggestion may be checked experimentally by measuring the photoeffects of the system as a function of temperature (Tien, 1984).

It is worth restating that one of the aims of pigmented BLM studies is to elucidate the mechanism of quantum conversion of photosynthetic thylakoid membranes. It seems that, in natural photosynthesis, the uniqueness of having an ultrathin lipid membrane, besides providing an appropriate environment for pigments and redox proteins, serves as a barrier to suppress the back reaction of the photoproducts (e.g., electrons and holes). With the photosystems (or pigments) located in different parts of the membrane (hydrophobic lipid interior and hydrophilic aqueous interface), these make possible for the upgrading of the energy of photons. As illustrated in Fig. 12, for two photons absorbed (one at each interface), only one electron and one hole are involved in the ensuing redox reactions. Previously, we have demonstrated in a system consisting of a chlorophyll-containing BLM with water-soluble chlorophyllin added to one side of the bathing solution, a very large photoresponse ($E_{op} > 200$ mV) was observed (Tien, 1974; Barber, 1979). This earlier work and the data obtained in the third type of experiments (MV^{2+}/ZnTPP-BLM/$Ru(bpy)_3^{2+}$, (EDTA) reported recently (Tien and Joshi, 1986), are consistent with the generally accepted Z-scheme of green plant photosynthesis (Akoyunoglou, 1981) in which Photosystem I and Photosystem II are believed to be spatially separated and located in different environments of the thylakoid membrane, thereby permitting the cooperation of two-photons to generate one powerful electron-hole pair for the reduction of $NADP^+$ and oxidation of water on opposite sides of the membrane.

In concluding this section, it should be mentioned that a pigmented BLM/aqueous interface has been likened to a double-Schottky barrier (or p-n type) cell, with the aqueous solution playing the role of metal. Bearing this in mind, one side of the membrane acts as a photocathode (p-type junction) and the other acts as a photoanode (n-type junction) as illustrated in Fig. 12. The photovoltages observed are influenced by the redox compounds (electron donors and acceptors) present in the bathing solution and/or externally applied voltage. In the scheme shown, two types of processes can be distinguished: photosynthetic and photocatalytic. In the former, the overall free energy is positive ($\Delta G >$ 0) in that external energy input is necessary to drive the uphill reaction. In photocatalytic processes which are thermodynamically favored ($\Delta G < 0$) the energy of light is not stored but merely used to overcome the energy barrier of the reaction.

C. BLM as a Model for the Photoreceptor Membrane of Vision

1. Background.

It seems appropriate to begin the discussion of the biophysics of vision with a rough sketch of the gross anatomy of the eye. From front to the back, the eye consists of cornea, anterior chamber, iris, lens, vitreous humor, retina, and optic nerve. Photons of light enter the eye at the cornea, passing through the pupil, focusing by the lens and finally reaching the retina. It is at the retina where the energy of photons are transduced into other forms of energy, which in turn result as electrical messages eventually decoded by the brain. The task of understanding how the eye transduces electromagnetic radiation into a signal that is ultimately perceived by the brain, requires an integrated knowledge of physics, photochemistry and neuroscience. As yet, the task of our understanding of vision at the molecular level is far from complete. We will describe what is known, concentrating on those aspects that are amenable to experimental approach with techniques and instruments now available.

In vertebrates, vision begins when the photons of light hit retina, where photoreceptors are located. There are rods and cones, so-called after their typical shape. Cones function in bright light for color vision, whereas rods operate in dim light, capable of detecting a single photon. Because of ease of isolation, rods have been extensively studied. A rod consists of two segments, the outer and the inner. The inner segment is concerned with cellular metabolism, which houses mitochondria, nucleus, etc. The outer segment, the light-absorbing body of the photoreceptor, is composed of 1000 or so disc-shaped sacs, neatly stacked (~300Å apart) which are enclosed by the plasma membrane. The sacs in a rod cell are specialized bilayer lipid membranes that contain the crucial light-sensitive molecule, rhodopsin. As for cones, the visual pigment appears to be located in the infoldings of the plasma membrane (Knowles and Dartnall, 1977; Bownds et al., 1985; Stieve, 1986).

Upon absorption of light, a transient hyperpolarization of the plasma membrane of photoreceptor is observed. It is generally assumed that an intracellular chemical transmitter exists that serves as a mediator between photoexcitation of rhodopsin in the sac membrane and gating of Na channels in the plasma membrane. Substances such as Ca and cyclic guanosine-3',5'-monophosphate (c-GMP) have been proposed as mediators (Stieve, 1986). Recent work in different laboratories has suggested that Ca^{2+} does not mediate visual excitation as evidenced by a decrease rather than increase of the free Ca^{2+} concentration after

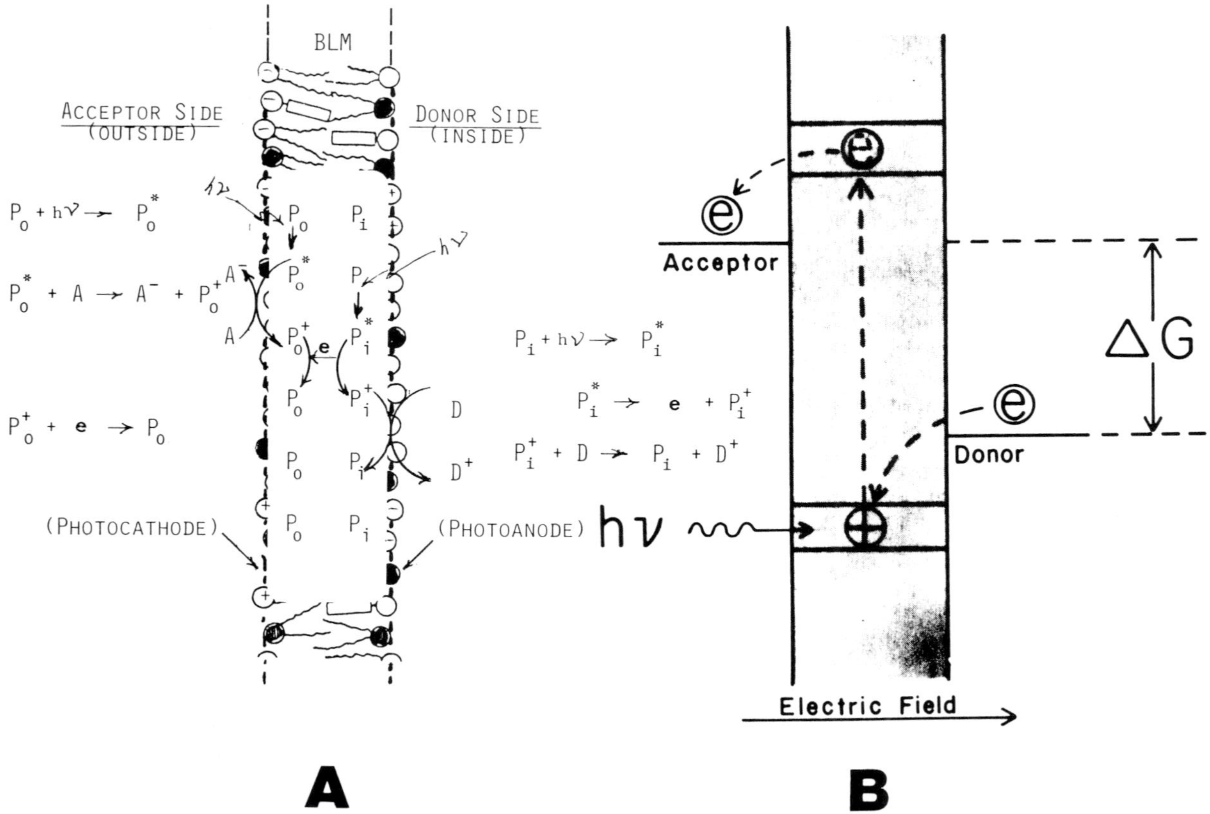

Fig. 12. Light-driven transmembrane charge separation and electron transfer in a pigmented BLM. (A) Diffusion of P, P^* and P^+ in the bilayer is restricted, owing to the viscosity of lipids. Intramembrane electron transfer is assumed by tunneling. Interfacial electron transfers to A (acceptor) and from D (donor) are rapid. To sustain steady-state photocurrent, transmembrane movement of electrons from the donor side to the acceptor side must occur. (B) Overall energy transduction in terms of free energy, G (see text).

injecting calcium buffers or cGMP into ROS cells. The Ca^{2+} hypothesis of Hagins predicts a sudden permeability change of sac membrane after light absorption by rhodospin, whereas the c-GMP hypothesis suggests that light leads to hydrolysis of c-GMP which in turn triggers the receptor potential. It is fair to state that, at present, little is known about the permeability of the sac membrane. Thus, questions like "How does excited rhodopsin alter the biophysics and biochemistry of its surroundings to generate an electrical signal eventually perceived by the brain?" "By what mechanism the energy of photons is transduced?" Rods are not only sensitive to a single photon, but also are able to adjust their sensitivity by light adaption over six orders of magnitude of light level. Facts like these are known. However, the details of amplification of light effect and questions posed above have not been resolved. Since rhodopsin and lipid bilayer are intimately involved, many investigators have resorted to artificial bilayer lipid membranes (BLMs and liposomes) to examine the rhodopsin (and its precursor, retinal) interactions with other components of the visual system (Tien, 1974). What follows is a brief summary of experiments using planar BLM as a model of the visual receptor membrane with emphasis of the work carried out in our laboratory.

2. BLM Studies.

The reconstitution of an artificial photoreceptor by incorporating rhodopsin into BLM was attempted as early as 1963 (see Burton, 1967). The early investigators of the BLM system extracted the rhodopsins from the retinal tissues of the eyes of bees, frogs and cows. These extracts were then added to one side of the BLM, but no photosensitivity was observed. One of the difficulties encountered was membrane instability, because of the presence of rhodopsin along with the detergent used in rhodopsin preparation. In contrast, it was discovered later that retinals and β-carotene, being oil soluble, could readily be incorporated in oxidized cholesterol and/or lecithin BLM (Anonymous, 1969; Tien and Kobamoto, 1969). Further, these carotenoid BLM generated appreciable photoelectric effects. The main findings were as follows. Depending on the carotenoid pigments used in the BLM and external factors, such as pH and applied potential, the voltage/time curves could vary from a simple monophasic response to biphasic modes (Kobamoto and Tien, 1972). In the latter, the photopotential waveform obtained with retinal-containing BLM was quite similar to that of early receptor potentials (Knowles and Dartnall, 1977). This early receptor potential (ERP) is observed when an intense light flash impinges on the retina, and is so-called because it precedes the full electrophysiological response of the retina. The ERP has essentially no latency (less that 0.5 μs) and its amplitude is a linear function of light intensity at lower levels, suggesting that the ERP is associated with initial events in transduction prior to the amplification mechanism between photochemical and neural excitation.

An earlier attempt to incorporate rhodopsin into BLM was reported. Takagi et al., in 1965, pioneered a different technique of BLM formation by adjoining the hydrocarbon chains of two lipid monolayers (see Section

III A 2; Tancrede et al., 1983). They observed that the BLM resistance as lowered to 10^5 to 10^3 ohm-cm^2 in 1.0 M KCL solution, and increase to 10^6 ohm-cm , when one side of the bathing solution was replaced with isotonic $CaCl_2$. Excitability, either electrical or light-induced, was not observed in their experiments, however. Using the classical system, Schadt (1973) also studied all-trans-retinal BLM, this time using $K_3Fe(CN)_6$ as the electron acceptor. Under the influence of light, excited retinals act as electron donors to $Fe(CN)_6^{3-}$, reducing Fe(III) to Fe(II). Further, Schadt found that the waveform of the photosignals depend on the pigments in the BLM, being pronouncedly biphasic when retinal was photo-oxidized to vitamin A acid.

Successful interaction of ROS with a BLM was reported by Fesenko and Lyubarskiy (1977), who observed a light-induced conductance increase. Depending on whether ROS was simply adsorbed at the BLM/solution interface or incorporated into the lipid bilayer itself, no specific mechanism could be chosen to explain the observed photoeffect. Similarly, exploratory experiments of forming rhodopsin BLM by apposing the two proteolipid monolayers (the Takagi technique) were also inclusive (Montal, 1975). More positive results were reported by Gambale et al. (see Gliozzi and Rolandi, 1984), who detected that the conductance of BLM containing honeybee pigment is increased in the dark over pigment-free BLM. They explained the observed photoeffect of a "conducting" dimer in the BLM. Related to this work is the experiment of Stillwell et al., (1982) who reported that retinol and retinoic acid, when incorporated into BLM and liposomes, increased the permeability to ions.

For investigating the ERP and its possible physiological role, many other model systems have been employed. Aizawa et al. (1977) employed a photosensitive membrane by casting a solution of 11-*cis*-retinal, phospholipid and triacetyl cellulose, whereas Kossi and LeBlanc (1981) succeeded in forming a rhodopsin-containing BLM on a porous disc. Mention should also be made of pigmented liposomes, reported by Schmidt (1984).

One problem common to all afore-mentioned membrane systems, is the orientation of chromophores in the lipid bilayer, which may be estimated from the polarized spectra of BLM (Antolini et al., 1982). New data have been obtained on the orientation of β-carotene and retinals in different types of lipid bilayer. Koyama et al., (1982) reported for the first time a resonance Raman spectrum of a β-carotene BLM. McIntyre and Kirschfeld (1981) reported the absolute absorption of an insect pigment that permits them to estimate its concentration and dichroic properties. They concluded that the photostable pigment is a carotene aligned parallel to the plane of the membrane. This view is contrasted in a brief report by Johansson et al. (1981) who suggested that carotene is oriented with its long axis perpendicular to the fatty acid chains in the membrane. Acuna and Gonzalez-Rodriguez (1979) reported that, in dark adapted bacteriorhodopsin, retinal makes a $20 \pm 3^\circ$ inclination angle relative to the plane of the membrane.

In the experiments reported recently, ROS fragments, from which soluble proteins had been depleted, or purified rhodopsin reconstituted with phospholipids in aqueous suspension, were incorporated into one side of the BLM (Liu and Tien, 1982; 1985). Electrical response to a white light flash was measured as pH, temperature and rhodopsin concentration varied. We observed a high amplitude electrical signal, apparently associated with conformational changes of rhodopsin. The action spectrum of the BLM-bound ROS photopotential arises from the bleaching of rhodopsin. A linear relationship was observed between a logarithmic plot of the amplitude of the photopotential rise and the number of flashes through the third. For later flashes, the decrease was slower. The relationship between photopotential rise and intensity of the flash shows that the bleaching rate was linearly related to flash intensity until higher intensities. Since every rhodopsin molecule absorbs but one photon, the effect upon bleaching of increasing intensities of light will eventually become negligible assuming light-scattering, absorption by bleached rhodopsin products and rhodopsin regeneration can be ignored. We also observed that photopotential amplitude increased with temperature and vanished when temperature exceeded 50° C or approached 0° C. From these findings, we infer that the photopotential rise in their experiments is identical with R_2 in the ERP of intact retinae of eyes. Both the amplitude and shape of the photopotential in our experiments were closer to the typical intracellular R_2 of the ERP than to the extracellular R_2. At 23°C, the amplitude of the initial photopotential rise was usually 3 mV or more and could increase to more than 8 mV at 37° , from which a fairly strong electric field (over 10^4 V/cm) was established across the BLM by the bleaching of rhodopsin molecules. The activation energies for the decay calculated was 125 $\pm$ 15 KJ/mol. The observation of reproducible high-amplitude photopotentials in stable, rhodopsin-binding BLMs reinforces the role that this system may play as a model of photoreceptor cell transduction.

Finally, relevant to the study described above, Bauer et al. (1984) in particular reported photoelectric signals generated by bovine rod outer segment sac membrane attached to a BLM. They interpret their results in terms of transient permeability increases after illumination but do not rule out a simultaneous contribution of several ionic mechanisms to the observed photosignal. In this connection, it is worth noting that exposure of the inner sac membrane surface to Ca does not radically affect the response of the ion channels, whereas very low level of cGMP rapidly opens them. It has been shown that bleached rhodopsin activiates the hydrolysis of cGMP. GTP is also required. Two distinct multi-subunits or arrangements with respect to the membrane is not yet clear. It is not surprising that none of the nucleotides nor cations had any effect on the rhodopsin photopotentials in our recent experiments, since all the evidence thus far favors a unidirectional chain of events triggered by rhodopsin bleaching. Nevertheless, study of activation by rhodopsin of one or both of the peripheral proteins would be advanced by a model system in which the location and orientation of rhodopsin in the membrane has been unequivocally established (Liu and Tien, 1985). Much less established are the light-sensitive channels in the sac membrane. What role does the rhodopsin molecule play? Perhaps with the advance in patch-clamping and recording techniques, future experiments with reconstituted rhodopsin-BLMs may provide molecular details of the interaction among Ca cyclic nucleotides and light-modulated channels in the visual receptor. The rhodopsin-containing sac membrane of the receptor and the outer plasma membrane of the rod (or cone) are isolated from each other. To convey the information that a photon has been absorbed by the receptor to the plasma membrane, a mediator (also called internal messenger) has been postulated. The first suggested mediator was calcium ions, the release of which from rhodopsin-sac membranes would result in the closing of ion channels in the plasma membrane, thereby causing a change in the membrane potential. Although the Ca hypothesis seemed plausible, cyclic nucleotides (e.g., c-GMP), the second candidate for the mediator, appears to be more likely. The results of recent patch-clamping experiments indicate that Ca-chelating agents do no alter the light response. Further evidence for the lack of Ca^{2+} involvement is that a lowering concentration of c-GMP closes the ion channels in the plasma membrane, thereby modulating the membrane current, whereas Ca ions do not (Bownds et al., 1986; Stieve, 1986).

D. BLM as a Model for the Purple Membrane of Halobacterium Halobium.

1. Background

To shed light on both photosynthesis and visual processes, one can also study the extraordinary and unusual photosynthetic bacterium, *H. halobium* (Schreckenbach, 1979; Oosawa and Imai, 1982). Extraordinary in that *H. halobium* requires an almost saturated NaCl solution for growth. What is unusual about *H. halobium* is that it can photosynthesize ATP when grown under anaerobic conditions. When it is grown in the presence of oxygen and nutrients without light, *H. halobium* thrives by oxidative phosphorylation. Therefore, *H. halobium* is an excellent experimental organism that is being extensively used for studying the structure and function of membranes, the bioenergetics of the cell, and the possible utilization of solar energy (Bolton and Hall, 1979).

The purple color of *H. halobium* is due to a rhodopsin-like complex termed bacteriorhodopsin (BR) in the plasma membrane, when the cells are grown in the absence of a suitable source of nutrients or at low oxygen tension. Bacteriorhodopsin, upon absorption of light, undergoes a photoreaction cycle which results eventually in the generation of an electrochemical gradient of hydrogen ions that can be used as the driving force for ATP synthesis in accordance with the chemiosmotic hypothesis of Mitchell. The purple membrane, of which BR is the main constituent, is made of patches covering as much as 50% of the surface of H. halobium cells. On the basis of electron microscopy and X-ray diffraction studies, the purple membrane may be considered as two-dimensional crystals comprising an array of BR in a lipid bilayer matrix. Chemically, the purple membrane consists of 25% lipids and 75% protein which has a MW of 26,000. This protein contains two chromophores, 13-*cis* and all-*trans* retinals, linked as a Schiff base to a lysine residue of the opsin. Naturally, this pigmented protein, spanning the plasma membrane, is presumed to be α-helical. This view is supported by circular dichroism studies. The amino acid sequence of various regions of BR has been determined, suggesting the retinal lysine is situated towards the inside surface of the purple membrane. Functionally, it is assumed that purple membrane functions as a light-driven hydrogen ion pump translocating hydrogen ions across the cell (Oosawa and Imai, 1982).

In the dark, the cells at pH 6.5 display pH gradient of about 0.5 units with inside alkaline, and in the light about 0.65 units. The pH gradient appears to be dependent on the external pH. Also of interest is the finding of transmembrane Na^+ and K^+ concentrations. The interior concentration of K^+ was found to be 3 to 5 M and that for Na^+, 0.5 to 1.2 M. Further, K^+ can be retained for days, even if the cells have been starved or poisoned. This observation strongly suggests the intracellular binding of K^+.

2. Model Membrane Systems

To aid the understanding of *H. halobium* at the molecular level, two types of artificial bilayer lipid membranes have been employed. As already discussed, the first consists of lipid microvesicles (liposomes). In this system, a bilayer lipid membrane of spherical configuration encloses a volume of aqueous solution. Since the liposomes are very stable and can be easily made in quantity, they are ideally suited for studies of permeability, ATP synthesis, chemical reactions, and oxygen evolution. The second system comprises of a planar bilayer lipid membrane (BLM) separating two aqueous solutions where photoeffects can be readily measured electrically, whereas it is not possible to do so in either intact cells or BR-containing liposomes (Antolini et al., 1982). Therefore, incorporation of BR into planar BLMs have been carried out successfully for electrical measurements by several groups of investigators. The generation of a photoelectric potential difference and a photocurrent by BR was measured in analogous manner as that of other pigmented BLMs described in the proceeding sections. Karvaly's group (Karvaly and Dancshazy, 1976; 1977; Ormos et al., 1978) appears to be the first to observe both photovoltaic effects and photoconductivity in BR-containing BLM, although the group headed by Skulachev had earlier reported similar effects in thick lipid membranes doped with BR (Drachev et al., 1984). In the experiments reported by Karvaly and his colleagues, the side of the BLM with sorbed pigment was negative in the light. Additionally, they have obtained photovoltaic action spectra of the pigmented BLM, which revealed some differences from the absorption spectrum of BR in aqueous solution. These differences coincide with absorption bands of carotenoids, indicating a possible role that the carotenoids may play in enhancing the photoelectric process. Further spectroscopy of BR-BLM has also been reported (Lopez and Tien, 1980). In a related experiment, it has been confirmed (Drachev et al., 1984) that the photoresponses of pigmented BLM are specific to the lipid bilayer region, and not to the Plateau-Gibbs border, as has been described earlier by (Hong, 1980; Mauzerall, 1979).

In our laboratory, we were successful in incorporating BR and purple membrane (PM) fragments into the BLM either directly or via liposomes (Higgins et al., 1979). The results obtained are as follows: (i) a photoresponse was detected when liposomes containing PM were fused by BLM. Low resistance membranes, or membranes shunted with an external resistor of 10^9 ohms, showed decay of the initial light response to some equilibrium value in the light in both presence and absence of octadecylamine. The light response could be abolished by the addition of a sufficient amount of triethylamine to either side of the membrane, (ii) the photo-voltage action spectra of BLM containing PM either directly or via liposomes were measured, and found to follow the absorption spectrum of bacteriorhodopsin, and (iii) liposomes containing bacteriorhodopsin (BR) were fused to one side of planar BLM. The photopotential resulting from flash excitation rose and fell as the sum of the three exponentials, with time constants for the leading edge of 30 $\pm$ 10 μs and 35 $\pm$ 10 ms. The decay time constant of the photopotential was 840 ms, a value consistent with the membrane time constant, given by the membrane dark resistance and capacitance (see Fig. 11). In this connection, Rayfield (1982) reported recently the short-circuit photoresponse of BLM containing bacteriorhodopsin to short (35 ms) and long (3.5 s) light pulses. It is shown that if the light pulse is short, compared with the charging and discharging times of the BLM capacitance, the temporal response of the light-driven proton pump can be measured. Experimental evidence indicated that the shape of the leading and trailing edges of the photoresponse curve for the pump deviates from simple exponential behavior. The short-circuit photoresponse of Chl-BLM was also studied for comparison purposes. More recently, Severina (1982) has shown that the photocurrent generated by BR-containing BLM can be increased by the presence of nystatin, whereas the photocycle of BR with modified retinals together with its photoelectric responses, ranging from less that 200 nsec to msec, has been investigated by Drachev and colleagues (1984). In connection with nystatin reported by Severina, a proton carrier, such as CCCP (carbonyl cyanide-m-chlorophenylhydrazone) can also dramatically alter the properties of BR-BLM. Bienvenue et al. (1984) reported that, in the presence of CCCP, the photoresponse of BR-BLM appears to be due to the pigment embedded in the lipid bilayer. Additionally, the direction of proton flow can be completely reversed under certain applied voltage conditions. Mirsky et al. (1984) have also studied the photoresponses of BLM modified with BR-containing liposomes and found that, in the presence of Ca^{2+}, short-circuit currents observed are not only due to BR in liposomes attached to BLM, but also to BR in the BLM itself. They also suggest that a BLM containing BR is the most convenient model system for BR investigations.

E. Miscellaneous Studies

In a series of papers, Huebner and co-workers have reported photoeffects in dye- and aromatic amino acid sensitized BLMs. Huebner and Varnadore (1982) reported that voltage transients are generated across BLMs by light flashes, as a result of photophysical processes in sorbed dyes, which displace electrical charges. Experimental results are presented for an azo dye, three carbo-cyanine dyes, an amino-pyridinium dye, and a xanthene dye (erythrosin, Red dye No. 3). Photoeffect quantum efficiency values, obtained by calculating optical absorption cross sections from the dyes' molar extinction coefficients in aqueous solutions are given. Huebner and associates (1982) present a hypothesis which suggests that: (i) the fraction of sorbed dye, which displaces charge from one flash, can be determined, providing the second flash occurs before dye excited by the first flash returns to its equilibrium condition, and (ii) the photoeffect quantum efficiency can be determined from the fraction of dye displacing charge, the light intensity and the dyes' optical absorption cross-section. However, the dyes' optical-absorption cross section values are uncertain, owing to solvent and orientational effects in BLMs. In connection with the cyanine dyes, mention should be made of the recent experiment of Takeguchi et al. (1985), who reported that in the presence of inorganic phosphate ions, the dyes induce ion pathway in BLM.

Owing to the importance of dyes, we have initiated a project to investigate systematically dye-sensitized BLMs, using the cyclic voltammetry technique (CV). 57 dyes of different chemical groups have thus been investigated. From the obtained voltammograms, 5 types of characteristic curves have been established (Tien and Kutnik; 1984). Electronic processes in, as well as across, the BLM are considered. Redox reactions at the BLM-solution interfaces are most conveniently studied by the CV technique (see Section II A 4). In the presence of light of known wavelengths, a number of dye-sensitized BLM have also been investigated by photoelectrospectrometry (Lopez and Tien, 1980). Ultraviolet light flashes can also induce voltage transients across BLMs, when aromatic amino acids are absorbed to one side of the membrane. These photoeffects varied with the chromophore structure, the aqueous solution salt concentration, pH and oxygen partial pressure. These photoeffects are attributed to the migration of electrically charged photochemical intermediates in the BLM, and provide a new method for studying the effect of UV light on membranes.

These effects of microwaves on BLMs have also been reported (Alekseev et al., 1978). The BLMs used by these workers were formed from lecithin with 10% of cholesterol and were oriented perpendicular to the electric vector of electric field. The power of high-frequency electromagnetic field was 0.9 GHz. Power absorption was monitored from the rate of heating of aqueous solution near the aperture in the Teflon cup. Alekseev et al. (1978) were interested in the effect on the conductance change of the BLM in the presence of tetraphenylborate (TFB), a well studied ion carrier (Milazzo, 1983). At a selected concentration of TFB, they found that the observed current was determined by the rate of diffusion of TFB through the aqueous solution/BLM interface, whose current saturation depended on temperature. A ten degree change in temperature caused about two-fold change in current, which was apparently due to a local heating of immobile layers of electrolyte adjacent to the BLM. The use of microwaves to probe the membrane, particularly the interfacial region between the bathing solution and the BLM, has a strong potential as an analytical tool for studying the mechanisms involved in the interaction of microwaves with membraneous structures of the living system. A paucity of data exist at present.

Besides light and microwaves, energy in other forms, such as heat and sound, can also affect BLMs. Scibona and colleagues (1978) reported that a temperature gradient at the two sides of BLMs can generate a thermal electrical potential, under isochemical conditions. BLM formed from sheep brain extracts behaved like a membrane with zero charge, while the phosphatidylcholine (from egg yolk) BLM acted like a charged membrane. The results obtained by Scibona and colleagues suggest that the measurment of electrical thermal potential is an interesting method to investigate the electrical behavior of bilayer lipid membranes. Alteration of BLM capacitance can also be induced by temperature during mechanical oscillation (Szekely and Morash, 1980).

In recent years, there has been a sharp increase in the availability and use of ultrasonic devices in medicine. As a result, there has been renewed interest in understanding the physical mechanisms involved in ultrasonically produced changes in living systems. Rohr and Ronney (1978) reported the effects of ultrasound of a frequency of 1 MHz on an oxidized cholesterol BLM for the intensity range from 0 to 1.5 W/cm . At an intensity of 1.5W/cm , the sound caused the BLM to rupture, whereas in the intensity range from 0.5 to 1.5 W/cm it accelerated the draining of the bulk lipid from the Teflon aperture during the formation of the BLM and, at all intensitites, the sound had no effect on the conductance, capacitance, or the dependence of each on the voltage applied across the membrane. This experiment of Rohr and Rooney (1978) is of particular interest, since details of the effect of ultrasound on membrane systems (cochlear inner hair cells) have not been extensively studied (Katsuki, 1982).

V. POTENTIAL APPLICATIONS

An experimental bilayer lipid membrane (BLM) less than 100 Å thick separating two aqueous solutions is a miraculous structure in itself. However, it has not been made into a practical device thus far, although there have been many attempts. In this last section, I shall summarize the past and current as well as future efforts of exploitation of the BLM system for practical applications.

A. Specific Electrodes, Biosensors, and Biomolecular Electronic Devices

1. Specific Electrodes

The most well-known specific electrode is the glass electrode used in pH measurements, where the glass can be thought of as a "membrane". As in the BLM system (solution/BLM/solution), the potential (E_m) developed is across the membrane separating a reference solution from the solution of interest. The first example of an ion specific BLM for K^+ was reported in 1967 by Mueller and Rudin (see Milazzo, 1983). They made a valinomycin-containing BLM which has a theoretical selectivity 10^4 to 1 for K^+ over Na^+ in the concentration range 10^{-5} to 10^{-1}. Beside valinomycin, many other ionophores (macrocyclic compounds, see Milazzo, 1983) have been incorporated into BLM. The second example is iodine-containing BLM which has been found to be specific to iodide (Tien, 1974). The presence of other ions such as CL^-, SO_4^{2-}, or F^- did not interfere with the "electrode" response to I^-. The behavior of iodine-containing BLM is reminiscent of metallic electrodes reversible to ions or their salts (e.g., Ag, Ag^+ and AgCl). Obviously, a variety of compounds can be and have been incorporated into BLM to make them useful in fundamental chemical and biomedical research (Koryta, 1982; Milazzo, 1983).

2. Biosensors

Essentially, a biosensor is a special class of devices that transduces a bioreaction into an electrical signal. In terms of membrane research, the heart of a biosensor discussed here is a modified BLM which serves as a model for the biomembrane. Thompson, Krull and associates (1982; 1983) have, in particular, exploited the BLM system for laboratory use. The most significant findings of their experiments are that modified BLMs can be used as detectors in a flow-injection system. The detection limit of sensors is in the range of 10^{-8} M.

We envision that, for detection in biological environments, the sensing element should be biocompatible and be biomembrane-like. Thus, the BLM system has great potential and is an ideal choice upon which to develop a new class of electrochemical biosensors. We have developed sophisticated techniques for measuring the electrical properties of BLMs such as the membrane potential, capacitance resistance, dielectric breakdown voltage, and other electrical parameters. With electrical methods, detection sensitivities less than 10^{-13} M can be easily achieved, which is quite comparable to what is now available using radioimmunoassay techniques. Biosensors based on BLMs, once successfully developed, should be very cheap with added advantage of ease of measurements.

For biosensing purposes, BLM formed by the conventional method, however, has a number of problems, the most serious of which is the extreme fragility of the BLM. This shortcoming greatly limits its usefulness as a practical device since it cannot be easily fabricated and will not sustain rugged laboratory use. One successful attempt to overcome this fragility problem has been achieved in our laboratory by forming BLMs in microporous polycarbonate filters (Nuclepore membranes, Nucleore Corp.). We have found that the electrical properties of micro-BLM in Nuclepore support are consistent with those of conventional BLMs (Mountz and Tien, 1978). Our work in this direction has been corroborated and extended by others (Thompson et al., 1982; Kocherginsky et al., 1984).

To illustrate our approach to the development of a biosensor for taste discrimination, for example, we would extract and isolate taste buds from bovine tongues for incorporation into Nuclepore-supported BLMs. Similarly, we would isolate receptor proteins from olfactory organs to construct biosensors for odor detection. Whether a new class of biosensors using Nuclepore-supported BLMs can be developed remains to be seen, but the advantages of specificity and biocompatibility of BLMs make such an endeavor exciting.

3. Biomolecular Electronic Devices (BED)

The basis for "biomolecular electronic device" (BED) concepts has been discussed at a MED Workshop organized by Carter. As a first step toward the designing and construction of "molecular" diode (or rectifier), the earlier article by Aviram and Ratner was intriguing to us. They suggested that the current would flow from the cathode to the anode via the well-known organic conductors TCNQ (tetracyano-p-quinodimethane) and TTF (tetrathiofulvalene) by a tunneling mechanism. Potember, Poehler, Cowan, Carter, and Brant have constructed switching and memory devices using TCNQ and other derivatives in the form of thin films (5-10 μM) sandwiched between two metal electrodes. A generalized approach to π-donor acceptor systems is excellently described by Sandman (see Carter, 1982), where again TCNQ and TTF among other derivatives are involved.

To construct a molecular diode, for example, based upon the aforementioned idea of Aviram and Ratner using a TCNQ-TTF complex, the classic monolayer technique of Langmuir and Blodgett has been used. Thus far, no positive results have been published. To date, one problem in relation to MED is the concern that molecules in Langmuir-Blodgett films have not exhibited any interesting electronic properties (e.g., non-linear I/V curves). We think one of the reasons, perhaps, lies in the substrate-film contact, which could be full of pin holes. This problem can be overcome by using the BLM techniques. In this connection, mention must be made of the work by Potember, Poehler, and Cowan (see Carter, 1982) who have been working with polycrystalline films of Cu or Ag salts of TCNQ. They have shown that such thin films 5-10μM thick sandwiched between metal electrodes can switch from a low to high level of conductance in response to an applied voltage. Similar effect could also be observed with a laser instead of a voltage. Thus, Potember et al. have constructed an optically driven switch. One should note that the films used in their switches are 5-10μM thick. When compared with a bilayer lipid membrane ($\sim$50 Å thick), a factor of 1000 times larger is involved. Thus, what new kind of phenomena could be observed as one approaches molecular sizes with TCNQ-containing BLMs? One way to find out is to form a BLM with a monolayer of TCNQ as the acceptor on one side and a monolayer of TTF on the other side using the Takagi technique (for details, see Tien, 1974, p. 477). The lipid bilayer phase (< 50 Å) can be modified by a number of additives such as the highly conjugated compounds beta-carotene, perylene and polyacetylenes (Zon and Tien, 1986).

As mentioned in Section IV E, only one isolated study has been reported on the effect of microwaves on BLMs. Now with the availability of TCNQ/TTF doped BLM exhibiting voltammograms (I/V curves) similar to the semiconductor diode, the possible presence of the Josephson effect in these BLMs is indicated. A typical Josephson junction is formed from two superconductors separated by a 30Å insulating barrier. In the absence of an applied voltage, a supercurrent could tunnel between the two superconductors. With an applied voltage (V) across the insulating barrier, the Josephson frequency (w) is given by: $w = 2eV/\hbar$, where $\hbar$ = Planck's constant($/2\pi$). For applied voltages of a few mV, frequencies (w) are in the microwave range. Thus, the study of the Josephson effect in BLM may offer an exciting perspective for microwave and for IR radiation device developments.

In the above paragraphs, we have stressed the development of BLM-based BED (or biochips). In the dimension perpendicular to the BLM, it is less than 100 Å. Thus, the principles governing the organic diode effect in the electron-conducting BLM can be investigated. I believe that on the basis of experimental results already accumulated (Tien, 1984), the BLM system should be a useful tool in the basic research area of MED and biochip design (Miyano, 1985).

B. BLM in Medicine

1. Immunology

The molecular basis of immunity is the recognition of what is "self" from "non-self". The destruction of foreign invading cells (non-self) depends on recognition sites on their membranes. Indeed, one of the principal defenses of our immune system is to make holes (lysis) in membranes of "non-self" cells, thereby causing their demise. However, it is not completely understood how the immune system achieves molecular recognition that a particular membrane should be attacked. Nor do we know the detailed mechanism of lysis. One thing is clear, nonetheless, that target cells and their membranes should be closely scrutinized for possible clues. Artificial lipid bilayers (BLMs and liposomes) for their simplicity have been chosen by several investigators as target membranes modified with recognition sites for other components of the immune system (Mountz and Tien, 1978a). Kinsky and Nicolotti (1977) have written an excellent review on the immunological properties of BLMs and liposomes. Here, we will only point out the highlights, with emphasis on recent experiments, on the use of BLM in immunology.

Using BLM as a model for the plasma membrane, aimed towards an understanding of membrane damages by complement, is the study initiated by del Castillo and his colleagues in 1966 (see O'Boyle et al., 1984). Since then, several groups of investigators have reported changes of ion flow across BLM after treatment with antigens, antibodies, and fresh blood serum as a source of complement. Analogous liposome experiments have also been reported (Kinsky and Nicolotti, 1977). Previously, it was believed that antibody and complement together form a complex entity, which attacked and destroyed living cells. The current view is that the enzymatic action of complement is not directed against the target cell, but represents a complex internal activation mechanism of the complement system. Mayer and his colleagues have clearly shown that, by parallel experiments with BLM and biomembranes, the conductance of membranes treated with C5b, C7 and C8 was affected markedly by the polarity of an applied field. Further, phospholipid is not required for the action of C5b-9; instead the only requirement is for a hydrophobic lipid bilayer for complement channel formation. From electrical measurements of complement-BLM, the diameter (or the effective size of the channel) has been estimated to be about 25 Å. O'Boyle, et al. (1984) have carried out the experiments of Mayer one step further. Owing to the instability of BLM, they used Nuclepore and Millipore membranes to support the BLM. They found these micro BLMs to exhibit stepwise reductions in electrical resistance in the presence of Forssman antigen, appropriate antiserum and complement. The results appear to support the "hydrophobic doughnut" or transmembrane channel hypothesis (see O'Boyle et al., 1984). Channel formation in these artificial membranes is believed to be due to the insertion of complement proteins. Concentrations and temperature studies were carried out to ascertain that the electrical responses were due to the generation of stable channels by complement across the membrane. The diameter of these channels was estimated to be of the order of 100 Å (O'Boyle et al., 1984).

2. Cancer

The direction of cancer research is increasingly molecular and membrane-bound. This is not surprising since the plasma membrane plays a crucial role in the processes of cell division, motility, and proliferation. Thus, to some extent, the problem is similar to immunity: how does the body deal with the invading (cancer) cells? Before suggesting what role can the BLM system play in the war on cancer, let me give a one-paragraph description of the progression of events on cancerous growth (i.e., metastasis).

The current consensus is that metastatic cells from a primary tumor somehow enter the bloodstream. Although the overwhelming majority of metastatic cells that enter the bloodstream are destroyed by natural killer cells, a few surviving ones that travel in the circulation as single cells or in clumps. After finding a suitable site for attachment, the metastatic tumor cells invade the region between the endothelium and the membrane. Once getting through the membrane, metastatic cells multiply and become new tumors. For a tumor to become 0.5 cm in diameter in size, about 10^9 cells are needed, a lot of which are potentially metastatic. By then, it may be too late to effect a cure. Thus, one of the problems in treating cancer is early detection so that effective drugs and other therapies can be applied.

From the above description, cancer may be thought of as a membrane-based disease. The BLM system may be used, therefore, as a tool for detecting the presence of metastatic cells in the bloodstream, for example. Also, it may be developed into a method for testing the influence of agents (known collectively as biological response modifiers or BRMs) capable of stimulating the body's defense systems. This is due to the fact that the electrical properties of BLM are highly susceptible to alteration as have been shown by many investigators (Milazzo, 1983). An example is the BLM experiment reported by Henkart and Blumenthal in 1975 (see Blumenthal and Klausner, 1982) on antibody-mediated cellular cytotoxicity, which is dependent upon recognition by the killer lymphocytes of an antibody directed to the antigen of the target cell. This situation was mimicked in the BLM with the correct specificity. More recently, Vassilev, Kanazirska, Dimitrov and Tien (1986, to be published) have found that natural killer cells (NK) can interact specifically with BLM made from lipids and proteolipids of K562 target cells. Interaction between NK cells and K562 cells causes a conductance change in BLM. The effects are much pronounced after pretreatment with interferon, one of the best known of the BRMs. Thus, this work may open a new vista to the study of the mechanism of the tumor killing in treating cancer.

3. Drugs, Anesthetics, and Toxic Compounds Testing

The possibility of using the BLM system as a tool for drug-membrane interaction studies was mentioned in the late 1960s (see Tien, 1974). Since that time, many investigations have been carried out (see Blank, 1970; Inui et al., 1976; Blumenthal and Klausner, 1982; Milazzo, 1983). Only recent experiments on the topic will be dealt with here.

Karadaghi, Tverdislov, Bucher, and co-workers (Karadaghi et al., 1984; Tverdislov et al., 1984) reported the incorporation of influenza virus proteins into BLM and the effects of ramantadine and amantadine on BLM conductance were studied. From the results, they suggest that these antiviral drugs may modify the cell membrane thus affecting virus penetration into the cell. For a different class of compounds, Arndt et al. (1982) have reported the effect of miconazole, a potent antimycotic agent with a wide spectrum of antimicrobial activity, on oxidized cholesterol BLM. According to Arndt and colleagues, the antimycotic drug induces a significant increase in BLM conductivity, which appears to be reversible after removal of miconazole by washing the BLM.

The high electrical resistance can also be reduced by a number of other interesting compounds. Rudnev and associates (1984) reported the effect of hemolysin on BLM-containing sphingomeytin. A single channel has an asymmetric current/voltage curve with maximal conductivity at positive potential on the side containing the toxin. Blumenthal and co-workers (Kafka et al., 1978; Steer et al., 1982; Blumenthal and Habig, 1984) have been very active in BLM research and have generated a wealth of information. In their most recent paper, they reported the ion selectivity of tetanolyin, a cytolytic toxin, in BLM and concluded that the permeability pathway induced by tetanolysin is caused by lipid perturbations rather than by formation of structural channels. Blumenthal's results appear to be consistent with our recent work on cardiac-active drugs described in the following paragraphs.

The class of drugs known as (Ca^{2+}) channel or Ca^{2+} antagonists has been a welcome addition of compounds such as the so-called β-adrenergic blockers used to combat cardiovascular diseases. However, the mechanisms and site of action as well as membrane selectivity of these drugs, are not fully understood. It is believed that these drugs are distributed to the periphery of the target cells through the blood circulation then enter the cells through the interaction with the plasma membranes. The binding of the drug to the membrane may influence the ionic channel states, which in turn can trigger physiological action. Because the action of these drugs is closely associated with the membrane, the insight into the molecular mechanism of interaction may be obtained by model membrane studies. This is owing to the complexity of biomembranes, which are not readily amenable to physico-chemical investigations at the molecular level. Recently, we have carried out a systematic study aimed at the above state goals (Shen et al., 1986). Our preliminary results can be summarized as follows: after the addition of the drug, a potential was usually developed across the BLM. The sign of potentials of the side, to which the drug was added, was negative. The magnitude of potentials was found to be dependent on the concentrations of the drug, pH, ionic species and strength of the bathing solution, the composition, and concentration of the membrane forming solution. Dose response curves show the difference in the intensity of interaction between the membrane and the drugs. The pH curve for verapamil shows clearly the potential change as a function of H^+ ion concentrations. In most cases, the potentials induced by the addition of drugs to solutions containing $CaCl_2$ were much less than those containing NaCl. Furthermore, in solutions containing both $CaCl_2$ and NaCl, the drug-induced potentials showed no significant change when the salt concentration was varied from 0 to 10^{-1} M, but it decreased dramatically when it was from 10^{-1} to 1M. The results reported recently were obtained with BMs formed from lipids only (Shen et al., 1986). The next step of our experiments is to isolate and incorporate specific protein channels and receptors into the membrane so that the drug-membrane interaction may be clearly delineated.

C. Solar Energy Transduction

Photoelectric effects in BLM discovered in 1968 have since been extensively investigated (see Chemical & Engineering News, Vol. 46 (No. 34, Aug. 12, 1968. pp. 36-37). Lately, photoactive BLMs have found their place in the field of photochemical conversion and storage of solar energy (for reviews, see Porter and Archer, 1976; Patsis and Seanor, 1976; Bolton, 1977; Bolton and Hall, 1979; Gerischer, 1979; Hautala et al., 1979; Gratzel, 1980; Lichtin, 1980). For instance, with suitable electron acceptors and donors in the bathing solution separated by a pigmented BLM, light-induced redox reactions can occur at opposite interfaces. It might be possible to couple BLMs of this type in such a way as to effect photolysis of water into hydrogen and oxygen (Tien, 1976). It should be stated, however, that ultrathin lipid films in the form of BLM, described in this chapter have not been developed into practical devices for energy transduction. The chief attraction of photoactive BLM to many scientists is that the system mimics, to some extent, the photosynthetic thylakoid membranes of green plants (see Section IV B), through which the most vital process of solar energy transduction is carried out. Investigations of photoactive BLM may provide insights on the mechanism of light-initiated redox reactions in photosynthesis, from which we may learn what are the basic conditions that must be fulfilled for efficient photoelectrochemical utilization of solar energy. Such knowledge may aid in the design and construction of practical devices for producing electricity and for splitting water to hydrogen by sunlight. A progress report of the state of the art will be given in the following paragraphs.

Insofar as we know, transduction of solar energy into electricity and/or chemical energy requires an ultrathin barrier less than 100 Å thick. Three such systems are known: (i) the p-n junction of the silicon photovoltaic cell, (ii) the thylakoid membrane of the chloroplast, and (iii) the pigmented bilayer lipid membrane (see Bolton, 1977). We shall only be concerned with the pigmented bilayer lipid membrane (PBLM) system.

The basic electronic process for the operation of a BLM containing pigments, such as chlorophyll (Chl) or Mg-meso-tetraphenyl-porphyrin (TPP), is as follows: when an electron acceptor (A) is added to one side and a donor (D) to the other side of a Chl-BLM, illumination causes the accetor side of the PBLM to become negatively charged relative to the donor side and a maximum open-circuit photovoltage of 60-350 mV to develop across, or a short-circuit photocurrent of up to 2 $\mu A/cm^2$ to flow through the BLM. The photoactive compound merely transfers electronic charges generated by light from one side of the membrane to the other side. There are at least three membrane processes which are important in the transduction of light to electrical/chemical energy. These are light absorption, energy transfer/charge separation, and charge translocation; they are closely dependent upon the molecular organization of the membrane and its constituent molecular orientation. Of these processes, the last one is most crucial, for without it there would be no energy conversion. The pigmented BLM is photoelectric in that it converts light into electricity. In this sense, the PBLM is regarded as an energy transducer (Kutnik and Tien, 1983). With this backgound in mind, we shall now describe two membrane-based solar cells and a scheme for the photolysis of water, using PBLM.

1. Photogalvanovoltaic Cells

Numerous photoelectrochemical cells (PEC) have been proposed for the direct conversion of solar energy to electricity or chemical fuel (Bolton, 1977; Lichtin, 1980). These cells, on the basis of two well established photoeffects, can be either a photovoltaic (PV) or photogalvanic (PG). The operation of a PV cell depends on the generation of an electromotive force as a result of the absorption of light, whereas the operations of a PG cell relies on the excitation by light of photoactive species in solution, which induces a faradaic process at the electrode. In the course of our investigation of photoelectric effects in PBLM, we came upon the idea that a new type of PEC might be developed. The operation of this novel cell would be based on the combined principles of the PG and PV phenomena mentioned above. It was, therefore, called the photogalvanovoltaic (PGV) cell. The key element responsible for the PV effect of the PGV cell was made from a Nuclepore membrane (Mountz and Tien, 1978), whereas the PG effect was due to the iron-thionine system (Lichtin, 1980). The photopotential of the PGV cell was equal to the sum of the photovoltage derived from PG and PV effects. Thus, our studies have demonstrated the validity of the concept of adding the voltage from each effect with no decrease in current. The PGV concept has many interesting and novel features, which are applicable to other types of PEC (Gerischer, 1979).

2. Nuclepore Membrane Supported Semiconductor Crystallite Solar Cells

From our analysis, it appears that an ultrathin barrier is the key for solar energy conversion (see Bolton, 1977). Thus, we have carried out the following experiment with interesting results (Bi and Tien, 1983-84, to be published). CdS crystallites were formed *in situ* in the pores of a polycarbonate (Nuclepore) membrane separating two aqueous solutions (Mountz and Tien, 1978). When both sides of the CdS-membrane were irradiated, open-circuit photopotentials up to 500 mV have been obtained. Although the initial results obtained from this novel system are intriguing, much efforts are needed to improve the design and optimize the photoresponse of the cell. (The preliminary results of this study were presented at the Beijing International Conference on Photochemistry, Oct. 21-26, 1985, Beijing, China).

3. Membrane-based System for Water Photolysis

One of the most important approaches for the photochemical conversion and storage of solar energy is the photolysis of water modeled on green plant photosynthesis (see Section IV B). The overall reaction is given by

$$2H_2O \xrightarrow{8h\nu} 2H_2 + O_2 \qquad (90)$$

The maximum conversion efficiency is about 30%, although only about one-third of which is realized by the green plant. To split water, a minimum energy of 1.23 eV is necessary, which could be supplied by a quantum of red light. However, various energy losses have to be considered, which are inherent in the process of quantum conversion. We have specifically assumed that the outer and inner membrane interface function, respectively, as a photocathode and photoanode. Thus, energy loss due to electrode overpotentials are unavoidable and subject to the usual considerations of electrochemistry. A more cogent question, that so far has not been raised, is "How can one hope to have BLM or for that matter thylakoid membranes about 70 Å thickness of sufficient dielectric strength to withstand 1230 mV, a minimum potential needed to electrolyze water, assuming such a large photovcltge could be generated?" Faced with this dilemma, it has been suggested (see Metzner, 1978) that, for the photooxidation of water in thylakoid membranes and in pigmented BLM, a light-induced potential different much less than 1.23 V should be sufficient if the membrane behaves as a semiconductor electrode as postulated earlier (Tien, 1976).

To put the reaction manifested in Eq. (90) to work, the hydrogen and oxygen produced should be separated from each other by a membrane as a first step. Since four photons are required for each molecule of H_2 generated and eight photons for O_2, the scheme of two coupled photosystems, as employed in natural photosynthesis, is a logical choice. The proposed scheme has been suggested since 1970 (see Tien, 1976; Bolton and Hall, 1979; Gerischer, 1979). The essential feature is a doped, pigmented BLM consisting a p-type semiconductor/membrane interface for hydrogen production and an n-type semiconductor/membrane interface for oxygen evolution (Metzner, 1978; Akoyunoglou, 1981; Mittal and Fendler, 1982). As in natural photosynthesis, all the transducers for water photolysis are packed into the ultrathin bilayer lipid membrane, where transfer of energy (excitons) and electronic charges (electrons and holes) can take place quickly and efficiently. With an ultrathin pigmented membrane, which in essence is an organic p-n junction, we can explore the properties of photoexcited particles in close proximity, as has been done in laminar semiconductor devices (Bastard et al., 1982). The findings of semiconducting BLM should, therefore, provide insights essential in the design of efficient membrane-based systems for water splitting by sunlight (see Lonsdale, 1983).

D. Molecular Design of Electro-active and Photoactive Compounds

With the advances in understanding the basics of electronic structure of molecules and solids on one hand, and the mechanism of photosynthesis and vision on the other, the time has come for a new endeavor, which may be called "molecular design." The aim of molecular design is to draft the structure of a compound with desired properties on paper before attempting actual synthesis. The approach differs from that which simply mimics natural products and improves upon them. Here, one is actively designing, developing and characterizing complex multi-molecular compounds which possess significant photo and/or electronic properties. Although the term olecular design has not been employed until now, the practice has been in operation for some time. Some of the compounds, for example such as TCNQ and TTF, are cases in point (Tien, 1984), which were synthesized by molecular design in the 1960s. Other compounds, such as covalently-linked porphyrin-quinone (TPP-Q) and porphyrin-carotene (TPP-carotene) complexes, mentioned in Section IV B, are also outstanding examples. TPP-Q and TPP-carotene are new compounds specifically designed to mimic the primary event of photosynthesis, the key step of which is light-driven charge separation in the form of electrons and holes. To prevent the separated charges from recombination (or back reaction) in photosynthesis, nature has evolved an ultrathin thylakoid membrane, providing an appropriate environment for the pigment systems (reaction centers) and redox proteins (electron transport chain). With the reaction centers located in different regions of the thylakoid membrane, the presence of appropriate electron acceptor (A) and donor (D) allow for solar energy transduction (see Fig. 5). Although our present knowledge of green plant photosynthesis is incomplete, especially in the domain of quantum conversion steps, many investigations have been undertaken to develop schemes for artificial photosynthesis (Mureramanzi and Tien, 1986). Towards this end, a number of supramolecular assemblies have been used, including BLMs as discussed in this paper.

Summarizing our research experience and that of others with pigmented BLM and liposomes (Antolini et al., 1982), an ultrathin barrier ($<$100 A) appears essential for efficient energy transduction. Fig. 11 shows a variety of compounds that have been tested in bilayer lipid membranes. Previously, we have shown that the photoelectric effects for TPP-Q and TPP-carotene BLMs are much larger than those obtained with simple TPP and quinone (or β-carotene) membranes (Wang et al., 1982). In the presence of suitable electron acceptor and donor species in the bathing solutions, a steady-state photocurrent was observed for TPP-Q containing BLM, clearly indicating the vectorial transfer of electrons

TABLE 2. Molecular Design of Electron-Conducting and/or Energy Transducing Compounds

Compound Type		Structure	Remarks
Tetraphenylporphyrin (TPP) based	X-1	(TPP)-$[CH{=}CH]_n$-(TPP)	[CH=CH]denoted unsaturated hydrocarbon n = 0,1,2,3...
	X-2	M_1(TPP)-$[CH{=}CH]_n$-(TPP)M_2	M_1 amd M_2 denote metal atom, Mg, Zn, Mn, Fe, Co, Cu, etc.
	X-3	A-(TPP)-$[CH{=}CH]_n$-(TPP)-D	$M_1{=}M_2$ or $M_1{\neq}M_2$
Phthalocyanine (PLC) based	X-4	(PLC)-$[CH{=}CH]_n$-(PLC)	A = electron acceptor such as quinone, methyl viologen, etc.
	X-5	M_1(PLC)-$[CH{=}CH]_n$-(PLC)M_2	D = electron donor such as β-carotene, EDTA, amines, etc.
	X-6	A-(PLC)-$[CH{=}CH]_n$-(PLC)-D	
Ruthenium-tris(2,2'-bipyridine)	X-7	$[Ru(bpy)_3^{2+}]$-$[CH{=}CH]_n$-$[Ru(bpy)_3^{2+}]$	(All above remarks apply)
$[Ru(bpy)_3^{2+}]$ based	X-8	A-$[Ru(bpy)_3^{2+}]$-$[CH{=}CH]_n$-$[Ru(bpy)_3^{2+}]$-D	
Dye-based	X-9	(p-type Dye)-$[CH{=}CH]_n$-(n-type Dye)	p-type Dye such as eosine, erythrosine, phenosafranine, etc.
	X-10	A-(p-type Dye)-$[CH{=}CH]_n$-(n-type Dye)-D	n-type Dye such as malachite green, crystal violet, coumarin, etc.

from the donor to the acceptor via pigmented membrane. Since, as already mentioned, in natural photosynthesis there are two photosystems connected together by an electron transfer chain, the next logical step toward artificial photosynthesis is to design molecules with requisite properties. A list of such compounds by molecular design are given in Table 2. In view of their ease of synthesis (backbone systems), stability and known properties, four classes of compounds are suggested. These are based on TPP (meso-tetraphenylporphyrin), PLC (phthalocyanine), Ru(bpy) (ruthenium tris-bipyridyl) and dyes. For example, Compound X-3 consists of two photochemically active TPPs connected by an unsaturated hydrocarbon chain. Further, appropriate electron acceptor (A) and donor (D) may also be attached to the molecule. Such compound(s), when incorporated into a bilayer lipid membrane (planar BLM or vesicular liposome), should permit detailed investigations of light-induced charge separation and transport leading to eventual redox reactions. Moreover, work with these suggested compounds offers a direct means for testing ideas and principles that may be applied later to more complex molecular-designed compounds.

Thus, the experimental work starting with organic synthesis of compounds by molecular design, suggested in Table , then to testing of these compounds in ultrathin lipid membranes, heralds a new approach towards artificial photosynthesis. Practical application of artificial photosynthesis by pigmented BLMs was predicted more than 15 years ago (see Chemical & Engineering News, 13 December 1971, P. 31), but its photoelectrochemistry, in a narrow confine of lipid bilayer less than 100 A separating two aqueous solutions, remains virtually unexplored. The proposed approach may be thought of as a unique way to investigate the diode effect of organic molecular junctions, free from the usual bulk phase effects. It could well become the method of choice for accomplishing the goal of artificial photosynthesis in the years to come(Mureramanzi and Tien, 1986).

ACKNOWLEDGEMENTS

The author's research programs have been supported in part by the National Institutes of Health (NIH-GM-14971), Department of Energy (DOE), the Office of Naval Research (ONR), and the Colleges of Natural Sciences and Human Medicine, Michigan State University. I am grateful to my colleagures, past and present, for their experimental work cited in this review. Special thanks are due to Ms. Carol Farmer for her excellent assistance in putting this manuscript in the final form.

REFERENCES

Acuna, A.U., and Gonzaliz-Rodriguez (1979). *J. Ann. Quim.*, **75**, 630.

Aizawa, M., Tomono, S., and Suzuki, S. (1977). *J. Memb. Sci.*, **2**, 289.

Akabas, M.H., Cohen, F.C., and Finkelstein, A. (1984). *J. Cell Biol.*, **98**, 1063-1071.

Akoyunoglou, G., ed. (1981). *Fifth International Congress on Photosynthesis*, Vol. I, Balaban International Science Services, Rehovot, Israel, pp. 254-262.

Alekseev, S.I., Tyazhelov, V., Faizava, L.K., and Chertishcher, V.V. (1978). *Biofisca*, **23**, 95.

Anonymous. (1969). *Nature*, **223**, 885.

Antolini, R., Gliozzi, A., and Gorio, A., eds. (1982). *Transport in Membranes: Model Systems and Reconstitution*, Raven Press, NY, pp. 57-75.

Antonov, V.F., Rovin, Y.G., and Trofimov, L.T. (1979). *A Bibliography of Bilayer Lipid Membranes*, All Union Institute for Scientific and Technical Information, Moscow.

Arndt, R., and Schulz-Harder, B.J. (1982). *Biochem. Pharm.*, **31**, 3120.

Bagaveev, L.A., Rovin, Y.G., and Starchk, N.V. (1979). *Biofizika*, **24**, 936.

Barber, J., ed. (1979). *Photosynthesis in Relation to Model Systems*, Topics in Photosynthesis, Vol. 3, Elsevier/North-Holland, NY, pp. 115-173.

Barltrop, J.A., and Coyle, J.D. (1978). *Principles of Photochemistry*, Wiley, New York.

Bastard, G., Mendex, E.E., Chang, L.I., and Esaki, L. (1982). *Phys. Rev.*, **B26**, 1974-1979.

Bauer, P.J., Bamberg, E., and Fahr, A. (1984). *Biophys. J.*, **46**, 111.

Bender, C., and Tien, H.T. (1987). "Dynamic polarization of lipid bilayers as a special case in the electrochemistry of liquid/liquid interfaces," *Anal. Chim. Acta*, **198**, 259.

Bender, C., and Tien, H.T. (1987). "Electrical properties of bilayer lipid membranes containing iodine and iodide, investigated by cyclic voltammetry," *Anal. Chim. Acta.*, **201**, 51-58 (987).

Benga, G. ed. (1985). *Structure and Properties of Cell Membranes*, Vol. III, CRC Press, Inc., Boca Raton, FL.

Benga, G. ed., (1986). *Water Transport in Biological Membranes*, CRC Press, Inc., Boca Raton, FL.

Benga, G. ed. (1988). *Water Transport in Biological Membranes*, CRC Press, Inc., Boca Raton, FL, pp. 41-75.

Benz, R., Prass, W., and Ringsdorf, H. (1982). *Angew. Chem. Suppl.*, 869.

Bienvenue, E., Seta, P., Hofmanova, A., Gavach, C., and Momenteau, M. (1984). *J. Electroanal. Chem.*, **162**, 275.

Blank, M., ed. (1970). *Surface Chemistry of Biological Systems*, Plenum Press, NY, pp. 135-154.

Blank, M., ed. (1980). *Bioelectrochemistry*, Am. Chem. Soc., Washington, D.C.

Blank, M., ed. (1986). "Electrical Double Layers in Biology," in *Proc. of Symposium of Electrochemical Society, Toronto, Canada, 1985,* Plenum Press, New York, pp. 149-166.

Blank, M., and Findl, E. ed. (1987). Mechanistic Approach to Interactions of Electric and Electromagnetic Fields with Living Systems, Plenum, NY, p. 301.

Blumenthal, R., and Klausner, R.D. (1982). In *Membrane Reconstitution,* G. Poste and G.L. Nicholson, eds., Elsevier, NY, p. 43.

Blumenthal, R., and Habig, W.H. (1984). *J. Bact.,* **157**, 321-323.

Bockris, J.O'M. (1980) In *Bioelectrochemistry,* H. Keyzer and F. Gutmann, eds., Plenum Press, New York, p. 5.

Bockris, J.O.'M., and Diniz, F.B. (1988). "The effect of electron transfer on the potential of certain electronically conducting membranes," *J. Electrochem. Soc.,* **135**, 1947-1954.

Bolton, J.R., ed. (1977). *Solar Energy and Fuels,* Academic Press, NY, pp. 167-225.

Bolton, J.R., and Hall, D.O. (1979). *Ann. R. Energ.,* **4**, 353.

Bownds et al. (1985). *Dahlem Konferenzen,* Berlin, Springer-Verlag, Germany.

Brasseur, R., Meutter, J.D., and Ruysschaert, J.-M. (1984). *Biochim. Biophys. Acta,* **764**, 295.

Brochard, F., De Gennes, P.G., and Pfeuty, P. (1976). *J. Physiol.,* **37**, 1699-1104.

Burton, R.M., ed. (1967). *J. Am. Oil Chemists' Soc.,* **45**, 201-246.

Carter, F.L., ed. (1982). *Molecular Electronic Devices,* Marcel Dekker, Inc., New York.

Carter, F.L., Siatkowski, R.E., and Wohltjen, H., eds. (1988). Molecular Electronic Devices, North-Holland, Amsterdam, 685 pp.

Chen, J.-W., and Tien, H.T. (1988). "A comparative study between a two-compartment cell and the standard electrode system," *J. Electroanal. Chem.,* **253**, 491-497.

Chen, Q.-S., and Li, S.-J. (1986). *Acta Phytophysiol. Sinica,* **12**, 9.

Connolly, J.S., and Bolton, J.R. (1989). "Artificial photosynthesis" in *Photoinduced Electron Transfer,* M.A. Fox and M. Chanon (eds.), Elsevier, Amsterdam, pp. 303-393.

Dancshazy, Z., Ormos, A., Drachev, L.A., and Skulachev, V.P. (1978). *Biophys. J.,* **24**(2), 423.

Davies, J.T., and Rideal, E.K. (1961). *Interfacial Phenomena,* Academic Press, NY.

Davison, S.G., ed. (1985). *Progress in Surface Science,* Vol. 19, No. 3, Pergamon Press, New York, pp. 169-274.

Dely, M., Prager, P., and Puppi, A. (1981). *Stud. Biophys.,* **85**, 115-123.

Derzhanski, A., Petrov, A.G., and Pavloff, Y.V. (1981). *J. Physique - LETTRES,* **42**, 119.

Disalvo, E.A., Siddiqi, F.A., and Tien, H.T. (1988). "Membrane transport with emphasis on water and non-electrolytes in experimental lipid bilayers and biomembranes," in *Water Transport in Biological Membranes,* G. Benga, ed., CRC Press, Boca Raton, FL.

Drachev, A.L., Evstigneeva, R.P., Kaulen, A.D., Lazarova, C.R.,

Laikhter, A.L., Mitsner, B.I., Skulachev, V.P., Khitrina, L.V., and Chekulayeva, L.N. (1984). *Biofisca,* **50**, 1142.

Fettiplace, R.L., and Haydon, D.A. (1980). *Physiol. Rev.,* **60**, 510.

Finkelstein, A. (1987). *Water Movement through Lipid Bilayers, Pores, and Plasma Membranes: Theory and Reality,* Wiley-Interscience, New York, 1987.

Finkelstein, A., and Cass, A. (1968). *J. Gen. Physiol.,* **52**, 145s.

Fesenko, E.E., and Lyubarskiy, A.L. (1977). *Stud. Biophys.,* **58**, 113. (1976; *Nature,* **268**, 562 (1977).

Gerischer, H. (1979). *T. Appl. Phys.,* **31**, 115.

Gliozzi, A., and Rolandi, R. (1984). In *Membranes and Sensory Transduction* (G. Colombetti and F. Lenci, eds.), Plenum Press, NY, pp. 1-69.

Goldman, D.E. (1967). *Ber. Bun. Ges. Physik. Chem.,* **71**, 799.

Gratzel, M. (1980). *Ber. Bun. Ges.,* **84**, 981.

Gruen, D.W.R., and Haydon, D.A. (1980). *Pur. A. Chem.,* **52**, 1229.

Guggenheim, E.A. (1950). *Thermodynamics,* North-Holland, Amsterdam, Ch. 10.

Harris, J.R., and Etemadi, A.H., eds. (1987). *Artificial and Reconstituted Systems, Subcellular Biochemistry,* Vol. 14, Plenum N.Y.

Hattenbach, A., Gundel, J., Hermann, G., Haroske, D., and Mueller, E. (1982). *Biochem. Physiol. Pflanzen.,* **177**, 611.

Hautala, R.R., King, R.B., and Kutal, C. eds. (1979). *Solar Energy: Chemical Conversion and Storage*, Human Press, Inc., NJ, pp. 203-235.

Higgins, J., Lopez, J.R., and Tien, H.T. (1979). *J. Electroanal. Chem.,* **104**, 3.

Hong, F.T. (1980). *Adv. Chem. Ser.,* **188**, 211.

Hong, F.T. ed. (1989). Molecular Electronics: Biosensors and Biocomputers, Plenum, NY.

Hotchandani, S., LeBlanc, R.M., Clarke, R.H., and Fragata, M. (1982). *Photochem. Photobiol.,* **36**, 235.

Hub, H.H., Hupfer, B., Kick, H., and Ringsdorf, H. (1980). *Angew. Chem.,* **19**, 938.

Huebner, J.S. (1979). *Photochem. Photobiol.,* **30**, 233.

Huebner, J.S., Arrieta, R.T., and Millar, D.B. (1982). *Photochem. Photobiol.,* **35**, 467.

Hughes, W. (1980). *Aspects of Biophysics,* Wiley, NY.

Inui, K.-I., Tabura, K., Hori, R., Kaneda, A., Muranishi, S., and Sezaki, H. (1976). *J. Pharm. Pharmac.,* **29**, 22-26.

Janas, T., and Tien, H.T. (1988). "Influence of dolichyl phosphate on permeability and stability of bilayer lipid membranes," in *Biochimica et Biophysica Acta,* **939**, 624-628.

Janas, T., Kotowski, J., and Tien, H.T. (1988). "Polymer modified BLM: Polypyrrolelecithin systen," *Bioelectrochem. Bioenerg.,* **19**, 405-412.

Ivanov, I. ed. (1988). *Thin Liquid Films*, Marcel Dekker, Inc., New York, pp. 181-242.

Jackowska, K., Krysinski, P., and Tien, H.T. (1987). "Modeling of BLM behavior by electrochemical cells," *Bioelectrochem. Bioenerg.,* **17**, 409.

Johansson, L.B.-A., Lindblom, G., Wieslander, A., and Arvidson, G. (1981). *FEBS Lett.,* **128**, 97.

Joshi, N.B., Lopez, J.R., and Tien, H.T. (1982). *J. Photochem.,* **20**, 139.

Kadoshnikov, S.I., and Stolovisky, Yu.M. (1982). *Bioelectrochem. Bioenerg.,* **9**, 79.

Kafka, M.S., Blumenthal, R., Walker, G.A., and Pollard, H.B. (1978). *Membr. Biochem.,* **1**, 279.

Karvaly, B., and Dancshazy, Z. (1977). *FEBS Lett.,* **76**, 45.

Karadaghi, S.E., Zakomirdin, J.A., Shimane, C., Bucher, D.J., Tverdislov, V.A., and Kharitonenkov, I.G. (1984). *Biochim. Biophys. Acta* **778**, 269.

Katsuki (1982). *Receptive Mechanisms of Sound in the Ear,* Cambridge Univ. Press, Cambridge, UK.

Kinsky, S.C., and Nicolotti, R.A. (1977). *Annu. Rev. Biochem.,* **46**, 49.

Knowles, A., and Dartnall, H.J.A. (1977). In *The Eye,* Vol. 2B, pp. 425-497.

Kocherginsky, N.M., Bromberg, L.K., Osak, I.S., Moshkovsky, Yu.S., Popkov, Yu.M., and Skuratnik, Yu.B. (1984). *Biofisca,* **29**, 963.

Koryta, J. (1982). *Ions, Electrodes and Membranes,* Wiley, NY, p. 197.

Kossi, C.N., and LeBlanc, R.M. (1981). *J. Cell. I. Sc.,* **80**, 426.

Kotowski, J., Janas, T., and Tien, H.T. (1988). "Immobilization of glucose oxidase on a polypyrrole-lecithin bilayer lipid membrane," *Bioelectrochemistry and Bioenergetics,* **19**, sec. of *J. Electroanal. Chem.,* **253**, 277-282.

Kotowski, J., Janas, T., and Tien, H.T. (1988). "Electrical oscillations in polypyrrole-lecithin bilayer lipid membranes," *Bioelectrochemistry and Bioenergetics,* **19**, sec. of *J. Electroanal. Chem.,* **253**, 283-289.

Kotowski, J., and Tien, H.T. (1989). "Sucrose influence on lecithin and polypyrrolelecithin bilayer membranes," *Bioelectrochem. and Bioenerg.,* **22**, 69-74.

Koyama, Y., Komori, M., and Shiomi, K. (1982). *J. Coll. Int. Sci.,* **90**, 293.

Krueger, B.K., Worley, J.F., and French, R.J. (1983). *Nature,* **303**, 172.

Krull, V.J., and Thompson, M. (1983). *Trend Anal. Chem.,* **2**, 6-7.

Krull, V.J., Thompson, M., and Wong, H.E. (1986). Langmuir-Blodgett technology and receptor action in stabilized lipid membranes in *Fundamentals and Applications of Chemical Sensors,* D. Schuetzle, R. Hammerle, J.W. Butler (eds.), Am. Chem. Soc., Washington, D.C., p. 351.

Krysinski, P., and Tien, H.T. (1988). "Some electrochemical studies on charge transfer of TCNQ modified BLMs," *Bioelectrochemistry and Bioenergetics,* **19**, *J. Electroanal. Chem.,* **253**, 227-233.

Kunitake, T., and Okahata, Y. (1980). *J. Am. Chem. S.,* **102**, 549.

Kutnik, J., and Lojewska, Z.K. (1983). *Ann. Univ. MCS,* **38**, 7-13.

Kutnik, J., and Tien, H.T. (1983). *Photobiochem. Photobiophys.,* **6**, 351.

Kutnik, J., and Tien, H.T. (1987). "Photoelectric effects in BLM-containing metalloporphyrins and dyes," *Photochem. Photobiol.,* **40**, 1009.

Kutnik, J., and Tien, H.T. (1987). "Depositions of metallic and semiconducting layers onto BLM," *Photochem. & Photobiol.,* **46**, 413.

Kutnik, J., and Tien, H.T. (1987a). *J. Electrophysiological Tech.,* **14**,

211.
Lakshminarayanaiah, N. (1969). *Transport Phenomena in Membranes,* Academic Press, New York.
Laprade, R., and Lapointe, J.-Y. (1982). *Rev. Can. Biol. Expt.,* **41**, 13.
de Levie, R. (1978). *Adv. Chem. Phys.,* **39**, 99-137.
Lewis, T.J. (1982). *Phys. Med. Biol.,* **27**, 335-52.
Lichtin, N.N. (1980). *Chemtech.,* **10**, 252.
Liu, Q.Y., and Tien, H.T. (1982). *Photobiochem. Photobiophys.,* **4**, 73.
Liu, Q.Y., and Tien, H.T. (1985). *Photobiochem. Photobiophys.,* **10**, 1.
Lonsdale, H. ed. (1983). *Membrane Technology in the 1980s,* Sunriver, Oregon, Sept. 19-21.
Lopez, J., and Tien, H.T. (1980). *Biochim. Biophys. Acta* **597**, 433.
Lopez, J.R., and Tien, H.T. (1984). *Photobiochem. Photobiophys.,* **7**, 25.
Marino, A.A. ed. (1988). Modern Bioelectricity, Dekker, NY.
Matynski, T., and Tien, H.T. (1990). Bioelectrochem. Bioenerg. in press.
Martonosi, A.N. ed. (1982). In *Membranes and Transport,* Plenum, NY, pp. 165-171.
Mauro, A., Blake, M., and Labarca, P. (1988). *PNAS,* **85**, 1071.
Mauzerall, D. (1979). In *Light-Induced Charge Separation in Biology and Chemistry,* H. Gerischer, ed., Verlag-Chemie, Berlin, pp. 241-254.
McIntosh, T.J., Simon, S.A., and MacDonald, R.C. (1980). *Bioc. Biop. A.,* **597**, 445; *Ibid,* **645**, 318.
McIntyre, P., and Kirschfeld, K. (1981). *J. Comp. Physiol.,* **143**, 3.
McLaughlin, S., and Eisenberg, M. (1975). *Ann. Rev. Biophys. Bioeng.,* **4**, 335.
Metzner, H. ed. (1978). *Photosynthetic Oxygen Evolution,* Academic Press, NY, pp. 411-438.
Milazzo, G. ed. (1983). *Bioelectrochemistry and Bioenergetics,* Vol. 5, Wiley, NY, pp. 157-224.
Miller, C. ed. (1986). *Ion Channel Reconstitution,* Plenum Press, New York.
Mirsky, V.M., Sokolov, V.S., Dyukova, T.V., and Melnik, E.I. (1984). *Bioelectrochem. Bioenerg.,* **12**, 327.
Mittal, K.L., and Fendler, E.J. eds. (1982). *International Symposium on Solution Behavior of Surfactants - Theoretical and Applied Aspects,* Plenum Press, NY, pp. 229-240.
Miyano, K. (1985). *Jap. J. Appl. Phys.,* **24**, 1379.
Montal, M. (1975). In *Molecular Aspects of Membrane Phenomena* (H.R. Kaback, G.K. Radda, and H. Neurath, eds.) Springer-Verlag, p. 316.
Moore, A.L., Dirks, G., Gust, D., and Moore, T.A. (1980). *Photochem. Photobiol.,* **32**, 691.
Mountz, J.M., and Tien, H.T. (1978). *Photochem. Photobiol.,* **28**, 395.
Murermanzi, S., and Tien, H.T. (1986). *Int. J. Ambient Energy,* **7**(1), 3.
O'Boyle, K., Siddiqi, F.A., and Tien, H.T. (1984). *Immunological. Commun.,* **13**, 85.
Olson, J.M., and Hind, G. eds. (1976). *Chlorophyll-Proteins, Reaction Centers and Photosynthetic Membranes,* Brookhaven Symposia Biology, Vol. 28, pp. 105-131.
Oosawa, F., and Imai, N. eds. (1982). *Dynamic Aspects of*

Biopolyelectrolytes and Biomembranes, Kodansha-Elsevier, Tokyo and Amsterdam, pp. 445-456.
Ormos, P., Dancshazy, Z., and Karvaly, B. (1978). *Biochim. Biophys. Acta*, **503**, 304.
Ostro, M.J. (1987). Liposomes: *From Biophysics to Therapeutics*, Dekker, New York.
Paphadjopoulos, D., and Watkins, J.C. (1967). *Biochim. Biophys. Acta*, **135**, 639.
Paphadjopoulos, D., and Kimelberg, H.K. (1974). *Prog. Surf. Sci.*, **4**, 141.
Pastushenko, V.F., and Chizmadzhev, Y.A. (1981). *Biofizika*, **26**, 458.
Patsis, A.V., and Seanor, D.A. eds. (1976). *Photoconductivity in Polymers*, Technomic Pub. Co., Westport, CT, pp. 309-328.
Pethig, R. (1979). *Dielectric and Electronic Properties of Biological Materials*, Wiley, NY, pp. 207-339.
Porter, G., and Archer, M.D. (1976). *Interdisc. Sci. Rev.*, **1**, 119.
Procopio, J., Varanda, W.A., and Fornes, J.A. (1982). *Biochim. Biophys. Acta*, **688**, 808.
Putvinskii, A.V. (1977). *Biotizika*, **22**, 725.
Rayfield, G.W. (1982). *Biophys. J.*, **38**, 79.
Rich, M., and Brody, S.S. (1982). *FEBS Lett.*, **143**(1), 45.
Robert, S., Tancrede, P., Houle, A., and LeBlanc, R.M. (1985). *Photochem. Photobiol.*, **41**, 101.
Rohr, K.R., and Rooney, J.A. (1978). *Biophys. J.*, **23**, 33.
Rudnev, V.S., Likhatskaya, G.N., Kozlovskaya, E.P., Monastyrnaya, M.M., and Elyakov, G.B. (1984). *Biofiska*, **29**, 1024.
Salamon, Z., and Tien, H.T. (1988). "Light dependence of the pigmented liquid crystal BLM capacitance." In *Mol. Cryst. Liq. Cryst.*, **154**, 195-207.
Salamon, Z., and Tien, H.T. (1988). "Bilayer lipid membranes formed from thermotropic liquid crystals." In *Liquid Crystals*, Vol. 3, No. 2, 169-174.
Salamon, Z., and Tien, H.T. (1988). "Light-induced electrical effects in a liquid crystal BLM containing TCNQ," *Photochemistry and Photobiology*, **48**, No. 3, Pergamon Press, 281-287.
Salamon, Z., Vitello, L.B., Erman, J.E., Butko, P., and Tien, H.T. (1989). "Electron transfer from cytochrome c to cytochrome c peroxidase across the bilayer lipid membrane (BLM)," *Bioelectrochem. Bioenerg.*, **21**, 213-221.
Salamon, and Tien, H.T. (1989). "Deposition of photosensitive complex within a lecithin bilayer lipid membrane," *J. Photochem. Photobiol., B. Biol.*, **3**, 351-357.
Sanderman, H. (1979). *Biochim. Biophys. Acta*, **87**, 789.
Schreckenbach, T. (1979). *Photosynthesis in Relation to Model Systems*, J. Barber, ed., Elsevier/North-Holland, Biomedical Press, New York, Ch. 6.
Scibona, G., Scuppa, B., Fabiani, C., and Pizzichi, M. (1978). *Biochim. Biophys. Acta*, **512**, 41.
Scott, A.C. (1977). *Neurophysics*, Wiley-Interscience, NY, Ch. 3.
Severina, I.I. (1982). *Biochem. Biophys. Acta*, **681**, 311.

Shchipunov, Y.A., and Drachev, G.Y. (1982). *Bioc. Biop. A.,* **691**, 353.
Shen, X.-C., Shi, B., and Tien, H.T. (1986). *Bioelectrochem. Bioenerg.,* **16**, 13.
Shen, X.-C., Shi, B., and Tien, H.T. (1986). "Transmembrane potentials in BLM induced by calcium channel and beta-adrenergic blockers," *Bioelectrochem. Bioenerg.,* **16**, 13-26.
Shi, B., and Tien, H.T. (1986). "Action of calcium channel and beta-adrenergic blocking agents in BLM," *Biochim. Biophys. Acta,* **859**, 125-134.
Sielewiesiuk, J. (1983). *Stud. Biophys.,* **96**, 117.
Steer, C.J., Klausner, R.D., and Blumenthal, R. (1982). *J. Biol. Chem.,* **257**, 8533-8540.
Stieve, H. ed. (1986). *The Molecular Mechanisms of Photoreception: Dahlem Konferenzen,* Springer, Berlin.
Stillwell, W., Ricketts, M., Hudson, H., and Nahmias, S. (1982). *Bioc. Biop. Acta,* **688**, 653.
Sugar, I.P., and Neuman, E. (1984). *Biophys. Chem.,* **19**, 211.
Szekely, J.G., and Morash, B.D. (1980). *Biochem. Biophys. Acta,* **599**, 73.
Tabushi, I., Koga, N., and Yakagita, M. (1979). *Tetrahedron Lett.,* **3**, 257.
Takeguchi, N., Saitoh, T., Morii, M., Yoshikawa, K., and Terada, H. (1985). *J. Biol. Chem.,* **260**, 9158.
Tancrede, P., Paquin, P., Houle, A., and LeBlanc, R.M. (1983). *J. Biochem. Biophys. Meth.,* **7**, 299.
Taylor, D.M., Mahboubian-Jones, M.G.B. (1982). *Thin Solid Films,* **87**, 167-79.
Thompson, M., Lennox, R.B., and McClellan, R.A. (1982). *Analyt. Chem.,* **54**, 76.
Tien, H.T., and Kabomoto, N. (1969). *Nature,* **224**, 690.
Tien, H.T. (1974). *Bilayer Lipid Membranes (BLM): Theory and Practice,* Marcel Dekker, Inc., NY, 672 pages.
Tien, H.T. ed. (1976). *Photoelectric BLM,* Photochemistry & Photobiology, Vol. 24, pp. 95-207.
Tien, H.T., and Karvaly, B. (1977). In Solar Energy and Fuels, J.R. Bolton, ed., Academic Press, New York, pp. 167-225.
Tien, H.T. (1984). *J. Phys. Chem.,* **88**, 3172.
Tien, H.T., and Kutnik, J. (1984). *Photobiochem. Photobiophys.,* **7**, 319.
Tien, H.T., and Lojewska, Z.K. (1984). *Biochem. Biophys. Res. Commun.,* **119**, 372-5.
Tien, H.T. (1984). *Bioelectrochem. Bioenerg.,* **13**, 299-316; *15,* 19-38.
Tien, H.T., and Joshi, N.B. (1986). *Photobiochem. Photobiophys.,* **10**, 241-251.
Tien, H.T. (1986). "Redox Reactions in Lipid Bilayers and Membrane Bioenergetics," *Bioelectrochem. Bioenerg.,* **15**, 19.
Tien, H.T. (1986). "Redox Reactions in Lipid Bilayers and Membrane Bioenergetics," *Bioelectrochem. Bioenerg.,* **15**, 19-38.
Tien, H.T., Kutnik, J., Krysinski, P., and Lojewska, Z.K. (1986). "Electronic Properties of Electroactive BLMS." In Electrical Double Layers in Biology, M. Blank, ed., Plenum Press, N.Y., pp. 149-166.

Tien, H.T., and Zon, J. (1987). "Field effects in experimental bilayer lipid membranes and biomembranes." In *Mechanistic Approaches to Interactions of Electromagnetic Fields with Living Systems*, Martin Bland ed., Plenum, New York.

Tien, H.T. (1987). "Reconstitution of reaction centers in planar BLM," *Progress in Photosynthesis Research*, Vol. 1, J. Biggins, ed., p. 209.

Tien, H.T. (1988). "Bilayer lipid membranes (BLM) in aqueous media: interfacial aspects and potential applications," *J. Surface Sci., Technol.*, **4**, 1, Jain-Interscience Press, Delhi 110007, India, pp. 1-21.

Tien, H.T. (1988). "Membrane bioenergetics as viewed from reconstitution experiments." In *Redox Chemistry and Interfacial Behavior of Biological Molecules*, G. Dryhurst and K. Niki, eds., Plenum Publishing Corp., New York, NY.

Tien, H.T. (1988) "Bilayer lipid membranes in aqueous media." in *Thin Liquid Films*, I. Iavanov, ed., Dekker, Inc., pp. 927-1057.

Tien, H.T. (1988). "BLM-based electrochemical biosensors," *J. Clinical Lab. Analysis,* **2**, 256-264.

Tien, H.T. (1988). "Membrane bioenergetics as viewed from reconstitution experiments." In Redox Chemistry of Biological Molecules, G. Dryhurst and K. Niki, eds., Plenum, NY, 529-556.

Tien, H.T., Salamon, Z., Kutnik, J., Krysnski, P., Kotowski, J., Ledermann, D., and Janas, T. (1988). BLMs: an experimental approach for biomolecular electronic device development. *J. Mol. Elec.*, **4**, 1-30.

Tien, H.T., Salamon, Z., Kutnik, J., Krysinski, P., Kotowski, J., Ledermann, D., and Janas, T. (1988) "Bilayer lipid membranes (BLM): an experimental system for biomolecular electronic device development," *J. of Molecular Electronics,* **4**, S1-S30.

Tien, H.T. (1989). *Prog. in Surf. Sci.,* **30**, 1.

Tien, H.T., and Salamon, Z. (1989). "Formation of self-assembled lipid bilayers on solid substrates," *Bioelectrochem. Bioenerg.*, **22**, 211.

Tripathy, A., Zviman, M., and Tien, H.T. (1989). "Reconstitution in brain microsomol calcium channels and olfactory epithelium cilia in planar BLM system," *J. Indian Chem. Soc.,* **66**, 674-651.

Tverdislov, V.A., Karadaghi, S., EL-1., Bucher, D.J., Zakomirdin, J.A., and Kharitonenkov, I.G. (1984). *Biochim. Biophys. Acta,* **778**, 269-275.

Vacek, K., Valent, O., and Skuta, A. (1982). *Gen. Physl. B.*, **1**, 135.

Vassilev, M., and Tien, H.T. (1985). In *Structure and Properties of Cell Membranes*, G. Benga, ed., CRC Press, Boca Raton, Chapter 3.

Vassilev, P.M., Kanazirska, M., and Tien, H.T. (1985). "Intermembrane linkage mediated by tubulin," *Biochem. Biophys. Res. Commun.*, **126**, 559.

Vassilev, P.M., Kanazirska, M.P., and Tien, H.T. (1986). "Microtube-dependent membrane interactions studied in two type of double bilayer membrane systems," *Bioelectrochem. Bioenerg.*, **15**, 395-406.

Vassilev, P.M., Kanazirska, M.P., Charamella, L.J., Dimitrov, N.V., and Tien, H.T. (1986). "Cell-mediated tumor-killing effect studied by using BLM," *Cancer: Biochemistry & BioPhysics,* **9**, 85-95.

Vassilev, P.M., Kanazirska, M.P., Charammella, L.J., Dimitrov, N.V., and Tien, H.T. (1987). "Changes in calcium channel activity related to the mechanism of cisplatin resistance of tumor cells," *Cancer Research,* **47**, 519.

Vassilev, P.M., Kanazirska, M.P., and Tien, H.T. (1987). *Biochim. Biophys. Acta,* **897**, 324-330.

Vassilev, P.M., and Tien, H.T. (1988). "Reconstitution of membrane molecular mechanisms in bilayer lipid membranes and patch-clamp bilayers." In *Subcellular Biochemistry,* Vol. 14, J.R. Harris and A.-H. Etemadi, eds., Plenum Press, New York.

Vassilev, P.M., and Tien, H.T., (1989) in *Artificial and Reconstituted Systems,* J.R. Harris and A.-H. Etemadi (eds.), Plenum Press, New York.

Vaz, W.L.C., and Schoellmann. (1976). *Biochim. Biophys. Acta,* **439**, 206-218.

Virysov, S.N., and Perelygin, V.V. (1984). *Biophysica,* 222.

Vodyanoy, V., Halverson, P., and Murphy, R.B. (1982). *J. Coll. Int. Sci.,* **88**, 247.

Weller, H., and Tien, H.T. (1973). *Biochim. Biophys. Acta,* **325**, 433.

Weller, P.F. ed. (1973). *Solid-State Physics and Chemistry,* Dekker, Inc., NY, pp. 847-903.

White, S. (1980). *Science,* **207**, 1075.

Worley, P.F., Baraban, J.M., Surachai, S., Wilson, V.S., and Snyder, S.H. (1987). *J. Biol. Chem.* **262**, 12132-12136.

Worthy, W. (1984). *Chemical & Eng. News,* **62**, No. 32, 19.

Xu, G., Shi, B., McGroarty, E.J., and Tien, H.T. (1986). "Channel-closing activity of porins from *Escherichia coli* in bilayer lipid membranes," *Biochim. Biophys. Acta,* **862**, 57-64.

Yoshida, T., and Okabayashi, H. (1982). *Chem. Scr.,* **19**, 42.

Zimmermann, U. (1982). *Biochim. Biophys. Acta,* **694**, 227-277.

Zon, J., and Tien, H.T. (1988). In *Modern Bioelectricity,* A.A. Marino, ed., Marcel Dekker, Inc., N.Y. and Basel, pp. 181-241.

ELECTROCHEMISTRY OF IONOPHORE FACILITATED ION TRANSPORT ACROSS LIQUID/LIQUID INTERFACES

J. Koryta
J. Heyrovský Institute of Physical Chemistry and Electrochemistry, Czechoslovak Academy of Sciences, Opletalova 25, CS-110 00 Prague 1
Czechoslovakia

ABSTRACT. Principles of ion transport across the interface of two immiscible electrolyte solutions (ITIES) are dealt with. The effect of macrocyclic ionophores on equilibria at ITIES and, particularly, on transport across ITIES under electric current flow will be described. Finally, the paper is concerned with applications to ionophore analysis, determination of selectivity coefficients of ion-selective electrodes, and extraction study.

1. ELECTROCHEMISTRY OF THE INTERFACE OF TWO IMMISCIBLE ELECTROLYTE SOLUTIONS (ITIES)

Ion transfer across surfaces of biological membranes, phenomena at liquid membranes of ion-selective electrodes, salt transfer in extraction processes and phase-transfer catalysis are important examples of processes connected with ion transfer across liquid/liquid interfaces. In many of these processes different species of the same sign of charge are exchanged across the interface, or cations and anions are transferred simultaneously in the same direction so that no net charge transfer takes place. A suitable approach to deeper understanding of these processes is the investigation of ion transfer during electric current flow or, as we put it, during electrolysis at the interface of two immiscible electrolyte solutions (ITIES).

In the present paper the equilibrium properties of ITIES will be discussed first, together with the interfacial structure. After that the kinetics of charge transfer across ITIES, the influence of diffusion to ITIES on the ion transport and the experimental methods will be described. The processes at liquid membranes under electric current flow will not be included into this paper. In succession, electrolysis at ITIES was investigated experimentally by Nernst and Riesenfeld (1), by Guastalla (2), by Gavach and coworkers (see, e.g. (3)), by the Prague group (4), by the Kyoto group (5), by Buck and coworkers (see, e.g. (6)) and by Girault and Schiffrin (7). Electrochemistry at ITIES has been reviewed (8-12).

T. Osa and J. L. Atwood (eds.), Inclusion Aspects of Membrane Chemistry, 275–286.

1.1. Equilibrium potential difference

Consider two contacting liquid phases, namely, an aqueous phase, w, and an organic oil phase, o, showing very low miscibility with water with each containing various ionic as well as non-charged components. In the phases that are in equilibrium the electrochemical potentials of each component in both phases must be equal

$$\tilde{\mu}_i(w) = \tilde{\mu}_i(o). \tag{1}$$

This condition only applies to the cases when the transfer rate across the phase boundary (ITIES) is sufficiently high so that equilibrium is established in a measurable time. When the chemical and electrical contribution to the electrochemical potential is separated, equation (1) gives

$$\mu_i^0(w) + RT \ln a_i(w) + z_i F\phi(w)$$
$$= \mu_i^0(o) + RT \ln a_i(o) + z_i F\phi(o). \tag{2}$$

Here, μ_i^0 terms are the standard chemical potentials (independent of the concentration of the i-th component), a_i terms are activities, z_i terms are charge numbers and $\phi(w)$ and $\phi(o)$ are the inner electric potentials of the phase w and o, respectively.

1.1.1. The Nernst potential. The difference between inner potentials of two phases is termed the Galvani potential difference; in the present discussion we shall call it the Nernst potential

$$\Delta_o^w\phi = \phi(w) - \phi(o) = \left(\mu_i^0(o) - \mu_i^0(w)\right)/z_i F +$$
$$+ RT/z_i F \ln(a_i(o)/a_i(w))$$
$$= \Delta_o^w\phi_i^0 + RT/z_i F \ln(a_i(o)/a_i(w)). \tag{3}$$

In general, the Galvani potential differences are not measurable under strict thermodynamic conditions. However, as will be shown, it is possible to circumvent this restriction by appropriate extrathermodynamic assumptions. The difference between standard chemical potentials in w and o is equal to the difference between standard Gibbs energies of solvation in w and o. In this respect, it is convenient to define the standard Gibbs energy of transfer of the i-th species from the phase o to the phase w (13) by the equation

$$\mu_i^0(w) - \mu_i^0(o) = \Delta G_{tr,i}^{0,o\to w}. \tag{4}$$

1.1.2. The distribution potential. Consider the case that a uni-univalent electrolyte is distributed among the phases, w and o. Because of different solvation of the ions B^+ and A^-, in the both media an electric double layer and, at the same time, an electric potential difference is formed at ITIES. This potential difference is termed the distribution potential (15). Combining equation (3) written once for B^+ and then for A^- with equation (4) gives

$$\Delta_o^w\phi = 1/2\left(\Delta G_{tr,-}^{0,o\to w} - \Delta G_{tr,+}^{0,o\to w}\right) \tag{5}$$

The Gibbs standard transfer energies are standardized with the help of Parker's TATB assumption (13) stating that the standard Gibbs transfer energies of tetraphenylarsonium cation and tetraphenylborate anion between an arbitrary pair of solvents are equal and, at the same time, equal to the standard Gibbs transfer energy of their salt. On the basis of this extrathermodynamic assumption Eqs. (3), (4) and (5) gain a good physical meaning. A detailed discussion of the distribution potential is found in Refs. (14, 15).

1.2. Electrical double layer at ITIES

In the same way as an ideal polarized electrode (see, e.g. (16)) ITIES becomes an ideal polarized interface if there are at least two salts with different both cations and anions present in the system, if the ions forming the aqueous phase are strongly hydrophilic and if the ions forming the electrolyte of the non-aqueous (oil) phase are strongly hydrophobic (4, 14). Under these circumstances there exists a definite "potential window" where determination of the electrical potential difference between the aqueous and the organic phase by the ion activities is rather fictitious. When a given amount of charge is passed across the interface only a small fraction of this charge is consumed in the ion transfer, the major part being used for charging the electrical double layer at the interface. In this situation it is the double layer charge, not the ion activities, that fixes the electrical potential difference.

According to the simple Verwey-Niessen model (17) the electric double layer at ITIES only consists of two potential differences in the diffuse electric layers, $\phi_2(w)$ in the aqueous phase and $\phi_2(o)$ in the oil phase,

$$\Delta_o^w\phi = \phi_2(w) + \phi_2(o) \tag{6}$$

which, in the first approximation, are described by the Gouy-Chapman theory. Recently, improvements of this model where on an ion-free layer on both sides of ITIES is assumed have been discussed (18, 19). As the surface charges in the both phases are equal in absolute value and only differ in their signs the potential differences $\phi_2(w)$ and $\phi_2(o)$ are connected by the relationship valid for a uni-univalent electrolyte

$$\frac{\sinh (F\phi_2(w)/2RT)}{\sinh (F\phi_2(o)/2RT)} = \sqrt{\frac{D(w)\ c(w)}{D(o)\ c(o)}} \tag{7}$$

where D(w) and D(o) are relative permittivities in the aqueous and in the oil phase and c(w) and c(o) are the concentrations in the aqueous and in the oil phase, respectively. For $|\Delta_o^w\phi| \gg RT/F$ we obtain

$$\phi_2(w) = \frac{1}{2}\Delta_o^w\phi \pm RT/2F\ \ln[D(w)c(w)/D(o)c(o)] \tag{8}$$

where the plus sign applies to $\Delta_o^w\phi > 0$ and the minus sign to $\Delta_o^w\phi < 0$.

1.3. Ion transfer under electric current flow

Ion transport across ITIES proceeds in three steps, electrodiffusion toward ITIES, interfacial reaction of ion transfer across ITIES and electrodiffusion into the bulk of the other phase. The ion transport is mostly studied in systems containing indifferent electrolytes with properties defined in Section 1.2. The only transport process occurring inside the phases in contact is then diffusion, like in electrode-kinetics investigations. The interfacial process equation represents the boundary condition of the partial differential equation of the 2nd Fick law.

In most cases the interfacial step is rapid and the concentrations of the ion concerned at the interface in both phases are governed by the Nernst potential Eq. (3).

When suitable experimental methods are used the kinetics of the interfacial step can be detected. It follows from the classical Volmer-Butler-Frumkin equation of current density (cf., e.g. (16), p. 225-256) that

$$\begin{aligned} j &= z_iFJ_i \\ &= z_iFk_i^0\Big\{c_i^!(w)\exp\Big[\alpha z_iF(\Delta_o^w\phi - \phi_2(w) - \phi_2(o) - \Delta_o^w\phi_i^0/RT\Big] - \\ &\quad - c_i^!(o)\exp\Big[-(1-\alpha)z_iF(\Delta_o^w\phi - \phi_2(w) - \phi_2(o) - \Delta_o^w\phi_i^0/RT\Big]\Big\} \end{aligned} \tag{9}$$

where z_i is the charge number of the ion transferred, J_i its flux across ITIES, k_i^o its standard transfer rate constant, $c_i^!(w)$ and $c_i^!(o)$ its concentrations at the interface in water and in the oil phase. In Eq. (9)

$\Delta_o^w\phi$ is the electric potential difference imposed on the system from an external source. This quantity is approximately given by Eq. (6). The concentrations $c_i^!(w)$ and $c_i^!(o)$ are connected with the concentrations $c_i^!(w)$ and $c_i^!(o)$ by the Boltzmann equation, for example, $c_i^!(w) = c_i(w)\exp(z_iF\phi_2(w)/RT)$. Assuming $|\Delta_o^w\phi| \gg RT/F$ and $\alpha \approx 1/2$ Eq. (16) gives in view of Eq. (8) (20, 21).

$$j = z_iFk_i^0\Big[D(w)c(w)/D(o)c(o)\Big]^{\pm z_i/2} \times$$
$$\times \left\{c_i(w)\exp\left[z_iF/(2RT)\left(\Delta_o^w\phi - \Delta_o^w\phi_i^0\right)\right] - \right.$$
$$\left. - c_i(o)\exp\left[-z_iF/(2RT)\left(\Delta_o^w\phi - \Delta_o^w\phi_i^o\right)\right]\right\} \quad (10)$$

where c(w) and c(o) are the concentrations of the uni-univalent indifferent electrolytes in water and in oil, respectively. A more advanced theory of ion transfer across ITIES is found, for example, in Ref. (22).

For investigation of ion transport across ITIES the chronopotentiometric method (3) and polarography with electrolyte dropping electrode (4, 23) were used but, at present, the favourite method is triangular sweep voltammetry (24). For elimination of the considerably large ohmic drop a positive feedback system (25) is used.

2. IONOPHORE FACILITATED ION TRANSFER ACROSS ITIES

The ion transfer is facilitated (that is, it takes place at lower potentials) when in the organic phase are present lipophilic substances which form associates with the ion transferred. These may be complex formers (usually belonging to the family of ionophores well known from biochemical membranology (26)) or, in the case of proton transfer, various proton acceptors. The term, facilitated ion transport, was introduced in Refs. (27-29).

2.1. Equilibrium situation

In the cases of ionophore-facilitated transport across ITIES studied so far the complexes with ligand: metal ion ratio 1:1 and 2:1 have been identified. In the equilibrium situation the aqueous phase only contains free metal ions while in the oil phase, in a general case, free metal ions M^{2+} and complexes MX^{2+} and MX_2^{2+} are present. Complex formation is described by consecutive stability constants defined as

$$K_{MX_i} = [MX_i]/([MX_{i-1}][X]). \quad (11)$$

In this equation the charge numbers have been ommitted and $[MX_o]$ denotes the concentration of free metal ions. The hydrophobic ligand X is practically confined to the oil phase. Thus, for equal overall concentrations of metal ions in water and in oil phase, Eq. (3) and (11) give (activities have been substituted by concentrations)

$$\Delta_o^w\phi(\text{fac}) = \Delta_o^w\phi_M^0 - RT/zF \ln\left(1 + K_{MX}[X] + K_{MX}K_{MX_2}[X]^2\right). \tag{12}$$

In the cases investigated so far either the term $K_{MX}[X]$ or the term $K_{MX}K_{MX_2}[X]$ were predominant in the logarithmic function in Eq. (20).

2.2. Facilitated transport in electrolysis at ITIES

The analysis of voltammetric curves of facilitated transport in the case that both the metal ions and the ligand are present in comparable concentrations is quite difficult. Thus, in order to eliminate the influence of diffusion of one of the components two versions of the experiment have been adopted.

i. When the concentration of the metal ion M^{z+} in water is considerably larger than the overall concentration of the ligand in the oil phase, the transport is controlled by diffusion of the ligand to ITIES, by diffusion of the complex formed into the oil phase and, possibly, by the rate of complex formation at the interface (Fig. 1) or by processes proceeding in the opposite direction.

ii. In this case the concentration of the ligand in the oil phase is much larger than the concentration of the metal ion in water. The process is then controlled by diffusion of the metal ion to ITIES and of the complex to the bulk of the oil phase and by the rate of the interfacial process.

If the interfacial process is rapid the concentrations in the vicinity of ITIES have equilibrium values and the voltammetric curve has a 'reversible' shape (see, e.g. (11)). The half-wave potential $\Delta\phi_{1/2}$ which is obtained as the arithmetic mean of the peak potentials (Fig. 1) is given in the case i. for the ligand: metal ratio 1:1 as

$$\Delta\phi_{1/2} = \Delta_o^w\phi_M^0 + RT/2zF \ln\left(D'_X(o)/D'_{MX}(o)\right) - RT/zF \ln K_{MX}c_M(w) \tag{13}$$

where $D'_X(o)$ and $D'_{MX}(o)$ are the diffusion coefficients of the ligand and of the complex in the oil phase and $c_M(w)$ the metal ion concentration in the aqueous phase. As the ligand and the complex are usually isosteric $D'_X(o) \approx D'_{MX}(o)$. In the case ii. the half-wave potential is

$$\Delta_o^w \phi_{1/2} = \Delta_o^w \phi_M^0 + RT/2zF \ln (D'_M(w)/D'_{MX}(o)) - - RT/zF \ln K_{MX} c_X(o) \qquad (14)$$

where $D'_M(w)$ is the diffusion coefficient of the metal ion in water and $c_X(o)$ the ligand concentrations in the oil phase. A simple criterion for reversible process follows from the fact that the current in any point of the voltammogram is proportional to the square root of polarization rate (Fig. 1).

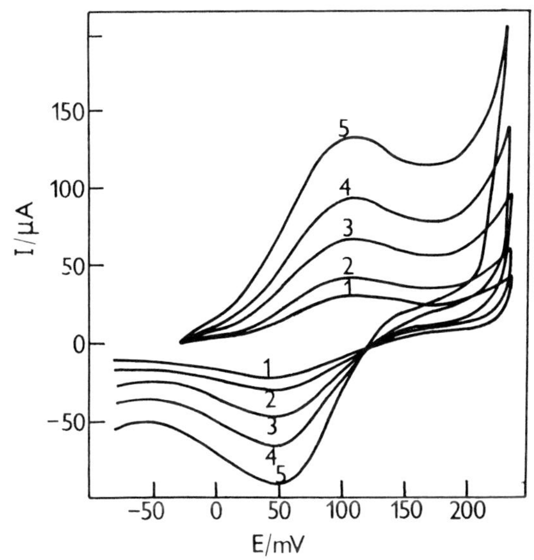

Fig. 1. Triangular sweep voltammograms of monensin. Aqueous electrolyte: 0.1 mol dm^{-3} sodium acetate, 3.2 mol dm^{-3} acetic acid, 0.01 mol dm^{-3} NaCl; nitrobenzene electrolyte: 0.01 mol dm^{-3} tetrabutylammonium tetraphenylborate, 0.001 mol dm^{-3} monensin. Polarization rates dE/dt: (1) 0.005 Vs^{-1}, (2) 0.01 Vs^{-1}, (3) 0.025 Vs^{-1}, (4) 0.05 Vs^{-1} and (5) 0.1 Vs^{-1}. The peak current is directly proportional to the square root of the polarization rate while the peak potential is constant.

If the interfacial process partially controls the transport process or if the ligand: metal ratio is 2:1 the voltammograms must be analyzed by means of the convolution method (30-32).

According to Kakutani et al. (33) the rate of the rapid facilitated cation transfer is directly proportional to the ligand concentration which supports the assumption of a heterogeneous reaction of complex formation resulting in the transfer of the complex across ITIES. Thus, Eq. (11) should be modified to the form

$$j = zFk^{0}_{MX}\left[D(w)c(w)/D(o)c(o)\right]^{\pm z/2} \times$$

$$\times \left\{c_M(w)c_X(o)\exp\left[zF/(2RT)\left(\Delta^w_o\phi - \Delta^w_o\phi^0_{MX}\right)\right] - \right.$$

$$\left. - c_{MX}(o)\exp\left[-zF/(2RT)\left(\Delta^w_o\phi - \Delta^w_o\phi^0_{MX}\right)\right]\right\} \qquad (15)$$

where k^{o}_{MX} is the standard transfer rate constant of the process $M^{z+}(w) + X(o) = MX(o)$, $D(w)$ and $D(o)$ are the relative permittitives and $c(w)$ and $c(o)$ the base electrolyte concentrations, and

$$\Delta^w_o\phi^0_{MX} = \Delta^w_o\phi^0_M - RT/zF \ln K_{MX}. \qquad (16)$$

In this way electrolysis at ITIES supports the theory of extraction kinetics where a surface reaction between the extracting ligand and the metal ion is assumed (34, 35).

2.3. Examples of individual ionophore effects

2.3.1. Cyclic alkali metal ion carriers. The transfer of alkali metal ions facilitated by valinomycin, nonactin and dibenzo-18-crown-6 was mainly studied by triangular-sweep voltammetry. The process which is 'reversible' is described by equation (13). In all cases the oil phase was nitrobenzene. The results are summarized in Table I.

TABLE I

Stabilities of cyclic ionophore complexes

Cation	Ligand	$\log K_{MX}$	Reference
K^+	Valinomycin	5.5	36
Na^+	Nonactin	7.3	37
K^+	Dibenzo-18-crown-6	6.0	37
Na^+		7.0	33

The kinetics of Na^+ transfer facilitated by dibenzo-18-crown-6 has been studied (33).

2.3.2. Acyclic alkali-metal ion and proton carriers. This property is shown by carboxylic antibiotics, monensin and nigericin. In contrast to the results obtained with bilayer lipid membranes where the acid anion acts as sodium and proton carrier (38, 39) the investigations at the nitrobenzene/water interface have shown that the undissociated form of the antibiotic (HX) functions as alkali metal ion carrier while the complex (MX) is the proton carrier (40, 42).

In a buffered solution containing much higher concentration of the metal ion than the ligand concentration in the oil phase the half-wave potential is governed by the equation

$$\Delta_o^w \phi_{1/2} = \text{const.} + RT/F \ln (k_1/c_M(w) + k_2/c_H(w)) \quad (17)$$

where $c_H(w)$ is the hydrogen ion concentration in water. The terms k_1 and k_2 are given by the equations

$$k_1 = \exp\left(F\Delta_o^w \phi^0_{M^+}/RT\right)K_1^{-1}$$

$$k_2 = \exp\left(F\Delta_o^w \phi^0_{H^+}/RT\right)K_2^{-1}$$

with K_1 and K_2 denotint equilibrium constants of MHX^+ formation in the organic phase

$$K_1 = \frac{[MHX^+]_o}{[M^+]_o[HX]_o}$$

$$K_2 = \frac{[MHX]_o}{[H^+]_o[MX]_o}$$

Table II contains the values of the constants K_1, K_2 and K_{exch} (for the reaction $M^+(w) + HX(o) \rightleftarrows MX(o) + H^+(w)$). Little difference in complexation of potassium and sodium by nigericin in nitrobenzene contrasts with their behaviour in methanol where the potassium complex, KHX^+ is much stronger than the sodium complex (42a).

2.3.3. Acyclic alkaline-earth ion carriers. Ligand: metal ion ratios 1:2 have been observed with the acyclic N,N'-di/(11'ethoxycarbonyl(undecyl/-N,N',4,5-tetramethyl-3,6-dioxaoctanediamide (43) and the cyclic ligand 7,19-dibenzyl-2,3-dimethyl-7,19-diaza-1,4,10,13,16-pentaoxacyclo-heneicosane-6,20-dione (44), in the case of divalent cations (Ca^{2+}, Ba^{2+}, Sr^{2+}, Mg^{2+}) while the ratio 1:1 has been found with univalent cations H_3O^+, Li^+, Na^+) (32). In the transfer of divalent metals a kinetic transfer step can also be detected.

TABLE II

Stability and exchange constants of carboxyclic ionophore complexes

Ligand	Cation	log K_1	log K_2	log K_{exch}	Reference
Monensin A	Na^+	5.9	10.4	-4.8	41
Monensin B	Na^+	6.05	10.4	-4.65	41
Monensin A+B mixture	Li^+	6.4	13.3	-5.16	41
Nigericin	Na^+	6.1	13.2	-7.1	42
	K^+	6.2	12.7	-4.9	42
	Rb^+	3.9			42
	Cs^+	2.9			42

3. APPLICATIONS

The peak currents appearing on the voltammograms are, in the case of excess metal ion (case i.), directly proportional to the concentration of the antibiotics. Thus, the straightforward application of this type of facilitated ion transport is the quantitative analysis of carrier-type antibiotics. The extraction of the antibiotic into nitrobenzene or 1,2-dichloroethane from a culture of an antibiotic is a simple procedure and the electrochemical measurement is quite rapid. Determination of monensin in cultures of Streptomyces fulvissimus has been described (45). However, it is not possible to determine individual components of a mixture of homologs like monensin A and B (41). A very simple determination of monensin in microbial cultures (45a) using the gel electrode (45b, 45c) has been described.

The data on stability constants of metal ion complexes of ionophores can be exploited for determination of selectivity coefficients of the liquid-membrane ion-selective electrodes based on these ionophores. For example, in the case of ligands mentioned in Section 2.3.3. the selectivity coefficients were determined with the help of voltammetry at ITIES and also potentiometrically using polymer-film based ion-selective electrodes (32). In spite of the fact that the nonaqueous solvent in the case of voltammetry was nitrobenzene and the platicizer of the membranes of ion-selective electrode was o-phenyl-octyl ether, the coincidence of the values was quite good.

As a rule the data on salt extraction kinetics are influenced by convective diffusion of the participating species (46) while only rarely the genuine interfacial process could be detected (34, 35). However, the data on standard transfer rate constants of both the cation and the anion of a salt can be combined to give the rate constant of salt extraction as described in Refs. (47) and (48). In the case of an extractant present in the oil phase the equation of cation transfer rate (15) and of the anion rate (10) can be employed for this purpose.

REFERENCES

1. W. Nernst and E.H. Riesenfeld, Ann. Phys. 8, 600 (1902).
2. J. Guastalla, J. Chim. Phys. 53, 470 (1956).
3. C. Gavach and F. Henry, J. Electroanal. Chem. 54, 361 (1974).
4. J. Koryta, P. Vanýsek and M. Březina, J. Electroanal. Chem. 67, 263 (1976).
5. M. Senda, T. Kakutani and T. Osakai, Denki Kagaku 49, 322 (1981).
6. O.R. Melroy and R.P. Buck, J. Electroanal. Chem. 136, 19 (1982).
7. H.H.J. Girault and D.J. Schiffrin, J. Electroanal. Chem. 170, 127 (1984).
8. J. Koryta, Electrochim. Acta 24, 293 (1979).
9. J. Koryta, Electrochim. Acta 29, 445 (1984).

9a. J. Koryta, Electrochim. Acta 33, 189 (1988).

10. J. Koryta and P. Vanýsek, 'Electrochemical phenomena at the interface of two immiscible electrolyte solutions', Advances in Electrochemistry and Electrochemical Engineering (H. Gerischer and C.A. Tobias, eds.), Vol. 12, John Wiley & Sons, New York, 1981, p. 113.
11. J. Koryta, Ion-Selective El. Reviews 5, 131 (1983).
12. V. Mareček, Z. Samec and J. Koryta, 'Charge transfer across the interface of two immiscible electrolyte solutions', Advances in Colloid and Interface Sciences 29, 1 (1988).
13. A.J. Parker, Electrochim. Acta 21, 671 (1976).
14. Le Q. Hung, J. Electroanal. Chem. 115, 159 (1980).
15. F.M. Karpfen and J.E.B. Randles, Trans. Faraday Soc. 49, 823 (1953).
16. J. Koryta and J. Dvořák, Principles of Electrochemistry, John Wiley & Sons, Chichester, 1987.
17. E.J.W. Verwey and K.F. Niessen, Phil. Mag. 28, 435 (1939).
18. Z. Samec, V. Mareček and D. Homolka, Faraday Disc. Chem. Soc. 77, 197 (1984).
19. T. Kakiuchi and M. Senda, Bull. Chem. Soc. Japan 56, 1753 (1983).
20. J. Koryta, Anal. Chim. Acta 139, 1 (1982).
21. M. Gross, S. Gromb and C. Gavach, J. Electroanal. Chem. 89, 29 (1978).
22. Z. Samec, Yu.Ya. Gurevich and Yu.I. Kharkats, J. Electroanal. Chem. 204, 257 (1986).
23. Z. Samec, V. Mareček, J. Weber and D. Homolka, J. Electroanal. Chem. 99, 385 (1979).
24. Z. Samec, V. Mareček, J. Koryta and W. Khalil, J. Electroanal. Chem. 83, 393 (1977).
25. Z. Samec, V. Mareček and J. Weber, J. Electroanal. Chem. 100, 841 (1979).
26. Yu.A. Ovchinnikov, V.T. Ivanov and M.M. Shkrob, Membrane-active Complexones, Elsevier, Amsterdam, 1974.
27. P.E. Scholander, Science 131, 585 (1960).
28. W.J. Ward, AIChE J. 16, 805 (1970).
29. C.F. Reusch and E.L. Cussler, AIChE J. 19, 736 (1973).

30. M. Greness and K.B. Oldham, Anal. Chem. 44, 1121 (1972).
31. Z. Samec, J. Electroanal. Chem. 111, 211 (1980).
32. Z. Samec, D. Homolka and V. Mareček, J. Electroanal. Chem. 135, 265 (1982).
33. T. Kakutani, Y. Nishiwaki, T. Osakai and M. Senda, Bull. Chem. Soc. Japan 59, 781 (1986).
34. W.J. Albery, R.A. Choudhery and P.R. Fisk, Faraday Disc. Chem. Soc. 77, 53 (1984).
35. W. Nitsch and K. Hillekamp, Chem.-Ztg. 96, 245 (1972).
36. P. Vanýsek, W. Ruth and J. Koryta, J. Electroanal. Chem. 148, 117 (1983).
37. A. Hofmanová, Le Q. Hung and W. Khalil, J. Electroanal. Chem. 135, 257 (1982).
38. B.C. Pressman, Fed. Proc. Fed. Am. Soc. Exp. Biol. 27, 1283 (1968).
39. R. Sandeux, J. Sandeux, C. Gavach and B. Brun, Biochim. Biophys. Acta 684, 127 (1982).
40. Guo Du, J. Koryta, W. Ruth and P. Vanýsek, J. Electroanal. Chem. 159, 413 (1983).
41. J. Koryta, D. Guo, W. Ruth and P. Vanýsek, Faraday Disc. Chem. Soc. 77, 209 (1984).
42. A. Sabela, J. Koryta and O. Valent, J. Electroanal. Chem. 204, 267 (1986).
42a. Y. Pointud, C. Tizzier and J. Juillard, J. Sol. Chem. 12, 473 (1983).
43. D. Ammann, M. Güggi, E. Pretsch and W. Simon, Anal. Lett. 8, 709 (1975).
44. J. Petránek and O. Ryba, Anal. Chim. Acta 128, 129 (1981).
45. J. Koryta, W. Ruth, P. Vanýsek and A. Hofmanová, Anal. Lett. B15, 1685 (1982).
45a. J. Koryta and V. Mareček, Mikrochim. Acta 1990I, 225.
45b. T. Osaki, T. Kakutani, and M. Senda, Bumseki Kagaku 33, E371 (1984).
45c. V. Mareček and M.P. Colombini, J. Electroanal. Chem. 241, 133 (1988).
46. G.J. Hanna and R.D. Noble, Chem. Revs. 85, 583 (1985).
47. J. Koryta and M. Skalický, J. Electroanal. Chem. 229, 265 (1987).
48. J. Koryta, Electrochim. Acta 32, 419 (1987).

SUBJECT INDEX

Inclusion Science

TOPICS IN INCLUSION SCIENCE

1. J. Szejtli: *Cyclodextrin Technology*. 1988 ISBN 90-277-2314-1
2. T. Osa and J.L. Atwood (eds.): *Inclusion Aspects of Membrane Chemistry*. 1991 ISBN 0-7923-1123-X
3. J. Vicens and V. Böhmer (eds.): *Calixarenes: A Versatile Class of Macrocyclic Compounds*. 1991 ISBN 0-7923-0714-3

ADVANCES IN INCLUSION SCIENCE

1. J. Szejtli (ed.): *Proceedings of the First International Symposium on Cyclodextrins*. Budapest, Hungary (1981). 1982 ISBN 90-277-1433-9
2. Alex I. Käiväräinen: *Solvent-Dependent Flexibility of Proteins and Principles of Their Function*. 1985 ISBN 90-277-1534-3
3. J. L. Atwood, J. E. D. Davies and T. Osa (eds.): *Clathrate Compounds, Molecular Inclusion Phenomena, and Cyclodextrins*. Proceedings of the 3rd International Symposium on Clathrate Compounds and Molecular Inclusion Phenomena and the 2nd International Symposium on Cyclodextrins, Tokyo, Japan (1984). 1985 ISBN 90-277-2010-X
4. J. L. Atwood and J. E. D. Davies (eds.): *Inclusion Phenomena in Inorganic, Organic, and Organometallic Hosts*. Proceedings of the 4th International Symposium on Inclusion Phenomena and the 3rd International Symposium on Cyclodextrins, Lancaster, U.K. (1986). 1987 ISBN 90-277-2601-9
5. O. Huber and J. Szejtli (eds.): *Proceedings of the Fourth International Symposium on Cyclodextrins*, Munich, F.R.G. (1988). 1988 ISBN 0-7923-0000-9
6. G. W. Gokel and K. Koga (eds.): *United States–Japan Seminar on Host–Guest Chemistry*. Proceedings of the U.S.–Japan Seminar on Host–Guest Chemistry, Miami, Florida, U.S.A. (1987). 1989 ISBN 0-7923-0262-1

KLUWER ACADEMIC PUBLISHERS – DORDRECHT / BOSTON / LONDON